T0358500

Earth Matters

Earth Matters

Earth Matters

INDIGENOUS PEOPLES, THE EXTRACTIVE INDUSTRIES AND CORPORATE SOCIAL RESPONSIBILITY

EDITED BY CIARAN O'FAIRCHEALLAIGH AND SALEEM ALI

LONDON AND NEW YORK

First published 2008 by Greenleaf Publishing Ltd

Published 2017 by Routledge
2 Park Square, Milton Park, Abingdon, Oxon OX14 4RN
711 Third Avenue, New York, NY 10017, USA

Routledge is an imprint of the Taylor & Francis Group, an informa business

Cover by LaliAbril.com

British Library Cataloguing in Publication Data:
Earth matters : indigenous peoples, the extractive
industries and corporate social responsibility
1. Social responsibility of business 2. Mineral industries
- Social aspects 3. Indigenous peoples - Social conditions
I. O'Faircheallaigh, Ciaran II. Ali, Saleem H. (Saleem
Hassan), 1973-
658.4'08

ISBN-13: 9781906093167 (hbk)

Contents

Foreword

Wayne Bergmann
Executive Director, Kimberley Land Council

The Kimberley Land Council is a grassroots community organisation founded 30 years ago, during one of a number of major confrontations between Kimberley Aboriginal people and resource developers in the late 1970s and early 1980s. This occurred when the Government of Western Australia supported a multinational company to drill for oil on sacred ground on Noonkanbah, a cattle station that Yungngora people had only recently secured a lease over, after walking off the station some years earlier in protest at their appalling work and living conditions. In the end the drilling went ahead. But as a result of Noonkanbah Kimberley Aboriginal people became organised and established the Kimberley Land Council in our struggle to protect the things most precious to us, our land and sea, the sites and stories connected with them, and our culture.

The way of doing business with Aboriginal people, back then, was to ride rough-shod right over us. Many things have changed over the last 30 years, a point evident from reading the chapters in this book, which examine relations between indigenous peoples and resource developers in many different parts of the world. Today nearly all major resource companies and their industry associations espouse principles of corporate social responsibility, stating their intention to engage with the communities affected by their operations and to ensure that those communities share in the benefits of resource development. In the Kimberley and in many other regions, companies negotiate with Aboriginal landowners, paying them compensation and involving them in managing the environment; establish employment, training and business development initiatives; and in some cases exceed their legal obligations in protecting sacred sites. These initiatives are important, and can make a valuable contribution in assisting indigenous people to overcome the serious social and economic disadvantages they face as a result of centuries of dispossession and political marginalisation.

But there is also need for caution in assessing these changes. As Kimberley Aboriginal people have discovered to their cost, some companies adopt the rhetoric of corporate social responsibility, but in practice show little respect for Aboriginal people, their land or their culture. Even where company intentions are good, adoption of corporate policies do not always translate into concrete benefits for Aboriginal people on the ground. Government policies are subject to change, and where they are hostile to indigenous interests, the enthusiasm of corporations for sharing the benefits of resource development with indigenous people can quickly evaporate. For these reasons there can be a large gap between corporate policies released in New York, London, Tokyo or Melbourne and the eventual outcome for Aboriginal peoples whose lands are used in resource exploitation.

Given this reality, it is essential to have available rigorous, critical analysis of the policies, motives and actions of multinational resource corporations in their dealings with indigenous peoples. It is equally important to appreciate the differing responses of indigenous groups, and to understand the impact of dominant political and legal systems on the choices open to them and on the strategies they pursue in dealing with corporations. This book makes a valuable contribution in this regard. It brings together information on the experiences of indigenous peoples around the world, of the impacts on them of corporate policies and actions, and of the successes and failures of indigenous peoples in engaging with resource companies and governments. By doing so it will help ensure that the principle of 'corporate social responsibility' becomes a reality in the Kimberley and in other indigenous homelands.

Introduction

Ciaran O'Faircheallaigh
Department of Politics and Public Policy, Griffith University, Australia

Indigenous peoples have historically gained little from large-scale resource development on their traditional lands and have suffered from its negative impacts on their cultures, economies and societies. During recent decades indigenous groups and their allies have fought hard to change this situation: in some cases by opposing development entirely; in many others by seeking a fundamental change in the distribution of benefits and costs from resource exploitation. In doing so they have utilised a range of approaches, including: efforts to win greater recognition of indigenous rights in international forums; pressure for passage of national and state or provincial legislation recognising indigenous land rights and protecting indigenous culture; litigation in national and international courts; and direct political action aimed at governments and developers, often in alliance with non-governmental organisations (NGOs).

At the same time, and partly in response to these initiatives, many of the corporations that undertake large-scale resource exploitation have sought to address concerns regarding the impact of their activities on indigenous peoples by adopting what are generally referred to as 'corporate social responsibility' (CSR) policies. This book focuses on such corporate initiatives. It does not treat them in isolation, recognising that their adoption and impact are contextual and are related both to the wider social and political framework in which they occur and to the activities and initiatives of indigenous peoples. It does not treat them uncritically, recognising that they may in some cases consist of little more than exercises in public relations. However, neither does it approach them cynically, recognising the possibility that, even if CSR policies and activities reflect hard-headed business decisions, and indeed perhaps *particularly* if they do so, they can generate significant benefits for indigenous peoples if appropriate accountability mechanisms are in place.

In undertaking an in-depth analysis of CSR and indigenous peoples in the extractive industries, the book seeks to answer the following questions. What is the nature and extent of CSR initiatives in the extractive industries and how should they be understood? What motivates companies to pursue CSR policies and activities? How do specific political, social and legal contexts shape corporate behaviour? What is the relationship between indigenous political action and CSR? How and to what extent can corporations be held accountable for their policies and actions? Can CSR help bring about a fundamental change in the distribution of benefits and costs from large-scale resource exploitation and, if so, under what conditions can this occur?

The next section provides a context for the individual contributions by analysing the concept of CSR and briefly outlining the circumstances that have lead to the widespread adoption of CSR initiatives by the world's major resource development companies. The focus of each chapter is then briefly outlined.

Corporate social responsibility

Three broad approaches are evident in writing on CSR (for a detailed discussion see Carroll 1999; O'Faircheallaigh 2007). The first sees it simply as a public relations exercise designed by companies to persuade governments and citizens that they are not only interested in maximising profits, but also have the public interest at heart. In this view CSR represents a cynical exercise designed to protect companies from public pressure to behave in a publicly spirited manner, and will involve them spending as little as possible and focusing what they do spend on publicity material such as glossy 'sustainability reports' and on highly visible and tangible projects such as construction of buildings or support for the arts or for sports (see for instance Ballard 2001).

A second approach sees CSR as a holistic and long-term view of what is required to allow a company to survive and continue to generate wealth into the future. CSR is thus not simply a public relations exercise, but will require a company to pay careful attention to societal values and to forgo profits in the short term in order to protect its 'social licence to operate'. However, it does so as part of a rational calculation of self-interest, as an integral part of a strategic approach to maximising profits over the longer term (see, for example, Cragg and Greenbaum 2002).

Others emphasise that CSR must involve activities that would not be dictated by a purely selfish calculation of corporate interests. In other words CSR refers to a duty or obligation on corporations to create benefits for society in ways that *go beyond* what they cannot avoid doing because of legal obligations, or what they would do in any case purely on the basis of economic self-interest. Carroll (1999), for example, states that society expects business to use economic resources efficiently and operate profitably and to obey all applicable laws, but also to behave ethically, which may require going well beyond what is demanded by the law, and to make voluntary donations to socially worthy causes (philanthropy), reducing corporate profits (Carroll 1999: 10-11). There are two major grounds on which business may be expected to behave in ways that do not simply reflect economic self-interest. The first involves the argument that corporations receive substantial benefits from society. For example, they have important legal privileges that are not enjoyed by individual citizens, including limited liability, and they benefit from public expenditures in areas such as education, infrastructure and scientific research. In return for these benefits, it is argued, they owe society a debt, which they should help to fulfil by taking societal interests into account in corporate decision-making (Donaldson and Preston 1995). The second involves the claim that corporations are not separate from the societies in which they operate, that they are run by human beings who are just as capable of making decisions on moral and ethical grounds as other individuals, and that ethics and social values as well as economic self-interest should therefore play a key role in corporate decision-making (Crawley and Sinclair 2003).

The argument that corporations owe a duty to society to behave in socially responsible ways raises the issue of whether such behaviour should be *required* of companies and enforced by public regulation or by other mechanisms that compel compliance, or whether relevant standards of corporate behaviour should be established only through voluntary mechanisms such as adherence to industry codes of conduct. For some, fundamental changes can be achieved only if companies internalise values that result in socially responsible behaviour, and such an outcome cannot be achieved through attempts by regulators or other actors external to corporations to impose standards. For others there is little prospect that CSR can result in fundamental change unless effective mechanisms exist to hold companies accountable for their behaviour, and voluntary mechanisms are incapable of providing this accountability (Coumans 2002).

Critics of CSR argue that, if it requires anything other than a focus on maximising profits, it involves real risks not just for corporations and shareholders but also for society. Milton Friedman, for example, identifies two such risks. The first is that productivity will decline as firms fail to focus on efficiency. The second involves a threat to the economic freedom of shareholders, as company managers and directors apply profits to their preferred social purposes rather than distributing them to shareholders and allowing the latter to decide how they should be applied (Friedman 1990).

Other critics of CSR focus not on the risks for corporations, but for society. They argue that large corporations already wield enormous economic power, and that as CSR practices become more widespread companies will start to accumulate social power as they use their wealth to intervene in social, political and cultural affairs. From this perspective it is preferable for governments, which unlike corporations can be held accountable by the electorate, to take responsibility for promoting the good of society (Cannon 1992: 47-48).

Whatever arguments can be made regarding CSR at a level of principle, a quick check of the websites of most major companies operating in the extractive industries will quickly reveal the widespread acceptance of CSR and the scale of activities and expenditures companies undertake to demonstrate their support for CSR principles. Multiple factors explain this phenomenon. Populations both in the home countries of corporations and in other countries in which they operate are better educated and better informed than in the past, in part because of the global reach of television and the internet. Global communication networks also assist in organising consumer boycotts across multiple markets, as happened with Shell after it was accused of causing widespread pollution in Nigeria and colluding with the Nigerian government to suppress popular protests against its operations (Mirvis 2000).

Corporations also face greater scrutiny and resistance at the local level from communities affected by their actions. Alliances between NGOs, which increasingly operate globally, and local groups mean that the latter also have a capacity to press their case internationally. During recent years people in countries as diverse as Papua New Guinea, Nigeria, Peru and Canada have taken direct action to delay the projects or disrupt the operations of companies they believed were failing to protect the environment, foster economic and social development or involve local communities in decision-making. It is important to note that in many cases local action is not based on accusations of illegality. Compliance with the law offers no guarantee whatsoever that a project will be allowed to proceed or operate smoothly. Local opposition can lead to major delays in developing industrial projects, to their temporary closure or to their abandonment, and the costs for companies and investors can be enormous (Gao *et al.* 2002). Conversely, the support of local communities can help facilitate early project completion and reliable operations, generating substantial economic benefits for the companies involved.

Another important development relates to the growing capacity of individuals or groups who have been adversely affected by a company's operations or products to take legal action against the company in its home country. Historically, such legal action tended to be restricted to the country in which the relevant events or activities occurred, and a company's liability was limited to the assets it held in that country. In recent years important legal precedents have been established that allow a parent company (and its assets) to be pursued in the courts of its home country. So, for instance, during the 1990s Papua New Guinea villagers affected by pollution of the Ok Tedi and Fly rivers, caused by BHP's Ok Tedi mine, took action in Australian courts to seek compensation. BHP sought a ruling in the Federal Court that Australian courts did not have jurisdiction but, having lost that action, reached a settlement with the villagers that included cash payments in the tens of millions of dollars, dredging of the Ok Tedi River and future containment of tailings. As the risks of being sued increase and now involve multiple legal systems, many companies feel it is better to use CSR policies to try to avoid problems that cause legal suits in the first place

The increasingly global nature of funding for major industrial projects represents another important factor in the trend towards CSR. Even the largest companies rarely fund major investments from their own resources. Funding typically comes from international banking consortia and international financial institutions (IFIs) such as the International Finance Corporation, the private investment arm of the World Bank. Particularly for projects in developing countries, banks and IFIs require political risk insurance for their investments, provided for example by the World Bank's Multilateral Investment Guarantee Agency or national organisations such as the US Overseas Private Investment Corporation (OPIC). International financiers and investment insurance agencies are increasingly reluctant to fund projects unless the companies developing them address environmental and social issues in the countries in which they operate (see for instance OPIC 2006). Here also, companies need to go beyond complying with local laws. If those laws fail to protect the environment or meet the aspirations and concerns of local communities, and projects are delayed or their operations disrupted, it is no comfort to banks or insurers that project operators complied with the law.

Focus of the chapters

The book begins with two chapters that consider aspects of the wider political and social context within which CSR policies are developed and resource development occurs. Trebeck focuses on the issue of whether corporations can be scrutinised and made responsive to those whose lives they affect ('civil regulation'), so contributing to the advancement of democracy in political systems where the responsiveness of state agencies is often imperfect. She uses empirical evidence from relationships between mining companies and indigenous Australians to assess the capacity of civil regulation to promote democratisation. Haley and Magdanz focus on another key contextual variable by examining the social ties that characterise indigenous societies affected by resource development. Drawing on the experience of Inupiat and Yupiit people living in areas of Alaska affected by oil development, they develop a theoretical framework and research agenda for understanding types of social ties, changes in social ties and implications of changes in social ties in relation to resource development.

As mentioned earlier, CSR initiatives affecting indigenous peoples cannot be analysed or understood in isolation, and in Chapter 3 Coumans examines a key aspect of the wider political context by examining relationships between indigenous peoples and rights-based non-indigenous NGOs. She illustrates the various means by which indigenous groups and NGOs seek to advance rights-based positions. She examines the interaction between these initiatives and the efforts of mining companies and their associations and home governments, and of international financial institutions, to determine standards for CSR, in ways that frequently seek to limit indigenous rights. The theme of indigenous–NGO alliances and their interactions with corporations is taken up later, in Chapter 10, by Cerretti, McAteer and Ali, who focus on what they term 'shareholder transnational advocacy networks' in the context of the foreign-dominated oil industry in Ecuador. These networks link indigenous communities in Ecuador, who have few avenues to influence corporate practice from the local level, with domestic and international NGOs. The NGOs in turn provide connections to corporate shareholder groups, including socially responsible investing firms, in attempts to influence the policies of corporate head offices in the USA and so, via a 'boomerang effect', company behaviour on the ground in Ecuador.

In 'settler' societies such as Australia and Canada it is now commonplace for CSR commitments to be enshrined in legally binding agreements between companies and indigenous landowners, a practice also increasingly common in some developing countries. O'Faircheallaigh argues in Chapter 4 that, while such agreements have important advantages, especially in establishing standards for company performance and corporate accountability, they can also bring about significant changes to the structural position of indigenous groups within wider political, legal and administrative frameworks. He explores the potential impact of agreements on indigenous peoples' relationships with the state, with NGOs and other political actors, and with the media, and on their existing rights under regulatory and legislative regimes. As Howitt and Lawrence note in Chapter 5, interpersonal as well as structural aspects of indigenous–company relations require attention, given that interaction on the ground is often dominated by personal relations between individual company employees and indigenous people they deal with on a day-to-day basis. Drawing on examples from mining and forestry in Australia and Scandinavia, they explore the tension between the potential benefits of, and the essential fragility of, such interpersonal relations.

Indigenous communities and peoples are not homogeneous in their experience of corporate policies and activities, a point highlighted by Gibson and Kemp in their discussion of corporate engagement with indigenous women in Chapter 6. They show that indigenous women have largely been excluded from negotiations about benefits from mineral development while disproportionately feeling the adverse impacts of mining. Literature about corporate responses to concerns articulated by indigenous women is virtually non-existent, and they argue that a remedy for this situation requires further empirical research informed by a number of theoretical lenses. Angelbeck's study of commercial forestry in British Columbia in Chapter 7 illustrates that the corporate sector is not monolithic, with most large forestry companies, unlike their counterparts in the mining industry, failing to adopt any significant CSR initiatives. This has left indigenous groups with no choice but to focus on litigation and direct action aimed at both companies and government, in attempts to protect their cultural heritage and forests in which to maintain contemporary cultural practices. Angelbeck notes the potential of CSR to deliver more positive outcomes for all involved.

In Chapter 8 Barker examines one arena for CSR activities, indigenous employment in mining, in the specific context of Australia. She focuses on an issue that has received little attention to date: the impact of employment on individual indigenous people and on their communities. She raises issues regarding the nature of employment effects, examines relevant evidence from a number of mines in north Australia, and indicates the sorts of research approaches that are required to enhance our understanding of various interlinking aspects of indigenous employment. Filer, Burton and Banks focus on the very different context of Melanesia. In Chapter 9 they examine the apparent paradox that a region that has experienced some of the most environmentally damaging mining projects of the modern era, including Bougainville, Ok Tedi and Freeport, is also regarded as containing, especially in Papua New Guinea, some of the world's most inventive and successful indigenous resistance to corporate domination.

In Chapter 11 Anguelovski examines 'dialogue tables' established at two mines in Peru as part of corporate responses to indigenous opposition to mining, which resulted from its damaging impact on indigenous territories and indigenous health. Using an environmental justice framework, she assesses the extent to which company practices matched corporate rhetoric and illustrates the critical role played by government policy and legislation in framing company responses to indigenous demands. In Chapter 12 Crate and Yakovleva examine another and quite different context for CSR: the recently privatised diamond-mining industry in Russia. They highlight the limited nature of corporate responses in a context where mining has, historically, caused widespread and serious environmental damage and where public officials are often reluctant to support indigenous efforts to achieve remediation and improve future environmental management. They consider the human rights and environmental justice implications of diamond mining in the Vilyuy regions, draw parallels with other indigenous contexts, and explore possible ways of better protecting indigenous lands and livelihoods.

The chapters show clearly the great diversity that characterises initiatives and policies represented as 'corporate social responsibility', the highly contingent and contextual nature of corporate responses to indigenous demands, and the complex and evolving nature of indigenous–corporate relations. They also reveal much about the conditions under which CSR can contribute to a redistribution of benefits and costs from large-scale resource development, a theme we return to in the Conclusion.

References

Ballard, C. (2001) *Human Rights and the Mining Sector in Indonesia: A Baseline Study* (MMSD Working Paper 182; London: Mining, Minerals and Sustainable Development Project of the International Institute for Environment and Development).

Cannon, T. (1992) *Corporate Responsibility* (London: Pitman Publishing).

Carroll, A.B. (1999) 'Corporate Social Responsibility', *Business and Society* 38.3: 268-95.

Coumans, C. (2002) 'Mining, Water, Survival and the Diavik Diamond Mine', in G. Evans, J. Goodman and N. Lansbery (eds.), *Moving Mountains: Communities Confront Mining and Globalisation* (London: Zed Books): 91-108.

Cragg, W., and A. Greenbaum (2002) 'Reasoning about Responsibilities: Mining Company Managers on What Stakeholders are Owed', *Journal of Business Ethics* 39: 319-35.

Crawley, A., and A. Sinclair (2003) 'Indigenous Human Resource Practices in Australian Mining Companies: Towards an Ethical Model', *Journal of Business Ethics* 45.4: 361-73.

Donaldson, T., and L.E. Preston (1995) 'The Stakeholder Theory of the Corporation: Concepts, Evidence, and Implications', *Academy of Management Review* 20.1: 65-91.

Friedman, M. (1990) *Free to Choose: A Personal Statement* (London: Pan Books).

Gao, Z., G. Akpan and J. Vanjik (2002) 'Public Participation in Mining and Petroleum in Asia and the Pacific: The Ok Tedi Case and its Implications', in D.N. Zillman, A.R. Lucas and G. Pring (eds.), *Human Rights in Natural Resource Development* (Oxford, UK: Oxford University Press): 679-93.

Mirvis, P.H. (2000) 'Transformation at Shell: Commerce and Citizenship', *Business and Society Review* 105.1: 63-84.

O'Faircheallaigh, C. (2007) 'Corporate Social Responsibility and Globalisation', in C. Curran and E. Van Acker (eds.), *Globalising Government Business Relations* (Frenchs Forest, NSW: Pearson Longman): 309-32.

OPIC (Overseas Private Investment Corporation) (2006) *Investment Policy* (Washington, DC: OPIC; www.opic.gov/doingbusiness/investment, accessed 20 August 2008).

1

Corporate social responsibility and democratisation

Opportunities and obstacles

Katherine Trebeck
Research and Policy Executive, the Wise Group, Glasgow, UK

Democracy seems to be accepted as a workable, albeit imperfect, mechanism of governing a polity and of political decision-making in Western liberal societies (Held 1987: 198; Young 2000: 4). Theories of electoral, representative democracy dominate many understandings of the practice of democracy. Capitalism, albeit in many variations, similarly seems to have prevailed over alternative methods of economic organisation. A capitalist system that delivers citizen satisfaction is therefore crucial. This chapter examines the need for broader conceptions of how democracy is attained, beyond representative processes. It then outlines why, considering democracy in a wider sense, companies should be scrutinised and accordingly how, given current configurations of capitalism, companies might be made responsive to those citizens whose lives they affect.

The chapter is largely theoretical, but draws on empirical evidence gleaned from research into relationships between mining companies and indigenous Australians. This research included almost 120 semi-structured interviews that took place between 2002 and 2005 with all levels of current and former mining company personnel, industry observers, bureaucrats, stock market participants, environmental and community activists (including indigenous communities) and academics. Additional data was used to contextualise personal anecdotes.[1]

1 Including company annual reports, corporate 'vision' statements and operating principles, websites, brochures and other publications, survey questions and results, speeches, corporate reports, media releases, media reports and company statistics.

It is argued that using the tool of 'civil regulation', CSR can lead to democratisation, but with several qualifications that limit the potential of civil regulation to bolster citizen sovereignty. To understand these limitations, the nature of corporate entities that influences their responsiveness to community demands is explored. The chapter concludes with an evaluation of the potential for democracy to be enhanced via CSR, and the role of organisations and sectors in this scenario.

Democracy

A fundamental democratic concept is the ability of people to shape policies and decisions relevant to them. Specifically, democracy seeks to structure policy decisions according to the preferences of those who will be affected (Eckersley 2000: 118; Shapiro 1999: 31). The extent to which a decision is democratically legitimate depends on the extent to which decision-making includes those impacted by the outcomes (Young 2000: 6; see also Dahl 1961; and Cohen 1989 cited in Dryzek 2001). In relation to mining, for example, how, where and when mining takes place should, in a democratic process, reflect the wishes of those affected by mining operations. Democratisation thus occurs as decision-making in a political community becomes more inclusive (Whitehead 2002: 27; see also Dahl 1961: 1; Shapiro 1999: 31; Young 2000: 6).

Processes of representation are necessary in modern times because no individual can ever be present when all decisions that affect their lives are taken. Representative democracy remains crucial in a wider democratic system, despite several shortcomings. These shortcomings, discussed in detail below, include difficulties some groups face in having their interests heard, and the state orientation of democracy and ostensible disregard for the possibility that many relevant decisions take place in non-state realms, such as companies.

Advocates of more participatory styles of democracy complain that current configurations based on representation are substandard in both substance and process (for example: Barber 1984: xiv; Hirst 1994: 3, 16; Cole 1920 quoted in Pateman 1970: 37; Walker 2002). Representative democracy entails relatively passive and infrequent citizen involvement, with most citizens participating only in regular elections. Minority interests are often disadvantaged by formal democratic processes, threatening democratic ideals of citizen sovereignty, self-determination and autonomy (for example, de Tocqueville 1945: 86). Indigenous communities, for example, can be impacted by decisions made by purportedly representative parliaments which are much removed, physically and culturally, from these communities and citizens, undermining any potential that outcomes will reflect the needs and values of those affected. Moreover, political decisions are often effectively taken by a small group of (not necessarily elected) citizens (Chandler 2002; Held 1987: 298; Hirst 1994: 70; Saward 2000: 67; Young 2000: 173). Skewed accessibility of formal political processes reinforces social and economic inequalities, and deepens political inequality, and vice versa (Held 1987: 14; Young 2000: 17).

Furthermore, government action sometimes does not reflect values held by a majority of its citizens (let alone those of minorities), constituting a democratic deficit. Societies have diverse needs, many beyond the scope of state-oriented democracy. Globalisation's impacts can exceed the responsive and management capacity of governments (Hirst 1994: 13; Pateman 1970: 38; Roßteutscher 2000: 172). The objective of being economically competitive often takes precedence, undermining democracy in capitalist states (Dryzek 1996: 3, 71-72). For example, concerted government efforts to facilitate mining, regardless of articulated opposition or concerns of local indigenous communities, deepen indigenous disenchantment with formal structures and processes of representative parliamentary democracy. In Australia, for instance, indigenous concerns were largely set aside by governments that actively pursued development of Hamersley Iron's Marandoo Mine, Century Mine in the Gulf of Carpentaria and the Jabiluka uranium mine in the Northern Territory (Trebeck 2005, 2007). Aligned state and corporate interest perpetuate indigenous marginalisation within representative democratic processes. This illustrates a cleavage between democratic ideals of sovereignty and participation by those affected in decision-making, and government developmentalism. It further highlights the necessity of alternative avenues by which affected communities might achieve sought outcomes.

Given deficiencies of representative democracy, and despite the ongoing influence of the state in the economy and people's lives, the state is hardly the most relevant and effective realm of decision-making for every process influencing the lives of citizens. Dryzek (2001) warns that tying evaluations of democracy to state-based scenarios limits its consideration to 'a needlessly thin conception of democracy'. Many key decisions are not state responsibilities or are beyond the state's capacity. Instead, civil society organisations and the private sector are increasingly influential in the direction and implementation of policy agendas. Consequently, new modes of thinking about the processes and realisation of democracy are necessary. Recognition of the inadequacies of state-based representative democracy enables evaluation to encompass entities beyond the state, including companies and civil society organisations.

The apparent ability of civil society organisations to elicit certain behaviour from powerful socioeconomic actors supports such reconceptualisation of governance and how democracy is realised. Various commentators have written about 'extra-parliamentary' modes of political decision-making that might enhance democratic ideals (Fisk 1989: 178–79 quoted in Dryzek 1996: 151; Moon *et al.* 2003; Pateman 1970: 106). Discussion now explores this widened terrain of democratic analysis, encompassing companies themselves, community efforts to change corporate behaviour and corporate characteristics that shape corporate responsiveness.

Companies and democracy

Given the variety of arenas for social decision-making and governance, analysis should similarly investigate social and economic entities that have an impact on the direction of society and the lives of individuals. As Menninger (1985) observes:

> the corporation reveals itself inevitably as part of the essential core of modern society. It stands next to the state as an institutional anchor of industrial polit-

ical economy and shares with the state, for better or worse, the rule and direction of social development (quoted in Galligan 1989: 8).[2]

Corporate decisions, however, impact on many who are not privy to internal company decision-making processes. Hence, how communities can influence these decisions needs to be considered if democratic outcomes are to be achieved.

While many communities may perceive that they possess little influence over a leviathan opposition that includes companies, capitalist institutions and government, when citizens form alliances internationally, nationally and locally, and act strategically to 'regulate' company actions, this imbalance can potentially be mitigated (see Coumans, Chapter 3 this volume, for numerous examples of such alliances). In such instances, a company changes its behaviour following pressure from civil society, rather than pressure from government. Corporate responsiveness to community pressure presents an opportunity for indigenous people to express themselves beyond formal political structures, using instead 'extra-parliamentary means' including physical protest, shareholder activism, reputational assault and tactical use of legislation.

CSR and civil regulation

Corporate social responsibility, in a practical sense, describes those activities, other than commercial outputs of a company and beyond legally required behaviour, that address social and environmental concerns of company 'stakeholders'. Stakeholders are anyone or any organisation affected by or able to affect a particular entity.[3] Stakeholder concerns, traditionally considered outside a company's remit, are important because the prerequisites of profit-making have broadened (Moon 1995; Warburton *et al.* 2004). Profit and shareholder interest remain the foremost corporate goals, but now certain stakeholders need to be satisfied if profit is to be sustained and shareholder interests served.

It is these key stakeholders who can attain desired corporate behaviour using informal means of setting standards with which companies comply. 'Civil regulation' describes the outcome of actions by civil society organisations and individuals to restrict the range of behaviours available to economic (or state) entities to achieve a specific response. Bendell (2000a: 8) defines civil regulations as those 'pressures exerted by processes in civil society to persuade, or even compel, organisations to act differently in relation to social and environmental concerns' (see also Murphy and Bendell 1999: vi).[4]

2 See also Bakan 2004: 5; Banerjee 2001: 40; Parker 2002: 4. Henderson (2001) and others, however, assert that processes of privatisation, deregulation and liberalisation of international capital flows have actually diminished the power of business through the increased openness and competitiveness of markets. Moreover, recently there has been a significant increase in controls placed on corporations in terms of environmental, safety, employee rights and other standards (R. Wilson 1999).

3 Including regulatory bodies, unions, other NGOs, international bodies, shareholder activists and local communities.

4 Corporate power has long been challenged by elements of society and social movements, such as the consumer movement in the 1960s and the environmental movement of the 1970s. Murphy and Bendell (1999: 1) argue that the history of business–civil society relations has been one of antagonism until the early 1990s when partnerships for sustainable development began to emerge.

Companies can be forced to address (invariably social and environmental) demands made by civil society entities. Methods of civil regulation range from campaigns to confront companies to supporting corporate effort to change through conciliatory engagement. Civil regulation accordingly operates via consumer preferences, employee and potential employee wishes, concerns of investors and regulators, provision of or withholding access to resources (where there is legal title or underpinned by physical intimidation), moral and reputational persuasion, and via costs incurred dealing with direct actions.

Civil regulation is often effective because corporate reputation among key sectors of society is extremely valuable and if damaged can result in financial loss (Goyder 1998: 51).[5] Examination of the mining industry's interaction with indigenous Australians confirms that reputation among key audiences is a crucial lever in civil regulation. If local indigenous communities view a prospective miner as an undesirable presence in their region, then the company will face a longer development process and perhaps even find it impossible to operate in a hostile context (Trebeck 2007). Operations, profit and shareholder value are thus threatened when key audiences hold poor perceptions of a company. Dobson (1989) explains that the reputation imperative acts as an 'invisible hand' to implicitly uphold contracts, however informal, between companies and their stakeholders that cannot be enforced explicitly (see also Walker 2002). CSR activities are accordingly undertaken by companies hoping to protect their reputation among key audiences.

Licence to operate

Civil regulation is often manifested via the leverage certain communities possess by virtue of a 'social (or community) licence to operate'. Corporations are subject to a variety of external controls, regulated by both government and civil society. Minimum levels of acceptable corporate behaviour are specified in laws and regulations. Historically, state regulations were interpreted as embodying societal expectations (Gunningham *et al.* 2002: 1; Rodgers 2000: 40, 43). Many companies, however, are recognising that their social obligations are no longer discharged simply by meeting legal duties. Government mandates remain necessary, but alone are insufficient basis for corporate legitimacy (Australian Minerals and Energy Environment Foundation 2002: 53; Edelman 2002; Gunningham *et al.* 2002: 1). In this context, companies are increasingly finding that they need to earn and maintain a social licence to operate.

The viability of a particular operation or business can be threatened if it is considered socially unacceptable. Companies face an ongoing danger that when their behaviour does not meet community expectations they will eventually be the subject of public censure, manifested in a number of possible direct ways, or via government and court actions. Community resistance to a project can be tacit in nature, such as intransigence

5 A study by Columbia University found that up to a third of shareholder value, depending on industry, is determined by the reputation of a corporation (cited in Bendell 2000b: 23). KPMG (quoted in Cohen 2002) estimates that for every $1 of lost reputation, $2 must be spent to regain that reputation.

during land-access negotiations, or it can be more palpable, such as public protest and physical occupation of a company operation, or even sabotage. Good corporate community relations, stakeholder engagement and consultation and efforts to meet particular community demands are therefore means by which companies seek to improve reputation among those with the ability to impact operations, and thereby attain a social licence to operate (Banerjee 2001: 46; Cragg and Greenbaum 2002: 325; Parker 2002: 81). Such CSR activities are essentially a tactic to win support from local communities, government, and key civil society organisations. Fundamentally, because companies require community support to exist over the long term, certain communities have potential ability to set parameters of corporate activity.

Civil society obstacles

Resulting civil regulation, however, brings significant caveats and dangers. For example, community organisations can turn to only a limited number of companies or corporate actions at a time, and consequently must prioritise where to deploy their effort. Confronting companies requires considerable sacrifice from would-be civil regulators, and engagement is equally time- and resource-intensive. Participating in negotiations, meetings and reviews all entails an opportunity cost. Pursuing enhanced democracy in one realm might therefore occur at the expense of participation in another.[6]

Similarly, apathy and indifference can leave civil regulation to people holding strong, but not necessarily widely held, positions (see, for example, Hirst 2002: 411; Sartori 1962 cited in Pateman 1970: 11; Whitehead 2002: 75). Individuals only mildly interested, or perhaps concerned, but unable to afford the time to participate, find it difficult to influence governance outcomes through civil regulation in such circumstances. Consequently, civil regulation that results from increased participation of the passionate may contradict passive or simply unmobilised sentiment.

Moreover, those conforming to norms of discourse are invariably advantaged, at the expense of those with less persuasive force or status (Eckersley 2000: 121; Young 2000: 38, 171). The exclusion of certain modes of communication means that outcomes from deliberation within civil society, and corporate response to demands emanating from civil society organisations, again reflect the interests of a few. This 'representativeness deficit' pertaining to those organisations seeking to 'regulate' companies means that the results of civil regulation will skew away from the interests of those affected.

Further still, communities themselves are not cohesive, seldom engage as one voice, nor advocate uncontested positions. Any notion of a common interest among or within communities is problematic because it is inherently difficult to ascertain (if such shared will and desire exists at all). Enforced or implied commonality is 'liable to narrow the possible agenda for deliberation' (Young 2000: 43), stifling views and interests divergent to the prevailing sense of common good (see also Eriksen 2000: 57; Trigger 1997, 1998, for discussion of division in indigenous communities). In addition, achieving cor-

6 Sirianni refers to this situation as the 'paradox of participatory pluralism' (quoted in Dryzek 1996: 7).

porate change through civil society action is ultimately predicated on certain conditions that enable such action, in particular civil rights and capacity to protest safely.

Increasing inclusion in corporate decisions via civil regulation could also be criticised as merely extending existing structures and institutions to marginalised groups, compelling them to acquiesce in the previously disenfranchising hegemony (Young 2000: 12). To achieve corporate change indigenous communities are forced to 'cross over' to the corporate 'world' in the tactics they use and the arguments they mount. For example, in seeking increased influence in how a mining project is run, indigenous communities were often compelled to act via mainstream forums, appealing to 'mainstream sensitivities' such as protection of the environment and human rights abuses and, of course, use of economic levers (Ballard and Banks 2003). This was evident in the campaign against development of Jabiluka uranium mine in Australia's Northern Territory. The anti-Jabiluka campaign not only appealed to concerns about the environment and world heritage values, but also highlighted the financial risks allegedly associated with Jabiluka in an effort to dissuade investors in Jabiluka's then owners, North Limited, from supporting development. Thus the North Ethical Shareholders group claimed there were unbudgeted costs of A$200 million: namely, those costs incurred if the company was compelled to build a mill at the Jabiluka site, rather than use an existing mill (discussed below). These financial issues were not the concern of most anti-Jabiluka campaigners, but they were used to appeal to the sensitivities of North's investors.

There is another quandary threatening the sovereignty of indigenous communities seeking to influence the decisions of mining companies: being forced to accept mining or risk a reduction of government services. Mining might be acquiesced to if communities view it as the only means to access necessary outcomes—for example, health and education services or employment. This concern was held by some indigenous people in communities near Century Mine in Queensland, Australia, where government and company both argued that, unless communities assented to Century's development, there would be a reduction of state services, and communities would forgo jobs and direct benefits stemming from the mine. Simultaneously, as governments seemingly decrease their service provision in some areas, often already disadvantaged indigenous communities increasingly look to mining companies to deliver infrastructure and services (Behrendt 2001; IIED 2002: 9-20; McMahon and Strongman 1999; UNEP 2002: 10).

Conceptions of corporate entities

At first glance companies might seem monolithic, but they are complex, and this complexity affects how community pressures for responsiveness are received and how comprehensively CSR is taken up. This section explores the characteristics of companies that determine when and why companies respond to community demands.

Many economists view each 'firm' as a nexus of contracts between respective stakeholders, including owners, managers, employees, suppliers, customers and government, and thus defined by a legal framework (see, for example, Butler 1997: 15, 391).[7]

7 Corporations, however, are given a legal identity as artificial people. In 1886 the US Supreme Court, e.g., declared that companies are 'natural' persons and are thus protected by the Constitution (White 1999).

This implies that companies are self-contained, profit-maximising unitary entities, and overlooks complex internal and external corporate realities. There are other interpretations of the company that alternatively advocate or reject the notion of corporations as 'moral agents' (for a comparison of the moral responsibilities of individuals and those of companies, including a discussion of corporate agency, see Hutton 1996: 24; Wilmot 2001). Understandings of corporate organisations that reject the idea of companies as moral agents assert that companies are social only in that they are made up of human beings, and that any ethics or morals espoused are only those formulated and implemented by individuals within the company.[8] Individuals who make up companies and give them any discernible moral character have their own spheres of influence and objectives, and are themselves complex and erratic. To a great extent, the behaviour of a company is an aggregation of the behaviour of individuals and sub-units within it, which in turn pursue diverse and possibly contradictory goals (Jensen 1983; Keskinen *et al.* 2003; O'Neill and Gibson-Graham 1999).[9]

Corporate culture and leadership also influence the behaviour of employees, reinforced by reward systems that provide incentives for a particular activity. Simultaneously companies, and individuals within them, function within a wider social context (Keskinen *et al.* 2003). Companies are porous entities. Individual employees have their own external networks that inform their actions and motivations. How an event, policy or directive is interpreted by these employees is often contingent on their personal background and position within the company. This diversity means that there is no singular culture within an organisation, with firms instead characterised by a variety of intersecting groups, such as committees, the board, friendships, work teams and so on, as well as individuals.

Internal complexity is an important factor in any traction that efforts to change company behaviour might gain. Which and how companies respond to community demands is determined by a combination of their financial, political and cultural characteristics. Uptake of CSR depends particularly on responsive individuals in the company—and especially senior managers—being sensitive to pressures for change, and then advocating the necessity of CSR for corporate benefit. The extent to which stakeholder demands are acknowledged and addressed can be considered a function of how managers perceive stakeholder power (ability to impact the firm), the legitimacy of the group or its claim, and urgency of a stakeholder's demand.[10] Individual recognition of the power, legitimacy and urgency of demands for CSR is, in turn, often shaped by personal values, which can be a function of experience of a crisis, notions of corporate performance or ethical beliefs (Agle *et al.* 1999; Parker 2002: 84).[11]

8 Korten (2001: 11), e.g., labels the notion that companies have the character of moral human beings as 'fiction' since the legal contracts that are the premise of companies have no 'conscience and no loyalty to people or place'. Jensen (1983) concurs, explaining that organisations do not have the preferences nor conscious decision-making capacities that people possess.

9 Thompson (1998) similarly argues that four important levels of organisational life can be considered—the individual level (micro), the level of team leadership ('macho'), the level of internal stakeholders (meso), and the level of external stakeholders (macro) (see also Chandler 2002: 9).

10 'Legitimacy' denotes that a stakeholder's activities are sought and appropriate according to social norms; the 'urgency' of a stakeholder encompasses importance and the unacceptability of delay (Agle *et al.* 1999; Mitchell *et al.* 1997; Wartick and Wood 1998: 111-12, quoted in King 2000).

11 Orlitzky and Swanson (2002) model incorporation of social expectations in corporate actions as dependent on how receptive executives are to the demands of stakeholders—how 'attuned' executives are to external demands will shape how likely they are to encourage adoption of CSR.

Habermas (cited in Llewellyn 2004) notes, however, that companies are more than simply the sum of individuals, and operate according to their own momentum. Change will thus only be realised when change is in the corporate interest. Zandvliet (2004) explains that if a stakeholder is relatively 'difficult' it will capture more CSR initiatives, and that companies often respond immediately to threats, sabotage and blockages, as opposed to written or verbal complaints.[12] For example, in the Gulf of Carpentaria in Queensland local indigenous communities were able to impel the prospective developer of Century Mine, CRA (Rio Tinto's predecessor in Australia), to the negotiating table. During a protracted negotiation period the company increased an initial offer to local indigenous communities of A$70,000 cash to a final agreement package of A$60 million over 20 years. The bargaining position exploited by indigenous interests, by virtue of native title provisions,[13] delay tactics and use of public forums, rendered communities and their demands of the miner visible to company managers and prominent in company decision-making equations (Trebeck 2005).

In the late 1980s, Hamersley Iron ('Hamersley'), a Rio Tinto subsidiary based in the Pilbara (Western Australia), sought to develop the Marandoo iron ore deposit located within Hamersley National Park (now Karijini National Park), claiming that Marandoo was imperative to maintaining supply (Hextall 1990; Stevens 1991). Hamersley could not proceed with development until the project obtained clearance under Western Australia's *Aboriginal Heritage Act* (AHA). Local indigenous communities demanded heritage surveys and compliance with the AHA. In doing so they were able to significantly stall Hamersley's intended development time-frame. Delaying Marandoo was a strategy to pressure the company to acknowledge indigenous interests, enabling local communities to make demands, such as increased indigenous employment in Hamersley's operations. Pertinently, seeking to improve relationships with local indigenous communities and prevent any future delay to its development plans, Hamersley established an Aboriginal Training and Liaison Unit in 1992 to increase indigenous employment and implement programmes to repair company relationships with indigenous people in the Pilbara.

Similarly, the campaign against the Jabiluka uranium mine in the Northern Territory encompassed legal action, efforts at education, mobilisation of national and international opposition, physical demonstrations, shareholder activism and parliamentary lobbying (Trebeck 2005). While Jabiluka's then owners, North Limited, impervious to many elements of the anti-Jabiluka campaign, stridently pursued Jabiluka's development, traditional owners were able to wield a lever that impacted the economics of the project. By virtue of a clause in a 1991 lease transfer agreement (transferring ownership from Pancontinental to Energy Resources of Australia [ERA], then majority-owned by

12 This accords with managerial theories of stakeholder relations that relate responsiveness to those stakeholders with capacity to impact a firm's operations (see, e.g., Wilmshurst 2004). Alternatively, normative prescriptions of stakeholder relations focus on the rights of all stakeholders. As seen in the case studies, and as Wilmshurst concludes, the two do not operate exclusively, but in conjunction.

13 The *Mabo* decision of 3 June 1992 was the first time that a court in Australia recognised that indigenous people had entitlements to their traditional lands according to traditional law (National Native Title Tribunal 2002). 'Native title' requires that an identifiable group or community has maintained, since colonisation, their traditional affiliation or occupation of the land according to indigenous laws and customs. The *Mabo* judgement thus held that other sources of law and governmental authority, pre-dating colonisation, remain in existence in Australia (Sanders 2002: 10; High Court of Australia, *Mabo v. Queensland (No. 2)*, 1992 175 CLR 1 at 5, quoted in Mantziaris and Martin 2000).

North), traditional owners held an effective veto over milling of Jabiluka ore at the existing nearby Ranger mill. The cost of building a mill to process Jabiluka ore at the Jabiluka site was over A$200 million, a cost ERA hoped to avoid through use of the Ranger mill. In the context of low uranium prices prevailing in the late 1990s, the veto over milling Jabiluka ore at the existing Ranger mill, which effectively necessitated construction of a new mill at Jabiluka, made the economics of the Jabiluka project far less attractive to developers.

The leverage necessary to capture corporate attention and response in these instances stemmed from community ability to impact the commercial position of the company, whether in the short term through economic levers, or over the long term via influence on company reputation and relationships with key audiences.

Concurrently, the importance of key employees in advocating the case for corporate adoption of CSR was evident at Rio Tinto headquarters where certain managers appreciated the imperative for improved indigenous relations and drove the initiation of new CSR policies and strategies. The leadership of Rio Tinto's then chief executive officer, Leon Davis, is considered by many as crucial in initiating change and enabling its implementation, as 'spearheading' the shift from legal wrangling to negotiations with indigenous people (Manning 2003). His senior position in the company hierarchy bestowed the ability to effect change quickly, while he held a personal commitment to a certain standard of behaviour towards indigenous Australians, as well as an appreciation of the business case for better indigenous relations. For example, his career included work at mine sites overseas where engagement and deals with local people were routine, and also at Bougainville in Papua New Guinea where hostility of local people forced early mine closure in 1989. He describes the 'loss of the [Bougainville] mine as a huge shock. [The Bougainville copper mine] was a huge investment . . . Bougainville was a lesson I learnt and which I took with me.'[14]

There was thus a confluence of a personal position and recognition of strategic imperative. Davis, and other key managers like him, combined the personal and the commercially prudent, attaching their own 'moral' stance to the business case, or simply recognising the commercial sense of addressing certain expectations of particular communities. Following advocacy of the need for CSR by Davis and others with similar perceptions, Rio Tinto's new position regarding indigenous relations was publicly stated in 1995. In a landmark speech, Davis (1995) outlined Rio Tinto's changed approach towards indigenous communities near its mines and exploration interests, away from a litigious stance towards development of mechanisms to share with and compensate indigenous people for mining activity on their land.

Corporate responsiveness obstacles

The nature of corporate responsiveness as demonstrated by the preceding examples entails possible obstacles to democracy through civil regulation of companies. These are

14 Personal Interview with L. Davis, Sydney, Australia, 14 February 2005.

presented to understand the parameters of CSR, where its potential ends and intervention of non-corporate entities is necessary.

Firms face a dilemma in allocating resources and responding to various community demands. CSR initiatives are constructed to utilise expenditure to maximum advantage, and expenditure will accordingly vary in relation to perceived leverage of communities over corporate interests. The implication is that democracy through civil regulation applies only to those who have something to offer companies: for example, withholding attack on public image or allowing access to land. Citizens with no such voice find themselves excluded from any franchise presented via civil regulation (Hertz 2002: 197; Murphy and Bendell 1999: vi, 42, 44). For example, on acquiring North in August 2000, and with it 68.4% ownership of ERA, Rio Tinto declared it wished to proceed with mining at Jabiluka, but only with traditional owner consent. Rio Tinto's priority was its reputation among local indigenous communities, rather than among environmental groups based in capital cities who opposed Jabiluka's development. This suggests that the benefits from civil regulation are segmented and are only attainable by those with something to offer companies: effectively, those able to threaten a company's social licence to operate and ultimately the company's commercial standing.

Responsiveness to community demands is, moreover, determined by the company, particularly the discretion of key managers. The way in which managers perceive community demands '[acts] as an important filter through which information about the external [social] licences is sifted, and guides their responsiveness to conflicting external pressures' (Gunningham *et al.* 2002: 17). The nature of the business is also important. Social pressures are more acute for companies that value their profile among key stakeholder audiences highly, for land access, regulatory or marketing reasons, for example (Moon *et al.* 2003). Moreover, the capacity of the company to respond ultimately determines whether, to what extent and how demands of certain communities are fulfilled (Gunningham *et al.* 2002: 18; Phillips 2001: 193). For example, Century Mine representatives explained to local indigenous communities that the dire financial situation of its parent company, Pasminco, meant it was unable to fulfil every community demand.[15] CSR thus stretches only so far, and the length of this stretch is determined by the company's financial position and internal culture and personalities, alongside the extent to which communities are prepared to enforce their demands. This reflects a perhaps unavoidable dilemma of CSR: that what companies are prepared to provide and what communities demand (or need) are unlikely to ever be completely reconciled.

The necessity of a business case for CSR requires that citizens, as consumers and investors and campaigners, take action against every corporation or corporate activity deemed unsatisfactory in order to 'regulate' desired corporate behaviour (see, for example, Korten 2001: 11). Most civil regulation, however, targets corporations individually, and, if effective, it will lead only to uneven and staggered progress in attaining broader goals, other than through a 'trickle-down effect', such as industry-wide codes of conduct or market leaders compelling laggards to follow.

At the individual company level, despite some successes in CSR and genuine responsiveness, CSR policies are inconsistently implemented. Policies are made, guidelines issued and publications distributed promoting 'good-news stories', but the extent of

15 Pasminco bought Century from CRA in 1997, but went into administration in 2002. In 2004 Pasminco was restructured as Zinifex.

change varies and uptake by employees is sporadic. Attempts to achieve community support and legitimacy are largely encapsulated by community programmes, especially employment-related initiatives, sophisticated and 'leading practice' as some of them may be. There is little effort to incorporate stakeholder views into corporate decision-making beyond consultation—in effect, 'tell us your views and we will decide what to do with them'. Much CSR from the mining companies considered appeared responsive, defensive and remote from operational decision-making. Corporate priorities are unchanged; what has changed is how these priorities are attained.

Equally, the external face of corporate CSR in the form of publicised programmes, speeches and launches masks internal dissent and scepticism among some employees about the value of CSR activities. For such employees, time and production constraints in pursuit of profit override other pressures. Management is charged with navigating the company's external environment for long-term shareholder value, whereas employees at mine sites have less cognisance of the geographical cross-subsidisation of reputational capital. Corporate headquarters are sometimes able to overlook tension between profit-maximisation and social responsibility, but are not required to honour obligations or incur the costs of CSR initiatives (Brereton 2002; Kapelus 2002). Many employees retain a narrow perception of what is required for corporate and personal success: namely, technical competence. For example, managers at mine sites often have strict targets that encompass quotas and performance reviews, directing their priorities towards immediate production tasks. Accordingly, if CSR initiatives are deemed expensive relative to other objectives, they are unlikely to be undertaken, unless specifically mandated by company headquarters.

This divergence might be overcome through implementation of incentives aligning objectives of employees with those ends sought by corporate headquarters. While employees responsible for delivery of CSR—by means of, for example, quotas for raising the number of indigenous employees or management of indigenous heritage—may not appreciate the necessity of such activities, they will appreciate that delivery in these areas impacts remuneration or promotion. Equally, structures, frameworks and programmes to deliver CSR are of little use if those implementing them lack enthusiasm, understanding and appreciation of the rationale behind their task. Education, such as cross-cultural training, might impart an understanding of the external context in which a particular company operates, fostering appreciation of the rationale driving CSR.

Conclusion: cause for hope?

The ability of certain sectors of society, in this case some indigenous Australians, to force companies to respond to their demands suggests that civil regulation can be effective. In broad terms, companies will be responsive to social concerns when two factors are present. The first is that it must be in the company's own interest to respond, even if this interest is considered over the long term. The second factor is that individuals within the company need to harness the 'business case' in order to effect change in corporate community engagement strategies, perhaps combining their own moral reasons for CSR with recognition of a prudent business rationale.

Much of the literature exploring CSR policy and practice seems to downplay, if not ignore, the necessity of this second factor in corporate responsiveness, failing to incorporate corporate complexity and neglecting the important role of internal advocates. In realising any corporate change in response to civil regulation, however, the business case remains paramount. Internal advocates are essentially the mechanism that links the business case to tangible outcomes in terms of changed corporate behaviour. Once corporate adoption of CSR has been announced and policies and structures developed, implementation of CSR then requires individuals who appreciate its prudence. Individuals who implement CSR need to recognise the business case, although they might be simultaneously driven by their own personal beliefs that support responsiveness to community demands.

Regulation of business by civil society, however, entails significant caveats and dangers. Current configurations of economic and legal frameworks necessitate an ever-present threat of the loss of social licence to operate in order to ensure that companies recognise and address demands of certain stakeholders. Without a persistent 'business case', predisposed individuals lose their commercial rationale for CSR, and the social prerequisites of profit-making diminish.

Necessity of the business case highlights a crucial role for civil society in maintaining vigilance and ongoing compulsion for companies to consider community demands. Companies and key employees need to be aware of the relevance of the social licence to operate to corporate performance. Without pressure from crucial stakeholders, any incentives to alter company behaviour (or maintain high social and environmental standards) wither away. Corporate memory of incidents illustrating the importance of the social licence to operate fades, and impetus for CSR abates. Thus civil society organisations need to maintain surveillance and pressure to ensure it is always in the corporate interest to respond to community demands. CSR as a means of democratisation accordingly stands or falls on the extent to which citizens are, first, concerned about an issue and, second, can mobilise through appropriate means to prompt desired corporate response. In CSR civil society shoulders a heavy obligation, but, as seen above, has only a partial capacity and legitimacy to perform this 'watchdog' role.

Moreover, civil regulation of companies and CSR is not going to achieve widespread improvement in the structures of society, or any overall increase in democratisation. With companies constituted as they are, because they are not omnipresent, even exemplary CSR inherently leaves gaps. This constitutes a vital role for the state, with its (as yet) unparalleled capacity for planning, redistribution and direct provision of citizenship rights.

Returning to the hypothesis of democratisation through civil regulation posed at the beginning of this chapter, it can be concluded that civil regulation offers, under some circumstances, positive results at the micro level, but only patchy democratisation on the macro scale. This is perhaps unsurprising given that it is within their local environs that most people are arguably most concerned and most prepared to take action (Pattie *et al.* 2004). In local areas, where communities instinctively turn to nearby companies rather than government to attain specific demands, as seen in the indigenous–mining company interface in remote Australia discussed above, CSR might be seen as a substitute for inadequate delivery of community demands by representative democratic structures.

Broadly, however, corporate social responsiveness offers more potential as a complement, rather than a substitute, for formal representative democracy and the mandate of the state. Voting and representative democracy will always be components of multifaceted democratic systems. At its best, civil regulation is an awkward form of community empowerment filled with inconsistencies and inequalities.

References

Agle, B., R. Mitchell and J. Sonnenfeld (1999) 'Who Matters to CEOs? An Investigation of Stakeholder Attributes and Salience, Corporate Performance, and CEO Values', *Academy of Management Journal* 42.5: 507-26.

Australian Minerals and Energy Environment Foundation (2002) *Facing the Future: The Report of the MMSD Australia Project* (Melbourne: Mining Minerals and Sustainable Development Project).

Bakan, J. (2004) *The Corporation* (London: Constable & Robinson).

Ballard, C., and G. Banks (2003) 'Resource Wars: The Anthropology of Mining', *Annual Review of Anthropology* 32 (October 2003): 287-313.

Banerjee, S.B. (2001) 'Corporate Citizenship and Indigenous Stakeholders: Exploring a New Dynamic of Organisational–Stakeholder Relationships', *Journal of Corporate Citizenship* 1: 39-55.

Barber, B. (1984) *Strong Democracy: Participatory Politics for a New Age* (Berkeley, CA: University of California Press).

Behrendt, L. (2001) 'Indigenous Self-determination in the Age of Globalisation', paper presented to *Human Rights and Global Challenges Conference*, Castan Centre for Human Rights Law, Monash University, Melbourne, 10–11 December 2001, www.law.monash.edu.au/castancentre/conference2001/papers/behrendt.html, accessed 20 August 2008.

Bendell, J. (2000a) 'Talking for Change? Reflections on Effective Stakeholder Dialogue', London: New Academy of Business Innovation Network; www.new-academy.ac.uk/publications/keypublications/documents/talkingforchange.pdf, accessed 20 August 2008.

—— (2000b) 'Introduction: Working with Stakeholder Pressure for Sustainable Development', in J. Bendell (ed.), *Terms for Endearment: Business, NGOs and Sustainable Development* (Sheffield, UK: Greenleaf Publishing): 14-29.

Brereton, D. (2002) 'The Role of Self-regulation in Improving Corporate Social Performance: The Case of the Mining Industry', in *Current Issues in Regulation: Enforcement and Compliance* (Melbourne: Australian Institute of Criminology).

Butler, K. (1997) *Multinational Finance* (Cincinnati, OH: South-Western College Publishing).

Chandler, G. (2002) 'Government and Corporate Social Responsibility: The Democratic Deficit', *Ethical Corporation*, 6 December 2002.

Cohen, S. (2002) 'Ethics and Business: What Can You Say? What Might Convince the Unconverted?', paper presented at *Centre for Applied Philosophy and Public Ethics Seminar*, Australian National University, Canberra, September 2002.

Cragg, W., and A. Greenbaum (2002) 'Reasoning about Responsibilities: Mining Company Managers on What Stakeholders are Owed', *Journal of Business Ethics* 39: 319-35.

Dahl, R. (1961) *Who Governs? Democracy and Power in an American City* (New Haven, CT: Yale University Press).

Davis, L. (1995) 'New Directions for CRA' (Melbourne: Securities Institute of Australia).

De Tocqueville, A. (1945) *Democracy in America* (New York: Random House).

Dobson, J. (1989) 'Corporate Reputation: A Free-Market Solution to Unethical Behavior', *Business and Society* 28.1: 1-5.

Dryzek, J. (1996) *Democracy in Capitalist Times* (New York: Oxford University Press).

—— (2001) 'Legitimacy and Economy in Deliberative Democracy', *Political Theory* 29.5: 651-69.

Eckersley, R. (2000) 'Deliberative Democracy, Ecological Representation and Risk: Towards a Democracy of the Affected', in M. Saward (ed.), *Democratic Innovation* (London: Routledge): 117-32.

Edelman, R. (2002) 'Corporate Responsibility, Is it Worth it and for Whom?', *Ethical Corporation*; www.ethicalcorp.com/content_print.asp?ContentID=373, accessed 20 August 2008.

Eriksen, E.O. (2000) 'The European Union's Democratic Deficit: A Deliberative Perspective', in M. Saward (ed.), *Democratic Innovation* (London: Routledge): 53-65.

Galligan, B. (1989) *Utah and Queensland Coal: A Study in the Micro Political Economy of Modern Capitalism and the State* (St Lucia, QLD: University of Queensland Press).

Goyder, M. (1998) *Living Tomorrow's Company* (Aldershot, UK: Gower Publishing).

Gunningham, N., R. Kagan and D. Thornton (2002) *Social Licence and Environmental Protection: Why Businesses Go Beyond Compliance* (London: Centre for Analysis of Risk and Regulation, London School of Economics and Political Science; www.lse.ac.uk/Depts/carr/Publications_folder_web_files/Disspaper8.pdf, accessed 20 August 2008).

Held, D. (1987) *Models of Democracy* (Cambridge, UK: Polity Press).

Henderson, D. (2001) 'Capitalism Needs Heart Bypass', *The Australian*, 19 June 2001.

Hertz, N. (2002) *The Silent Takeover: Global Capitalism and the Death of Democracy* (London: Arrow Books).

Hextall, B. (1990) 'CRA Clear to Mine in National Park', *Sydney Morning Herald,* 16 November 1990.

Hirst, P. (1994) *Associative Democracy* (Cambridge, UK: Polity Press).

—— (2002) 'Renewing Democracy through Associations', *The Political Quarterly* 73.4: 409-21.

Hutton, W. (1996) *The State We're In* (London: Vintage).

IIED (International Institute for Environment and Development) (2002) 'Mining, Minerals and Sustainable Development: Project Draft Report for Comment' (London: World Business Council for Sustainable Development).

Jensen, M. (1983) 'Organization Theory and Methodology', *Accounting Review* 58.2: 319-39.

Kapelus, P. (2002) 'Mining, Corporate Social Responsibility and the "Community": The Case of Rio Tinto, Richards Bay Minerals and the Mbonambi', *Journal of Business Ethics* 39.3: 275-96.

Keskinen, A., M. Aaltonen and E. Mitleton-Kelly (eds.) (2003) *Organisational Complexity* (Scientific Papers 1/2003; Helsinki: TUTU Publications, Finland Futures Research Centre).

King, D. (2000) 'Stakeholders and Spin Doctors: The Politicisation of Corporate Reputations' (Hawke Institute Working Paper Series No. 5; Magill: Hawke Institute, University of South Australia; www.hawkcentre.unisa.edu.au/institute/resources/Working%20paper%205.pdf, accessed 25 November 2003).

Korten, D. (2001) 'Predatory Corporations', in G. Evans, J. Goodman and N. Lansbury (eds.), *Moving Mountains: Communities Confront Mining and Globalisation* (Sydney: Minerals Policy Institute): 1-18.

Llewellyn, S. (2004) 'Meeting Responsibilities "on the Stage" and Claiming Rights "behind the Scenes": The Re-casting of Companies', paper presented at *Fourth Asia Pacific Interdisciplinary Research in Accounting Conference*, Singapore, 4–6 July 2004; www.accountancy.smu.edu.sg/Apira/Final%20 Papers/1192.Llewllyn.pdf.

Manning, P. (2003) 'Poor Fellow Mining Country', *Sydney Morning Herald*, 18 April 2003.

Mantziaris, C., and D. Martin (2000) *Native Title Corporation: A Legal and Anthropological Analysis* (Leichhardt, NSW: Federation Press).

McMahon, G., and J. Strongman (1999) 'The Rise of the Community', *Mining Environmental Management* 7.1.

Mitchell, R., B. Agle and D. Wood (1997) 'Toward a Theory of Stakeholder Identification and Salience: Defining the Principle of Who and What Really Counts', *Academy of Management Review* 22.4: 853-86.

Moon, J. (1995) 'The Firm as Citizen? Social Responsibility of Business in Australia', *Australian Journal of Political Science* 30: 1-17.

——, A. Crane and D. Matten (2003) *Can Corporations be Citizens? Corporate Citizenship as a Metaphor for Business Participation in Society* (Nottingham, UK: International Centre for Corporate Social Responsibility, Nottingham University).

Murphy, D., and J. Bendell (1999) *Partners in Time? Business, NGOs and Sustainable Development* (Geneva: United Nations Research Institute for Social Development).

National Native Title Tribunal (2002) *10 Years of Native Title: Information Kit* (Perth, WA: National Native Title Tribunal).
O'Neill, P., and J.K. Gibson-Graham (1999) 'Enterprise Discourse and Executive Talk: Stories that Destabilize the Company', *Transactions of the Institute of British Geographers* 24.1: 11-22.
Orlitzky, M., and D. Swanson (2002) 'Value Attunement: Toward a Theory of Socially Responsible Executive Decision-making', *Australian Journal of Management* 27 (Special Issue): 119-28.
Parker, C. (2002) *The Open Corporation* (Cambridge, UK: Cambridge University Press).
Pateman, C. (1970) *Participation and Democratic Theory* (Cambridge, UK: Cambridge University Press).
Pattie, C., P. Seyd and P. Whiteley (2004) *Citizenship in Britain: Values, Participation and Democracy* (Cambridge, UK: Cambridge University Press).
Phillips, R. (2001) 'Confrontation to Partnership: Corporations and NGOs', in J. Goodman (ed.), *Protest and Globalisation* (Sydney: Pluto Press).
Rodgers, C. (2000) 'Making it Legitimate: New Ways of Generating Corporate Legitimacy in a Globalising World', in J. Bendell (ed.), *Terms for Endearment: Business, NGOs and Sustainable Development* (Sheffield: Greenleaf Publishing): 40-48.
Roßteutscher, S. (2000) 'Associative Democracy: Fashionable Slogan or Constructive Innovation?', in M. Saward (ed.), *Democratic Innovation* (London: Routledge): 172-83.
Sanders, W. (2002) 'Towards an Indigenous Order of Australian Government: Rethinking Self-Determination as Indigenous Affairs Policy' (Canberra: Centre for Aboriginal Economic Policy Research).
Saward, M. (2000) 'Less than Meets the Eye: Democratic Legitimacy and Deliberative Theory', in M. Saward (ed.), *Democratic Innovation* (London: Routledge):66-77.
Shapiro, I. (1999) *Democratic Justice* (New Haven, CT: Yale University Press).
Stevens, M. (1991) 'Showdown at Marandoo', *Business Review Weekly*, 6 December 1991.
Thompson, I. (1998) 'Inducing Change: Can Ethics be Taught?', *Legislative Studies* 13.1: 40-61.
Trebeck, K. (2005) 'Democratisation through Corporate Social Responsibility? The Case of Miners and Indigenous Australians' (PhD thesis, Australian National University, Canberra).
—— (2007) 'Tools for the Disempowered? Indigenous Leverage Over Mining Companies', *Australian Journal of Political Science* 42.4: 541-62.
Trigger, D. (1997) 'Reflections on Century Mine: Preliminary Thoughts on the Politics of Indigenous Responses', in D. Smith and J. Finlayson (eds.), *Fighting over Country: Anthropological Perspectives* (Canberra: Centre for Aboriginal Economic Policy Research): 110-28.
—— (1998) 'Citizenship and Indigenous Responses to Mining in the Gulf Country', in N. Peterson and W. Sanders (eds.), *Citizenship and Indigenous Australians* (Cambridge, UK: Cambridge University Press): 154-68.
UNEP (United Nations Environment Programme) (2002) *Finance, Mining and Sustainability: Exploring Sound Investment Decision Processes (2001–2002)* (Paris: UNEP; Washington, DC: World Bank Group; London: Mining Minerals and Sustainable Development; www.mineralresourcesforum.org/workshops/finance_wp/docs_slides/financereport.pdf, accessed 20 August 2008).
Walker, P. (2002) *We, The People: Developing a New Democracy* (London: New Economics Foundation).
Warburton, J., M. Shapiro, A. Buckley and Y. van Gellecum (2004) 'A Nice Thing to do but is it Critical for Business?', *Australian Journal of Social Issues* 39.2: 117-27.
White, A. (1999) 'Sustainability and the Accountable Corporation: Society's Rising Expectations of Business', *Environment* 41.8: 30-43.
Whitehead, L. (2002) *Democratization Theory and Experience* (Oxford, UK: Oxford University Press).
Wilmot, S. (2001) 'Corporate Moral Responsibility: What Can we Infer from our Understanding of Organisations?', *Journal of Business Ethics* 30.2: 161-70.
Wilmshurst, T. (2004) 'Stakeholder Theory: One Grand Theory or Multiple Perspectives?', paper presented to *Fourth Asia Pacific Interdisciplinary Conference on Research in Accounting*, Singapore, 4–6 July 2004.
Wilson, R. (1999) 'Corporate Citizenship in a Multinational Business', paper presented to *Global Values, Global Difference Conference*, London, 1 July 1999 (London: Centre for Tomorrow's Company).
Young, I. (2000) *Inclusion and Democracy* (New York: Oxford University Press).
Zandvliet, L. (2004) *Redefining Corporate Social Risk Mitigation Strategies* (Social Development Notes 16; Washington, DC: World Bank).

2

The impact of resource development on social ties

Theory and methods for assessment

Sharman Haley
Institute of Social and Economic Research, University of Alaska Anchorage, USA

James Magdanz
Alaska Department of Fish and Game, Kotzebue Alaska, USA

The concept of corporate social responsibility (CSR) matured in the context of debates over sustainable development, embedding the principle that current economic activities must not compromise environmental and social resources for future generations. While there are various contending theories of CSR, most of them recognise corporate responsibilities to the communities in which they operate, responsibilities to align their activities to some degree with the long-term interests of these communities.

Usually, providing local jobs with decent wages and working conditions is counted as a benefit to be weighed against whatever negative environmental or social impacts might accompany them. But, when the communities in question are indigenous, with predominantly non-market social and cultural systems, more complex trade-offs arise about the impact of jobs and wages on community well-being. Jobs and wages may have negative as well as positive social effects. In this chapter we address the question: How does resource development and the associated increase in the cash economy affect social relationships and well-being in indigenous communities rooted in traditional relations of subsistence production?

The chapter reviews empirical findings regarding social impacts of development in Arctic Alaska; develops a theoretical framework for understanding types of social ties, changes in social ties and implications of changes in social ties; explores methodological strategies for measuring social ties; and concludes with a research design for measuring changes in social ties that may result from resource development in Arctic Alaska. It uses Woolcock's (1998) definitions of social capital at the micro level—intra-community ties and extra-community networks—and explores Granovetter's and Burt's ideas that 'weak ties' (i.e. acquaintances) that bridge network clusters may be more beneficial than 'strong ties' (i.e. friends and relatives) in a market economy context.

Well-being in Inupiat communities

The Survey of Living Conditions in the Arctic (SLiCA) (Poppel *et al.* 2007) collected data on well-being in a sample of 700 Inupiat and Yupiit households in the Bering Straits, Northwest and North Slope regions of Alaska. More than three-quarters of the residents are Inupiat (all three regions) or Yupiit (Bering Straits), living in three regional centres and 32 small villages. All the communities are remote, with no road access. The economy is a mix of wage employment and traditional subsistence hunting and gathering activities. The largest employer by far is government. Employment on the North Slope is fuelled by tax revenues from the oil and gas industry. The Northwest Arctic (NWA) is home to the Red Dog Mine, the world's largest zinc and lead mine and the second largest employer in the region. The Bering Straits region has no significant resource development or basic industry. All three regions are characterised by high unemployment and low household incomes.

Survey results show that, despite widespread (income) poverty, 90% of respondents are satisfied with their life as a whole. The primary factors predicting life satisfaction are:

- Family ties
- Social support networks
- Income and employment
- Subsistence activities
- Local control of resources

The biggest problem, cited by 83% of respondents, is unemployment; 42% have considered moving to another community, and the most frequently cited motive is better job opportunities. Yet 77% of households prefer to combine jobs with subsistence. Subsistence and social relationships are the most important reasons people choose to remain in small communities, despite the lower (cash-based) standard of living there (Poppel *et al.* 2007).

Alaskan Inupiat and Yupiit households tend to be closely connected to other households socially and spatially, thereby forming extended families (Magdanz and Utermohle 2002; Sumida 1988; Usher 1992). As Craver notes (2001: 19) 'the socio-economic functioning of Inupiat households is seldom accomplished by a single household'. Research throughout the Arctic has shown that participation in social networks and sharing is associated with increased subsistence activity (Berman 1998; Craver 2001; Magdanz and Utermohle 2002; Usher 1992). Sharing of subsistence harvests is integral to traditional Inupiat values and social relationships. Subsistence activities are the medium for the reproduction of core Inupiat values and identity.

Ritchie and Gill (2004) analyse subsistence as a form of social capital. They have assembled a series of quotations from subsistence communities in Prince William Sound in south central Alaska to illustrate the social and cultural functions of subsistence activities. The interviews were conducted some years after the *Exxon Valdez* oil spill polluted 44,000 km^2 of the Gulf of Alaska waters and 1,900 km of shoreline. The impact on marine life and subsistence resources was devastating. These local voices explain that subsistence is about kinship and sharing, reciprocity, security, identity, reproduction of cultural values and skills, heritage, spirituality and social cohesion, as well as a unique lifestyle with its own rhythms and relationships with the natural world.

> Subsistence is part of rural economy, but it has little or no relation to western views of economic value. Subsistence is about eating, but wild foods can't simple be replaced by a processed substitute. Subsistence is about kinship and social cohesion, but it is not a ritual or ceremony (Piper 1993: 107; see also Jorgensen 1990).

> I hunted 19 days one year . . . I shot 21 deer and 20 of them were bucks . . . I hang [the meat], cut it, then deliver it to people all around town. I give it away. I know the limit's only 5 or whatever but . . . out of those 21 deer, I probably gave 11 of them away and only used what I needed (Ritchie and Gill 2004: 169-70).

> It is during the cycles of subsistence that bonding is strengthened and expanded. The sense of worth is solidified and new skills are learned. It is during these bonding times that our individual value is placed within our community, and we are able to understand what we must do to preserve our lives and to live in harmony (Ott 1994: 47).

> We could always fall back on the land, whatever happened. Whatever they [non-Natives] did . . . to us, you always have the land or sea . . . This is what you always have to fall back on. It is always there . . . You take care of it; it will take care of you . . . You identify with it. That is what you are (Ritchie and Gill 2004: 163).

> We have . . . a different attachment to [the environment]. As long as it is there . . . life is going to be okay . . . You have the Father Sky and there is the Mother Earth. It is like she got damaged. It is not just economically you are going to get deprived, but psychologically . . . you have lost something too (Ritchie and Gill 2004: 164).

> Social networks, associations, norms of reciprocity, social cohesion, and trust among family, extended kinships, friendships, and community are intrinsic in a subsistence lifestyle (Ritchie and Gill 2004: 23).

To summarise, subsistence serves a wide range of economic, social and cultural functions in Inupiat society, including: food and nutrition; economic production, consumption, cost of living and economic security; sharing, social ties and cultural identity; values and spiritual resilience; social capital in the form of reciprocity, trust, cooperation and leadership; and physical and mental health. Time on the land promotes observation-based knowledge, skills, experience and judgement; hunting provides a positive outlet and valued social role for young men; and self-reliance promotes a sense of efficacy and fate control.

Jobs and income are also important factors for well-being. Closer analysis, however, shows that the effects of employment and income on subjective well-being (SWB) are mixed. Consistent with the findings of Lane (2000), the benefits of increasing income are concentrated at the low end of the income distribution, with diminishing returns to well-being as income rises. SLiCA finds the threshold is around 60% of the median income; below that level, increasing income correlates with increasing SWB, but the correlation largely disappears above that amount (Poppel *et al.* 2007). And, while the raw correlation in SLiCA data between employment and SWB is positive, when Martin

FIGURE 2.1 Subsistence harvest-sharing network in an Inupiat village

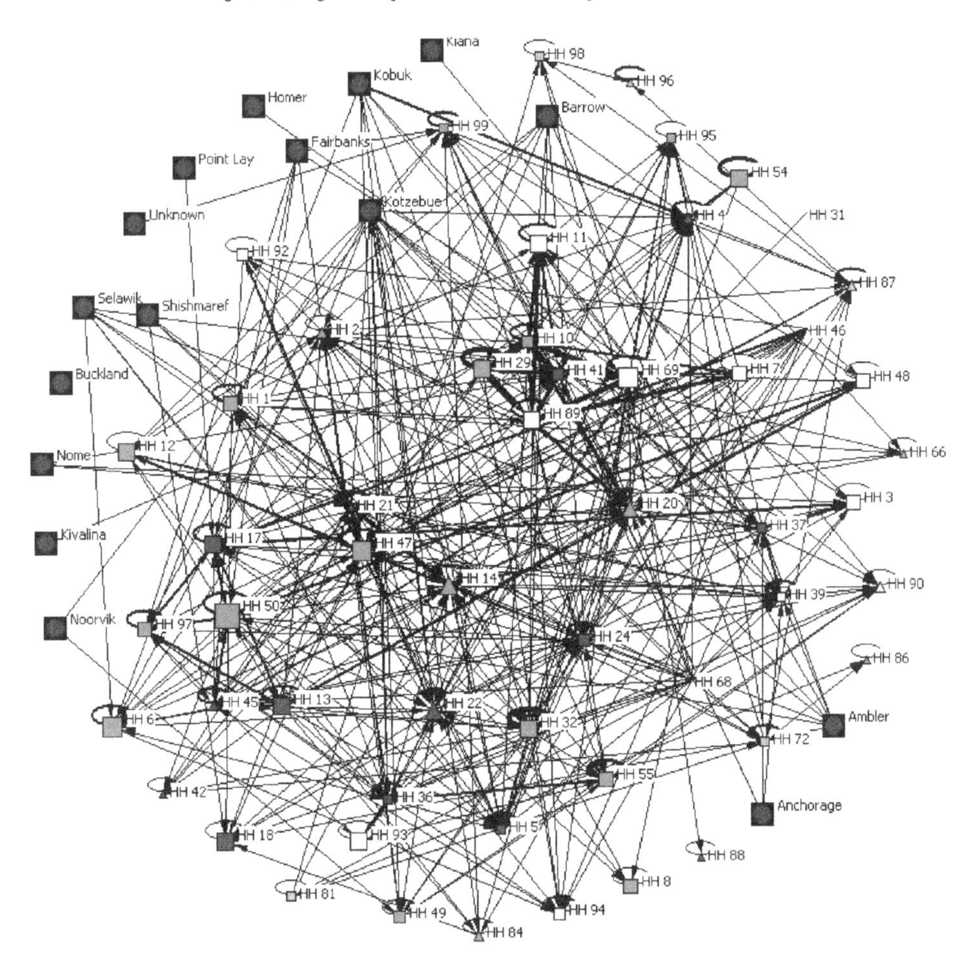

FIGURE 2.2 Cash-sharing network in an Inupiat village

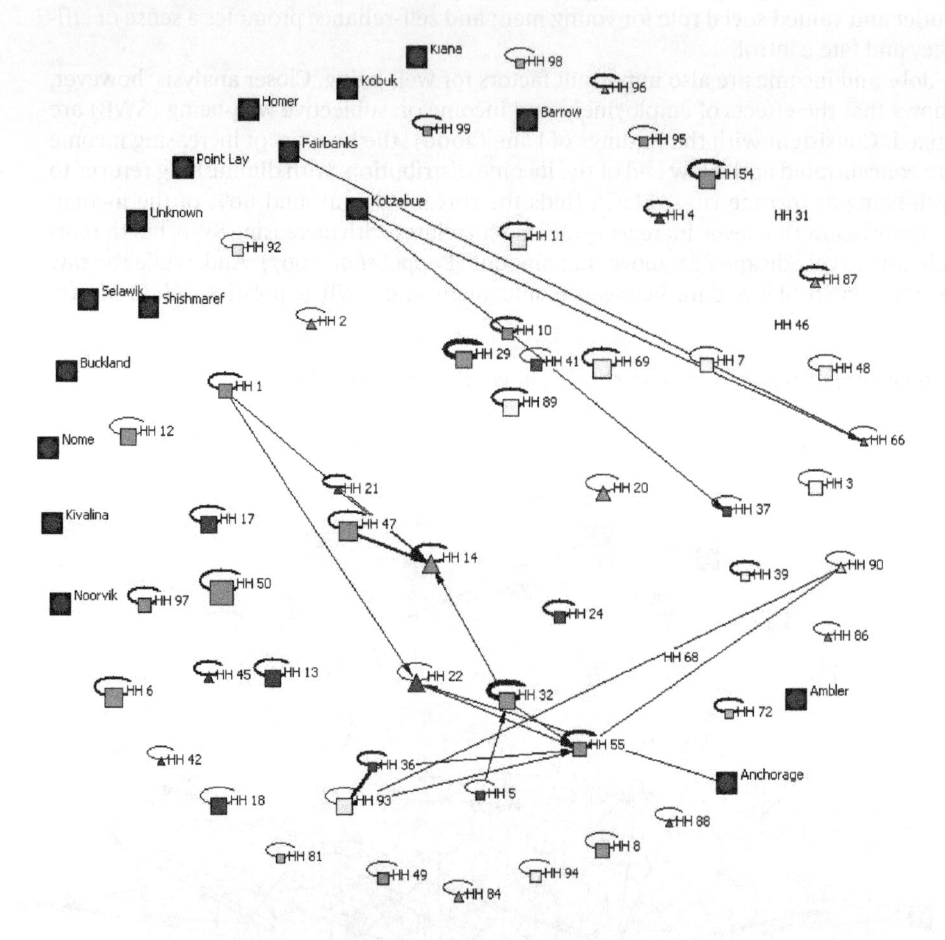

(2005) introduced more control variables (using a three-equation ordered probit model to estimate the relationship between overall life satisfaction and the probability of employment, along with a range of personal and community variables) the relationship she finds is negative. 'The negative relationship may be because jobs take time away from participating in family, social and community activities that are [more] important for satisfaction' (Martin 2005: 142).

This tension between the cash and subsistence economy may be further illustrated with subsistence-sharing network data. Figure 2.1 shows the subsistence harvest-sharing network between households in one Inupiat village. The survey data set reports 'Who caught, cut, & gave away the [subsistence resource] your household used?' for more than 60 different species of fish, shell fish, marine mammals, large land mammals, small mammals, birds and bird eggs, and wild edible plants. It is visually dense. The measured density—that is, the number of ties expressed as a percentage of the number of ordered pairs within the community—is also high.

Similar network maps and indices can be constructed for other types of ties covered in the survey instrument, including information and advice about hunting, sharing hunting equipment, providing household labour (housework, childcare, cutting firewood, building or vehicle maintenance and repair), social support, financial ties, and, as a possible new addition to the subsistence survey instrument, information and advice about job search.

Figure 2.2 shows the cash-sharing network in the same community. In marked contrast to the subsistence economy, relationships in the cash economy are disintegrated. Even in Inupiat communities with strong ties, cash is not widely shared between households.

This dichotomy between strong, dense ties in the subsistence economy and weak, disintegrated ties in the market economy raises the question: What are the social effects of increasing integration in the cash economy?

Economic development, social capital and well-being

The social capital literature to date has primarily focused on how networks, norms and trust shape economic outcomes (Borgatti and Foster 2003; Granovetter 2005; Woolcock 1998). Here, however, we focus on the converse: how changes in economic activities change social relationships and values. We place this in a larger normative context: economic development per se is not the goal of human endeavour: the goal is human development and well-being. Economic activities and social relations are both means to this higher-order goal.

Human development and well-being are expanding arenas for academic and policy research. 'Human development' usually refers to aggregate indicators representing constituent components of human welfare, as in the United Nations Human Development Index, which combines measures of per capita income, life expectancy and education. The *Arctic Human Development Report* (AHDR 2004) expands on this to include fate control, cultural integrity and ties to nature, but stops short of defining or identifying indicators to measure these.

'Well-being' usually refers to individual-level measures, and the research generally focuses on identifying the underlying determinants of well-being. Subjective and objective measures of well-being are highly congruent (Helliwell 2005). Many studies have found that social relationships are a source of happiness (Myers and Diener 1995) and a factor predicting a range of life outcomes, including finding employment (O'Regan and Quigley 1996), socioeconomic status (Smith 1998) and health status (Larsen 1993).

Helliwell (2005) assesses the state of the science linking social capital to well-being. He finds that subjective well-being (SWB) rises with:

- Frequency of contact with family, friends and neighbours
- Neighbourhood norms of trust and trust in the public realm
- Mastery or command of personal or social circumstances, which is correlated with and amplified by education

- Job satisfaction and workplace trust, and
- Good government (effectiveness, efficiency, lack of corruption; voice, accountability and political stability), especially at the community level

So the linkage between social capital and human well-being is well established. But social capital is an amorphous concept. Woolcock (1998) reviewed the several classical and contemporary research traditions on social capital and proposed his own definitional synthesis. First, social capital is defined by structure (for example, ties), not effects (for example, trust). He then identifies and names four types of social capital. At the group or community (micro) level, 'integration' refers to the strength and density of internal relationships, while 'linkage' refers to the network of contacts outside the group. At the national or governmental level, 'integrity' means the competence and effectiveness of governmental institutions, while 'synergy' refers to the breadth and quality of relationships between the government and civil society. He argues that all four dimensions working in concert are key to economic development. Furthermore, social capital is dynamic, and the component forms and functions change as development progresses.

Granovetter (1973, 1983) distinguishes between 'weak ties' (i.e. acquaintances) and 'strong ties' (i.e. friends and relatives). Strong ties exhibit greater time, intensity, intimacy and reciprocal services. The networks of strong ties are also more dense, forming clusters of closely connected people. Weak ties tend to be more diverse and diffused, and therefore more far-reaching. 'Weak-tie' acquaintances are typified by co-workers, old school chums, and members of civic organisations. They provide access to information and resources beyond that available in one's own social circle. Strong ties are especially important in poor communities where mutual helping relationships, exchanging all manner of goods and services, provide a measure of economic security (see also Lomnitz 1977; Stack 1974).

Ties that bridge different social groups have particular value. Because of the transitive nature of strong ties—my relatives are also related to each other, and my friends tend to be friends with each other—weak ties are much more likely to bridge social differences than strong ties. Diverse sources of information are the key advantage of weak ties that bridge. 'Individuals with few weak ties will be deprived of information from distant parts of the social system and will be confined to the provincial news and views of their close friends' (Granovetter 1983: 202).

Granovetter (1983) elaborates on work by Coser (1975) to argue that bridging weak ties are associated with complex role sets and cognitive flexibility. People concentrated in strong-tie relationships have less opportunity to explore their roles in relation to the complexities of the outside world. On the other side, negotiating a wide variety of different viewpoints and activities requires intellectual flexibility and self-direction. Indeed, it is prerequisite to the social construction of individualism.

Bridging weak ties also have macrosocial effects. Without weak ties to link them, a society composed of closely knit groups will be fragmented and incoherent. 'Weak social ties extend beyond intimate circles and establish the intergroup connections on which macrosocial integration rests' (Blau 1974: 623, quoted in Granovetter 1983: 220). These links enable social movements, the wide diffusion of new ideas, and effective mobilisation for collective action (Granovetter 1983). Links between diverse constituent groups and political leaders improves information flow both ways, improves decision-making and increases the likelihood of civic engagement (Granovetter 1973).

Putnam (2000) picks up on Granovetter's distinction and explains the difference between 'bonding social capital' and 'bridging social capital':

> Bonding social capital constitutes a kind of sociological superglue, whereas bridging social capital provides a sociological WD-40. Bonding social capital, by creating strong in-group loyalty, may also create strong out-group antagonism . . . nevertheless, under many circumstances both bridging and bonding social capital can have powerfully positive social effects (Putnam 2000: 23 quoted in Ritchie and Gill 2004).

Burt's (2001) discussion of bridging versus closure in the topography of network ties provides a particularly useful model that integrates both forms of social capital in one framework. The elements in Burt's model are essentially identical to Woolcock's (1998) typology, and so we will take the liberty of using Woolcock's terminology for them. Figure 2.3 illustrates the Burt–Woolcock synthesis. Internal cohesion enhances social capital because dense networks facilitate trust and shared norms by facilitating effective sanctions. External integration enhances social capital by increasing access to diverse information. Burt argues that groups with both forms of social capital have the highest performance:

> Performance is highest [in the fourth quadrant] where in-group closure is high (one clear leader, or a dense network connecting people in the group) and there are many non-redundant contacts beyond the group (member networks into the surrounding organisation are rich in disconnected perspectives, skills and resources). Performance is lowest [in the first quadrant] where in-group closure is low (members spend their time bickering with one another about

FIGURE 2.3 Burt–Woolcock synthesis model of social capital

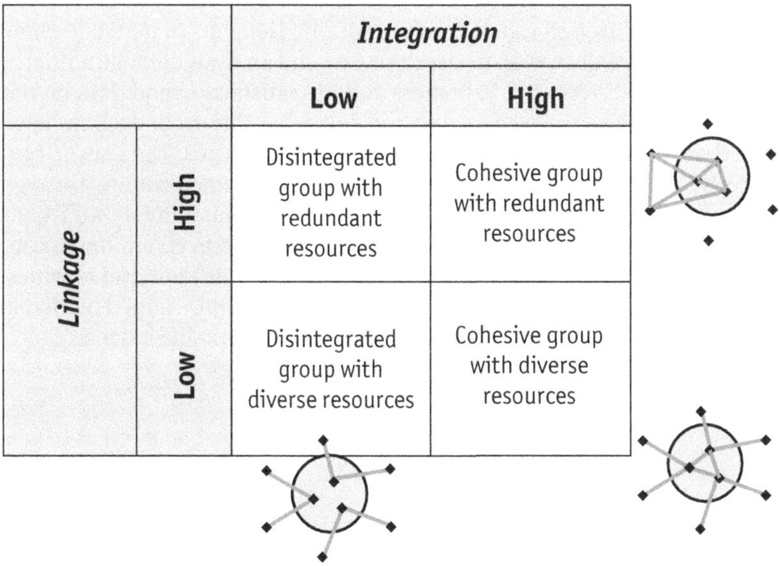

		Integration	
		Low	**High**
Linkage	**High**	Disintegrated group with redundant resources	Cohesive group with redundant resources
	Low	Disintegrated group with diverse resources	Cohesive group with diverse resources

> what to do and how to proceed) and there are few non-redundant contacts beyond the group (members are limited to similar perspectives, skills and resources) (Burt 2001: 41).

This thesis is empirically supported not only in the management literature discussed by Burt, but also in evaluation research in rural Alaska on the effectiveness of community initiatives to improve the operation of water and sanitation facilities:

> Overall, we want to emphasise that strong community leadership, a broad base of support for improving utility management [cohesion], and cooperation among local officials and outside agencies [integration] are crucial to improving operation and maintenance of water and sewer systems in rural Alaska (Haley 2000: E-13).

These theories and empirical studies have been elaborated mostly in the context of urban communities in market economies. Social capital researchers typically measure levels of trust, organisational membership and societal participation to explain outcomes such as social cohesion, social solidarity and economic achievement (Ritchie and Gill 2004). The thesis about strong and weak ties has been explored mostly in relation to employment search, focusing on information and referral and expanded choice sets. The structural holes hypothesis has been tested in terms of individual and team performance in corporate environments. Woolcock (1998) reviews how social capital theory has been applied in studies of economic development, education, organisations, democracy and governance. But what concerns us here is the converse: how economic development restructures social relations.

The transition from traditional relations in a subsistence hunting and gathering society to market relations in a cash economy has significant implications for social capital and well-being. The traditional economy is based on household production and kinship-based sharing networks that create strong social ties and cultivate trust. The market economy is based on individual, wage labour and impersonal and often transient exchange relationships.

Lane (2000) and Layard (2005) analyse specific elements of market economies that erode SWB. Lane conducted an in-depth review and analysis of longitudinal and cross-national data on self-reported happiness and life satisfaction and data on depression. The striking finding he seeks to explain is a pervasive pattern of declines in well-being associated with increases in affluence in developed countries. The reason, he argues, is the corrosive effect of the market economy on social connectedness, through smaller and more fragmented families and greater geographic mobility as well as increasing materialism and consumerism. Layard finds that short-term commitments and linking monetary rewards to individual performance both corrode trust and loyalties and create unhappiness. Helliwell (2005) notes that SWB by age is U-shaped, with one-third of the mid-life decline in SWB explained by the stress of work–life balance.

Social ties and market integration: empirical evidence from SLiCA

The strong ties so abundantly manifest in the subsistence economy are the most important source of well-being reported in Inupiat communities (Kruse *et al.* 2007). But jobs and income are also important, and what the market economy does incomparably well is to improve material standards of living. Granovetter (1973, 1983) and Burt (2001) argue that bridging ties that link individuals or groups to diverse network resources may be more important for success in the market economy. Bridging ties have been shown to be more effective for diffusion of information, including activities such as job search, innovation or adaptation.

So what are the social effects of increasing integration in the cash economy? In terms of the Burt–Woolcock model we would expect community integration to decrease and linkage to increase. This and the foregoing discussion raise three sets of hypotheses:

1. Increases in market activities are associated with decreases in strong, dense ties and increases in weak, bridging ties
2. Decreases in strong, dense ties and increases in weak, bridging ties will:
   - Decrease trust, reciprocity and common values for the community
   - Decrease social support and well-being for the individual
   - Increase diversity of information resources and choices for the individual
3. Increases in income will, on average, yield no net increase in well-being for individuals:
   - Increases in well-being due to improved material standards of living, expanded opportunities and choice sets, and improved health will be offset by declines in family ties and social support

The first step in exploring these hypotheses is to mine existing data. The SLiCA includes information on approximately 3,000 individuals from 663 randomly selected households in the northern regions of Alaska. The survey collects both subjective assessments and objective measures of well-being, including income, education, housing, social relationships, mental and physical health and cultural practices.[1] Some of this analysis, as discussed above, has already been done by Martin (2005) and Kruse *et al.* (2007). The factors the authors found that explain overall life satisfaction include family ties, social support, personal income (at levels below 60% of median income only), working at least part of the year, subsistence activities, satisfaction with the amount of fish and game available locally, and satisfaction with the number of job opportunities in the community.

Here we are interested in how social capital varies with degree of market integration. For this we use three proxies for market integration: whether the survey respondent is employed full-time; the household income for the respondent; and whether the respondent lives in a village or a regional centre.

1 www.iser.uaa.alaska.edu/projects/Living_Conditions, accessed 15 August 2008.

In Arctic Alaska more than two-thirds of indigenous people live in villages with a population of less than 1,000. The villages have lower levels of market economic activity than the regional centres. The regional centres are hubs for regional commerce, offering more jobs, employment, income and a wider range of goods and services. They are also more socially and ethnically diverse.

The villages are more culturally homogeneous and more traditional than the regional centres. Some 87% of villagers report that both parents are Inupiat and only 13% have a non-Inupiat parent, while in the regional centres 26% report one parent is non-Inupiat. In the villages more than 30% of households include three or more generations, while in the regional centres households of three or more generations make up only 17% of households. Overall, 40% have lived in their community all their lives, and 60% have lived outside the community for a year or more. In the regional centres, 70% have lived outside the community for a year or more sometime in their life, while in the villages this drops to 55%. Interestingly, the Northwest Arctic (NWA) region has a much higher percentage than the other two regions: over 80% for the regional hub Kotzebue and 70% for the NWA villages. Kotzebue also leads the pack for the percentage of respondents (57%) who have considered moving away in the past five years, while NWA villagers are much more content to stay put, with only 34% having considered moving.

Overall, 61% learned Inupiat as a child, but, while 65% of villagers spoke with their parents at home in Inupiat, only 47% of respondents in the regional centre were spoken to in Inupiat by their parents when they were young. While 50% of the adults in the NWA villages speak Inupiat some or most of the time, only 26% of adults in Kotzebue do so. Over 90% of adults read, write, speak and understand English well; the percentage is higher in regional centres than in villages.

Respondents living in the regional centres report higher levels of formal education and employment. Seventy-one per cent of Inupiat adults have completed high school, with 44% completing some or all of it outside the community; 35% of adults in regional centres and 21% in villages have some college or vocational education beyond high school.

Among Inupiat adults 77% prefer a mix of wage employment and subsistence activities for their livelihood; 15% prefer paid work as their primary occupation, while 8% prefer a traditional subsistence lifestyle. Unsurprisingly, those who prefer wage employment are more likely to live in the regional centre, and are also more likely to be young: 26% of respondents aged 16–24 prefer working on a wage job. Eighty-seven per cent report being somewhat or very satisfied with the combination of activities they have.

Regarding traditional values, 78% of respondents are satisfied or very satisfied with their community's promotion of traditional values, with a smaller percentage in Kotzebue and Nome than in the villages; 92% of Alaska respondents apply traditional values in their personal life, and this does not vary between the regional centres and the villages (Poppel *et al.* 2007: tables 157 and 158) More than three-quarters of Inupiat adults are educated in traditional skills including hunting and fishing, preserving meat and fish, knowing when berries are ripe and where to find them, cooking and preparing traditional native foods, and over-nighting on the land. Knowledge of traditional skills is across the board more prevalent in villages than in regional centres, and skills are more likely to be used. Some 88% of respondents said children today are learning these skills, and there were no significant differences between villages and regional centres in perceptions of learning traditional skills.

People in Kotzebue follow politics more than villagers, with 51% compared to 42% considering themselves somewhat or very knowledgeable in this area. Yet the higher the proportion of non-natives living in the community, the less satisfied they feel with their control over environmental problems in their area. Perceived social problems, such as alcohol and drug abuse, sexual abuse and family violence, and suicide, are also consistently higher in regional centres than in villages.

We extend this description of survey findings with bivariate analysis of social characteristics by extent of market activity. Table 2.1 shows three sets of comparisons: villages versus regional centres; high-income versus low-income households; and whether or not the respondent is employed full-time. As we would expect, people in regional centres, those with higher household incomes, and those who are employed full-time are highly correlated and are also more likely to have high levels of educational attainment. Inupiats in communities of all sizes and income levels report strong family ties.

Yet the three domains reveal important differences. For example, high-income households enjoy more social support than low-income households. The difference in social support is weakly significant by community size, and disappears altogether by employment status.

For this analysis we constructed an indicator of strong ties, defined as frequency of contact with family not living in the household plus self-rated strength of family ties, and an index of bridging ties, defined as having a non-native parent or spouse, having parents born in another place, having lived for a year or more in another place (greatest weight), having been away for a month or more for work or education, and memberships of regional or state-level boards and organisations. Both measures range from 0 to 10. We find no difference in strong ties by size of community, by household income, or by employment status.

We do find significant differences in bridging ties exactly as expected. Bridging ties are higher in regional centres, higher in high-income households, and higher for those who are employed full-time. We checked the relationship between bridging ties and community size in a multivariate linear regression controlling for age, gender and full-time employment. Only age was significant, and, while the relationship to community size weakened slightly, it remained significant.

We constructed an index of variation in opinion, defined as the sum of the squared differences in the respondent's opinion compared to the mean opinion overall across 17 opinion questions in the survey. As expected, we find greater variation in opinion in regional centres and less variance in villages. There is no significant variance by income, but people employed full-time are somewhat more likely to hold deviant opinions than those with more tenuous ties to the labour market. We checked the relationship between opinion diversity and community size in a multivariate linear regression controlling for age and bridging ties: age was positively correlated with deviant opinions and bridging ties were insignificant. While the relationship with community size weakened slightly, it remained significant.

Lastly, we analysed the sources of Inupiat identity. The six most important are, ranked by their importance to village Inupiat: (1) eating traditional foods, (2) hunting, fishing, gathering and preserving traditional foods, (3) childhood upbringing, (4) naming kinship relationships, (5) occupation or profession, and (6) personal contacts with other Inupiat. All of these markers of identity are more important to village Inupiat than to their counterparts, and the comparisons by income and employment status follow suit.

TABLE 2.1 Mean social characteristics by degree of market integration

Variable	Village	Regional centre	Pr. [a]	Sig.	Low HH income	High HH income	Pr. [a]	Sig.	PT or not employed	FT employed	Pr. [a]	Sig.
Household income	$49,460	$72,089	0.000	**	$28,562	$88,693	na		$45,295	$69,152	0.000	**
Education: % with post-secondary education[a]	33.86%	19.33%	0.016	**	17.92%	30.98%	0.000	**	10.58%	39.04%	0.000	**
Percent of HH adults employed full-time	42.3%	59.8%	0.000	**	37.8%	59.6%	0.000	**	25.6%	72.5%	0.000	**
Family ties index	3.97	4.03	0.364		4.01	3.95	0.356		3.98	3.99	0.972	
Social support index	4.04	4.18	0.075	*	4.00	4.18	0.016	**	4.04	4.13	0.283	
Strong ties index	7.93	8.06	0.364		8.03	7.91	0.356		7.97	7.97	0.972	
Bridging ties index	2.66	3.35	0.001	**	2.54	3.28	0.000	**	2.51	3.29	0.000	**
Variation in opinion	19.33	22.80	0.031	**	20.45	20.79	0.821		19.19	21.88	0.068	*
Participation in civic activities	3.03	2.51	0.053	*	2.51	3.27	0.002	**	2.29	3.51	0.000	**
Sources of Inupiat identity *(5= very important, 1= not at all important)*												
Inupiat food I eat	3.97	3.65	0.000	**	3.99	3.73	0.005	**	3.96	3.75	0.016	**
Hunting, fishing, gathering and preserving traditional foods	3.88	3.54	0.001	**	3.88	3.65	0.021	**	3.87	3.65	0.027	**
Childhood upbringing	3.81	3.56	0.024	**	3.80	3.65	0.152		3.79	3.65	0.195	**
Naming kinship relationships	3.68	3.58	0.383		3.76	3.51	0.033	**	3.71	3.56	0.205	
Occupation or profession	3.67	3.44	0.080	*	3.60	3.59	0.963		3.62	3.55	0.616	
Personal contacts with Inupiat	3.64	3.60	0.746		3.78	3.45	0.002	**	3.69	3.53	0.137	

[a] All are t-tests, except 'Education' which is an ordered nominal variable analysed with a chi-square test. HH = household; PT = part time; FT = full time

* Weakly significant at the <10% level ** Statistically significant at the <5% level Pr. = probability that there is no difference between the two categories Sig. = Statistical significance

Childhood upbringing is a more important source of identity to village Inupiat and to those without full-time employment, but the difference is not significant by income. Naming kinship relations is more important in low-income households and less important in high-income households. Occupation or profession is less important to hub dwellers (regional centres), presumably because comparatively less of their identity is tied to subsistence activities. Low-income households put the most weight on personal contacts with other Inupiat, while high-income households put the least weight on that marker.

Our regression confirmed earlier findings that strong ties and health are important to well-being. We found no significance to income or employment variables, community size or our bridging ties variable. Other research, however, adds information. As part of her study of nutrition security among the elderly, J. Smith (2007) found that, despite the fact that elders living in urban Alaska have better indicators of health status due to better medical care, their functioning in daily activities and their social integration are better when they live in their home villages, presumably because they have more social support and their participation in community life is more highly valued.

While analysis of the SLiCA data is suggestive, it is limited in several respects. Information on social networks, market activities and other variables we would like to analyse to better understand the effects of economic development on Inupiat society is missing; the dataset does not include Inupiats living in urban areas; and the sample size is too small to isolate the effects of different types of job or development activity. To empirically evaluate these hypotheses about the effects of the expanding market economy on social capital and well-being we must collect new primary data.

Research design for measuring changes in social ties

Our research model entails two levels of analysis: individual and community. The decision agents are individuals, yet the phenomena of interest are the collective patterns that result. There is an externality here. Individual choices take existing social relations as a given; yet the aggregate of individual choices may transform social connectedness and social capital, and therefore well-being. When Martin (2005) found that family ties, social support and living in a long-inhabited whaling community were the most important determinants of well-being, these were exogenous community attributes, not individual choice variables in her model. The market economy in particular promotes individual choices to pursue opportunities for personal gain.

Our (as yet hypothetical) design is a random sample survey of Inupiat shareholders in a regional native corporation, organised under the Alaska Native Claims Settlement Act of 1971. This will provide a coherent set of individuals with strong cultural and family relations living in diverse economic settings. About one-third of the shareholders live in villages, 20% in the regional hub, and 45% live outside the region. This will allow us to analyse in cross-section the patterns of social ties among shareholders by residence and degree of market integration. We can also examine how their social ties vary with employment status (locally employed, remotely employed or unemployed). Because a sample survey provides ego network data and not complete network data, we are limited to ego- or micro-level analysis and the incomplete inferences we can make to macro- or community-level patterns.

We propose to measure strong and weak social ties and bridging ties with survey questions on contacts for: food sharing, social support, employment information, resource and referral, and political process or 'fate control'. Our proposed survey instrument is based on the subsistence harvest and sharing surveys conducted by fish and game managers in rural Alaska. We will incorporate questions from SLiCA and instruments used in rural Alaska so we can benchmark and leverage the pooled data to extend the analysis in geography and in time.

How will we measure the strength of ties? Granovetter (1995) and Langlois (1977) both define weak ties as infrequent recent contact. Ericksen and Yancey (1980) define strong ties as a relative or friend of the respondent, while people identified as acquaintances are classed as weak ties. Lin *et al.* (1981) add neighbours as strong ties and friends of friends as weak ties. Murray *et al.* (1981) studied academic job search and classified as strong ties mentors and dissertation advisers who know the respondents and their work well. Friedkin (1980) also studied academics and defined a strong tie as one where professional consultation went both ways, while in a weak tie the consultation went one way. Weimann (1980) measured the strength of ties between members of a kibbutz by the tenure and importance of the relationship and frequency of contact. Wellman (1979) asked his subjects to rank order the intimacy of their various relationships. Lin *et al.* (1978) utilised two measures, frequency of contact and the type of relationship named, and got similar results with both. We propose to collect and analyse both measures.

Bridging weak ties are those that connect to social circles different from one's own. 'Difference' is not well defined. Lin *et al.* (1981) found that, for individuals of low socioeconomic status, only the links to persons of higher socioeconomic status have value for occupational mobility. Steinberg (1980) measured differences in organisational memberships. She found that the group with strong ties

> was linked to the fewest organizations and individual memberships were concentrated in the same organizations which formed a dense network . . . Groups formed on the basis of weak ties, on the other hand, were linked to more organizations that were loosely knit and individual memberships tended to be scattered throughout these organizations (Steinberg 1980: 19, quoted in Granovetter 1983: 225).

The index that Burt suggests for measuring bridging ties at the individual level is network betweenness (Freeman 1977), an index of the extent to which a person brokers indirect connections between all other people in a network. We cannot calculate this from a sample survey because it requires complete information on the whole network. Burt's purpose for this index is to measure opportunities for information entrepreneurship and individual success. In our application we seek an aggregate measure of linkage between the community—our unit of analysis for understanding the persistence of cultural systems—and diverse external resources. Network constraint (Burt 1992) is an index that measures the extent to which the household's ties are redundant, each to the others. We could use this with subsistence harvest survey data where the entire community is surveyed—the mean index for the whole community would serve as a measure of integration—but not for a sample survey. For our analysis we propose using the proportion of households in each type of community that have ties to households outside their community. In doing so we assume that these external ties are non-redundant. Even if different households have ties to the same external community, it is likely to be to a different household in that community.

This proposed methodology provides social network data on people of a common indigenous cultural heritage, representing the full spectrum of degrees of market integration. It targets data on strong and weak ties, bridging ties, information exchange, employment status, socioeconomic status, food security, social support, civic participation and subjective well-being to analyse the effects of development and increasing integration in the cash economy on the social capital and well-being of indigenous people who practise and value traditional social relations of subsistence production.

Conclusion

How do resource development and the associated increase in the cash economy affect social relationships and well-being in indigenous communities? Theory suggests that increases in market activities are associated with decreases in strong, dense ties and increases in weak, bridging ties, which increase material standards of living, opportunities and choice sets for the individual, yet decrease social support and well-being. Our review of existing empirical data shows that living in a regional centre, full-time employment and higher household income are each associated with higher income and education, more bridging ties, more participation in civic activities and differences in the sources of cultural identity, yet showed no significant difference in family ties or strong ties and mixed relationship with social support and diversity of opinion. New primary data on social networks and well-being for indigenous people of common cultural heritage representing the full spectrum of degrees of market integration will be required to tease out these hypothesised relationships.

References

AHDR (*Arctic Human Development Report*) (2004) (Akureyri, Iceland: Stefansson Arctic Institute).

Berman, M.D. (1998) 'Sustainability and Subsistence in Arctic Communities', paper presented at the *37th Annual Meeting of the Western Regional Science Association*, Monterey, CA, 18–22 February 1998.

Blau, P. (1974) 'Parameters of Social Structure', *American Sociological Review* 39.5: 615-35.

Borgatti, S.P., and P.C. Foster (2003) 'The Network Paradigm in Organizational Research: A Review and Typology', *Journal of Management* 29.6: 991-1,013.

Burt, R.S. (1992) *Structural Holes* (Cambridge, MA: Harvard University Press).

—— (2001) 'Structural Holes versus Network Closure as Social Capital', in N. Lin, K.S. Cook and R.S. Burt (eds.), *Social Capital: Theory and Research* (New York: Aldine de Gruyter): 31-56.

Coser, R. (1975) 'The Complexity of Roles as Seedbed of Individual Autonomy', in L. Coser (ed.), *The Idea of Social Structure: Essays in Honor of Robert Merton* (New York: Harcourt Brace Jovanovich).

Craver, A. (2001) *Complex Inupiaq Eskimo Households and Relationships in Two Northwest Alaska Rural Communities* (report prepared for the US Bureau of the Census by the Alaska Native Science Commission; Anchorage, AK: University of Alaska Anchorage).

Ericksen, E., and W. Yancey (1980) 'Class, Sector and Income Determination', unpublished manuscript, Temple University, Philadelphia, PA.

Freeman, L.C. (1977) 'A Set of Measures of Centrality Based on Betweenness', *Sociometry* 40: 35-40.

Friedkin, N. (1980) 'A Test of the Structural Features of Granovetter's "Strength of Weak Ties" Theory', *Social Networks* 2.4: 411-22.

Granovetter, M.S. (1973) 'The Strength of Weak Ties', *American Journal of Sociology* 78.6:1,360-80.

—— (1983) 'The Strength of Weak Ties: A Network Theory Revisited', *Sociological Theory* 1: 201-33.

—— (1995) *Getting a Job: A Study of Contacts and Careers* (Chicago: University of Chicago Press).

—— (2005) 'The Impact of Social Structure on Economic Outcomes', *Journal of Economic Perspectives* 19.1: 33-50.

Haley, S. (2000) *Evaluation of the Alaska Native Health Board's Sanitation Facility Operation and Maintenance Program: Final Report on Phase II Projects. Vol. I: Summary and Analysis* (Anchorage, AK: University of Alaska, Institute of Social and Economic Research).

Helliwell, J.F. (2005) 'Well-being, Social Capital and Public Policy: What's New?' (Working Paper 11,807; Cambridge, MA: National Bureau of Economic Research; www.nber.org/papers/w11807, accessed 20 August 2008).

Jorgensen, J.G. (1990) *Oil Age Eskimos* (Berkeley, CA: University of California Press).

Kruse, J.A., B. Poppel, L. Abryutina, G. Duhaime, S. Martin, M. Poppel, M. Kruse, E. Ward, P. Cochran and V. Hanna (2007) 'Survey of Living Conditions in the Arctic (SLiCA)', unpublished manuscript, February 2007.

Lane, R.E. (2000) *The Loss of Happiness in Market Democracies* (New Haven, CT: Yale University Press).

Langlois, S. (1977) 'Les reseaux personnels et la diffusion des informations sur les emplois', *Recherches Sociographiques* 2: 213-45.

Larsen, J.S. (1993) 'Measurement of Social Well-being', *Social Indicators Research* 28: 285-96.

Layard, R. (2005) *Happiness: Lessons from a New Science* (London: Penguin).

Lin, N., P. Dayton and P. Greenwald (1978) 'Analyzing the Instrumental Use of Relations in the Context of Social Structure', *Sociological Methods and Research* 7.2: 149-66.

——, W. Ensel and J. Vaughn (1981) 'Social Resources, Strength of Ties and Occupational Status Attainment', *American Sociological Review* 46.4: 393-405.

Lomnitz, L. (1977) *Networks and Marginality* (New York: Academic Press).

Magdanz, J., and C. Utermohle (2002) 'The Production and Distribution of Wild Food in Wales and Deering, Alaska' (Technical Paper Number 259; Alaska Department of Fish and Game).

Martin, S. (2005) 'Determinants of Well-being in Inupiat and Yupiit Eskimos: Do Communities Matter?' (PhD dissertation, University of Texas).

Murray, S., J. Rankin and D. Magill (1981) 'Strong Ties and Job Information', *Sociology of Work and Occupations* 8.1: 19-136.

Myers, D.G., and E. Diener (1995) 'Who is Happy?', *Psychological Science* 6: 10-19.

O'Regan, K.M., and J.M. Quigley (1996) 'Spatial Effects upon Employment Outcomes: The Case of New Jersey Teenagers', *New England Economic Review*, May/June 1996: 41-58.

Ott, R. (1994) 'Sound Truth: Exxon's Manipulation of Science and the Significance of the *Exxon Valdez* Oil Spill' (Anchorage, AK: Greenpeace).

Piper, E. (1993) 'The *Exxon Valdez* Oil Spill: Final Report, State of Alaska Response' (Anchorage, AK: Alaska Department of Environmental Conservation).

Poppel, B., J.A. Kruse, G. Duhaime, L. Abryutina (2007) *SLiCA Results* (Anchorage, AK: University of Alaska Anchorage, Institute of Social and Economic Research; www.arcticlivingconditions.org).

Putnam, R.D. (2000) *Bowling Alone: The Collapse and Revival of American Community* (New York: Simon & Schuster).

Ritchie, L.A. (2004) 'Voices of Cordova: Social Capital in the Wake of the *Exxon Valdez* Oil Spill' (PhD Dissertation; Department of Sociology, Anthropology, and Social Work; Mississippi State University).

—— and D.A. Gill (2004) 'Social Capital and Subsistence in the Wake of the *Exxon Valdez* Oil Spill', paper presented to the *Annual Meeting of the Rural Sociological Society*, Social Science Research Center, Mississippi State University, 2004.

Smith, J. (2007) 'Food Customs of Rural and Urban Inupiaq Elders and their Relationships to Select Nutrition Parameters, Food Insecurity, Health, and Physical and Mental Functioning' (doctoral dissertation, Florida International University/Institute for Circumpolar Health Studies, University of Alaska Anchorage).
Smith, J.P. (1998) 'Socio-economic Status and Health', *American Economic Review* 88.2: 192-96.
Stack, C. (1974) *All Our Kin* (New York: Harper & Row).
Steinberg, L. (1980) 'Preexisting Social Ties and Conflict Group Formation', paper presented to the 1980 meeting of the American Sociological Association, New York.
Sumida, V.A. (1988) 'Land and Resource Use Patterns in Stevens Village, Alaska' (Technical Paper 129; Fairbanks, AK: Alaska Department of Fish and Game, Division of Subsistence).
Usher, P. (1992) 'Modeling Subsistence Systems for Social Impact Assessment' (report prepared for the Grand Council of the Cree of Quebec; Ottawa: Grand Council of the Cree of Quebec).
Weimann, G. (1980) 'Conversation Networks as Communication Networks' (PhD dissertation abstract, University of Haifa).
Wellmann, B. (1979) 'The Community Question: The Intimate Networks of East Yorkers', *American Journal of Sociology* 84.5: 1,201-31.
Woolcock, M. (1998) 'Social Capital and Economic Development: Toward a Theoretical Synthesis and Policy Framework', *Theory and Society* 27.1: 151-208.

3

Realising solidarity

Indigenous peoples and NGOs in the contested terrains of mining and corporate accountability

Catherine Coumans
MiningWatch Canada

Mining exploration and exploitation is rapidly expanding in remote regions of the world where large-scale mining has not taken place traditionally.[1] As mining companies move into areas such as the highlands of Papua New Guinea or the Arctic regions of Canada they frequently encroach on lands and territories of indigenous peoples. These peoples find themselves on the front lines of both local struggles and global debates about economic, environmental and human rights impacts and benefits of mining. Non-governmental organisations (NGOs) that work in solidarity relationships with communities affected by mining increasingly find themselves working with indigenous peoples and their organisations. Some of these NGOs are based in countries that are exporting mining companies around the world and are not primarily staffed by indigenous peoples.

The participation of non-indigenous NGOs in the mining struggles of indigenous peoples takes place in a context in which local and international indigenous organisations have developed rights-based positions with respect to their lands, territories and natural resources and with respect to the right to determine the kind of development that takes place on their lands. The UN Declaration on the Rights of Indigenous Peoples,

1 The factors behind this rapid expansion include: sustained high demand and concomitant high prices for many mineral commodities; declining numbers of readily accessible lucrative deposits; advances in technology and communications; increased accessibility to equity and debt financing through public and private sources and international financial institutions; regulatory revisions in many countries that provide favourable fiscal and regulatory regimes, privileged access to land, and favourable ownership and profit repatriation provisions.

which was adopted in a vote by the UN General Assembly in September 2007, exemplifies indigenous organisation at an international level and provides a clear articulation of indigenous rights.[2] Where indigenous rights have gained recognition through local struggles or in national and international forums, the texts, agreements, rulings and legislation that reflect this recognition become a resource for non-indigenous NGOs in their solidarity work with indigenous communities.

Mining companies and their associations are actively developing their own positions, approaches, and resources for interactions with indigenous peoples.[3] Home governments of some of the world's largest mining companies, in particular those of Canada, the UK, the USA and Australia, are also increasingly focused on indigenous issues with respect to mining, as are the host governments of countries where mining expansion is rapidly progressing. NGO critics assert that home-government involvement in mining overseas through, among others, embassies, export credit agencies and aid agencies takes place in consultation with domestic mining companies, but is insufficiently transparent and accountable to the public that funds these activities or to the citizens of host countries.

Struggles related to mine expansion on indigenous land are both driving and informing corporate social responsibility (CSR) standard-setting for mining that is engaged in by NGOs,[4] financial institutions,[5] industry associations and mining companies[6] and some governments,[7] sometimes as multi-stakeholder efforts involving two or more of these sectors, usually in some form of consultation with indigenous peoples' representatives. Key normative anchors for these international CSR frameworks are the human rights principles of the UN and the risk-based policies of the World Bank Group.

This chapter explores some of the roles played by rights-based non-indigenous NGOs from mining exporting countries as agents that act in, and create linkages between, various terrains of struggle. These NGOs frequently develop long-term solidarity relationships with local indigenous communities and facilitate meetings and information-sharing between mining-affected indigenous communities. They engage mining companies

2 It is perhaps not a coincidence that three of the four countries that voted against adoption of the UN Declaration on the Rights of Indigenous Peoples, the United States, Canada and Australia, not only have substantial indigenous populations, but are also countries that are major players in resource extraction at home and abroad. On 8 April 2008 Canada's Federal Parliament passed a resolution calling on the Government of Canada to endorse the Declaration and fully implement the standards it contains.

3 In September 2006, the Government of Canada, Prospectors and Developers Association of Canada, Mining Association of Canada and the Canadian Aboriginal Mining Association launched the *Mining Information Kit for Aboriginal Communities* (Canada Government *et al.* 2006). This publication is now being used by the industry to promote mining among indigenous communities and is also being translated and promoted abroad through Canada's embassies.

4 See, e.g., Miranda *et al.* 2005; 'The Golden Rules', www.nodirtygold.org/goldenrules.cfm; developing certification systems such as the Initiative for Responsible Mining Assurance, www.responsiblemining.net; Association for Responsible Mining, www.communitymining.org (all accessed 15 August 2008).

5 See, e.g., the IFC Performance Standards, www.ifc.org/ifcext/sustainability.nsf/content/performanceStandards (accessed 6 October 2008) and the Equator Principles, www.equator-principles.com/index.shtml (accessed 6 October 2008).

6 See, e.g., ICMM 2003; Mining Association of Canada 2006; PDAC 2008.

7 See, e.g., the National Roundtables on CSR and the Canadian Extractive Industry in Developing Counties and its final report (Canada Government 2007); the Australian Government's Leading Practice Sustainable Development Program for the Mining Industry, www.ret.gov.au/sdmining (accessed 9 August 2008).

directly, sometimes together with indigenous partners. They lobby their governments, often together with indigenous partners, and expose their governments' roles in mining struggles involving indigenous peoples. These NGOs may help facilitate interventions by indigenous people in international forums, such as at the World Bank and the UN, where international standards that impact on mining-related struggles of indigenous peoples are being developed. This chapter also provides examples of the ways in which NGO actors that work in solidarity with indigenous communities may introduce or support indigenous positions in multi-stakeholder CSR standard-setting processes.

It is clear that some indigenous peoples and their organisations welcome collaboration with rights-based NGOs and are adept at providing direction for NGO activity at all levels, even as they continue to evolve their own local and international linkages and responses to the challenges, and opportunities, posed by global mining expansion. Close communications and joint actions with indigenous peoples have provided some NGOs with a chance to deepen their understanding of power relations and human rights in the complex political contexts in which these mining struggles are taking place.

Indigenous peoples organise locally, nationally and internationally

Struggle by indigenous peoples over resource extraction is part of a broader struggle to assert their right to self-determination and control over their lands, territories and resources. This struggle is globally organised through local, national,[8] regional[9] and international indigenous organisations[10] and through networks that may include non-indigenous organisations (see 'Rights-based civil society organisation' opposite.)

Local interactions between indigenous peoples and resource extraction companies are increasingly characterised by indigenous peoples' use of rights-based language, principles—such as free, prior and informed consent—and related expectations regarding appropriate process and acceptable outcomes for these interactions. Indigenous peoples' rights to their traditional lands, territories and resources are recognised in some

8 For example: Organización Nacional Indigena de Colombia; Defensoría Maya de Guatemala; Russian Association of Indigenous Peoples of the North (RAIPON); Indonesia-Aliansi Masyarakat Adat Nusantara/Alliance of Indigenous Peoples in the Archipelago (AMAN); National Coalition of Indigenous Peoples, Malaysia (JOAS); Kalipunan ng Katutubong Mamamayan ng Pilipinas/National Council of Indigenous Peoples and Tebtebba Foundation in the Philippines (KAMP); Nepal Federation of Indigenous Nationalities (NEFIN); Indian Confederation of Indigenous and Tribal Peoples (ICITP); Assembly of Indigenous/Tribal Peoples of Thailand; Indigenous Information Network, Kenya; Communauté des Autochtones Rwandais; Na Koa Ikaika Kalahui, Hawai'i; Rhéébù Nùù Committee, Kanaky-New Caledonia; National Aboriginal and Torres Strait Islander Legal Services Secretariat, Australia (NAILSS). Sources: Colchester *et al.* 2003; Tauli-Corpuz 2005; Tauli-Corpuz and Carino 2004.

9 For example: Inuit Circumpolar Conference; the Asian Indigenous Peoples' Pact (AIPP) and the Asian Indigenous Women's Network (AIWN); Indigenous Environmental Network (IEN).

10 For example: UN Permanent Forum on Indigenous Issues; International Indian Treaty Council (IITC).

countries' laws and policies,[11] as well as 'in international law as reflected in the jurisprudence of the United Nations and the regional human rights treaty bodies'.[12] UN instruments, International Labour Organisation Convention 169, the Convention on Biological Diversity, the Convention on the Rights of the Child, findings by UN treaty bodies, reports from the UN Special Rapporteur on indigenous land rights, and other sources reflect rights-based principles informed by and supportive of indigenous struggles.[13] The UN Declaration on the Rights of Indigenous Peoples, which was adopted by the UN General Assembly in 2007, is the seminal document that reflects both decades of evolving recognition and definition of indigenous rights, and decades of international organising, research and lobbying by indigenous peoples.

The key principle of free, prior and informed consent is referenced in numerous places in the UN Declaration on the Rights of Indigenous Peoples as a condition for undertaking activities that will affect indigenous lands or resources. Free, prior and informed consent is also recognised in UN instruments and human rights law.[14] Process elements that are important to indigenous peoples relate to fulfilling expectations of appropriate consultation on resource extraction projects; mechanisms for rejecting a resource extraction project; and processes by which to negotiate legally binding agreements that set out the conditions under which a project may proceed.

Rights-based civil society organisation

The USA, Australia and Canada, as 'mining exporting' countries, have NGOs whose mandate, both domestically and internationally, is to counter the negative environmental and social impacts of mining.[15] These organisations are, respectively, Earthworks,[16] the

11 For example, the Philippines introduced the Indigenous Peoples' Rights Act in 1997, which recognises the principle of free, prior and informed consent. Colombia's 1991 Constitution recognises Colombia as a multi-ethnic and multicultural state and includes a series of rights with respect to indigenous peoples recognising cultural integrity and territoriality (including collective ownership) and the right to participate in decision-making with respect to natural resource exploitation. Colombia has also adopted ILO 169 in Law 21. The Fifth Schedule of the Indian Constitution guarantees indigenous peoples (*Adivasis*) protection for their lands and prohibits the grating of mining leases to non-*Adivasis*. While this aspect of the Constitution has been widely violated, the 'Samatha Judgement' of 1997 saw India's Supreme Court rule that all leases to private mining companies on tribal lands (Fifth Schedule Areas) were illegal, affecting 100 million *Adivasis* in eight states.

12 Forest Peoples Programme and Tebtebba Foundation 2006.

13 See Miranda *et al.* 2005: Appendix A.2 International Instruments and Legal Precedents Regarding Indigenous Surface and Subsurface Rights to Land and Natural Resources; Forest Peoples Programme and Tebtebba Foundation 2006: 26-44.

14 See Miranda *et al.* 2005: Appendix A.3 International Instruments, Multistakeholder Processes, and United Nations Positions Supporting Free, Prior, and Informed Consent; Forest Peoples Programme and Tebtebba Foundation 2006: 45-54.

15 In addition to these national organisations that work both domestically and internationally, there are other mining-focused organisations in these countries that tend to have a more localised focus, such as Great Basin Resource Watch in Nevada, USA, whose slogan is 'Working with communities to protect their land, air, water': www.gbrw.org, accessed 20 August 2008.

16 www.mineralpolicy.org, accessed 15 August 2008.

Mineral Policy Institute[17] and MiningWatch Canada.[18] Each of these organisations works with mining-affected indigenous peoples, but is not staffed primarily by indigenous peoples. Additionally, the Mining Ombudsman Project[19] under the umbrella of Oxfam Australia focuses on mining and works with indigenous peoples internationally. In the UK, another mining exporting country, a major focus for Philippine Indigenous Peoples Links (PIPLinks)[20] is to provide support for Philippine indigenous peoples affected by mining projects. In addition, the Forest Peoples Programme[21] focuses extensively on issues related to indigenous rights and resource extraction (Colchester *et al.* 2003; Colchester and MacKay 2004).

These organisations explicitly support the principle of free, prior and informed consent for indigenous communities. In their publications and campaigns with indigenous peoples whose rights are threatened by mining projects, these organisations commonly reference international law as reflected in UN rulings, as well as jurisprudence of the regional human rights treaty bodies, to support campaign demands. In addition to organisations that focus explicitly on mining, many civil society organisations with mandates related to development, research, human rights or faith focus on mining as part of their work. Many of these organisations also explicitly support the principle of free, prior and informed consent.

Faith-based organisations such as Kairos[22] and Development and Peace[23] in Canada and Christian Aid[24] and CAFOD[25] in the UK provide both policy and campaign support for communities and indigenous peoples affected by mining. Development and Peace is currently running a two-year campaign on mining with the slogan 'Life Before Profit' which promotes the principle of free, prior and informed consent and features the struggles of indigenous peoples in Guatemala and the Philippines against Canadian mining companies. Kairos presented a brief before the UN Committee on the Elimination of Racial Discrimination in 2007 that chastised the Government of Canada for its opposition to the adoption of the UN Declaration on the Rights of Indigenous Peoples. The brief pointed out the fallacies in the government's arguments for opposing the principle of free, prior and informed consent, and further berated the government for its treatment of Aboriginal peoples in Canada, questioning why the government continues 'to grant licenses to resource extraction on the territory of the Lubicon Cree First Nation without their free, prior and informed consent?'[26]

The North-South Institute,[27] an organisation that provides 'policy-relevant research', has been engaged in a series of research projects in conjunction with southern and Canadian indigenous partners that aim to better understand the necessary conditions and mechanisms for rights-based consultation with indigenous peoples regarding mining

17 www.mpi.org.au, accessed 15 August 2008.

18 www.miningwatch.ca, accessed 15 August 2008.

19 www.oxfam.org.au/campaigns/mining/ombudsman, accessed 15 August 2008.

20 www.piplinks.org, accessed 15 August 2008.

21 www.forestpeoples.org, accessed 15 August 2008.

22 www.kairoscanada.org, accessed 15 August 2008.

23 www.devp.org, accessed 15 August 2008.

24 www.christianaid.org.uk, accessed 15 August 2008.

25 www.cafod.org.uk, accessed 15 August 2008.

26 www.kairoscanada.org/e/media/letters/aboriginalrightsCERD0702.pdf, accessed 13 October 2008. See also Assembly of First Nations 2006: 21-23.

27 www.nsi-ins.ca, accessed 15 August 2008.

activities on or near ancestral lands. Rights and Democracy,[28] a human rights organisation, has been developing a participatory Human Rights Impact Assessment methodology to 'improve the capacity of civil society organisations to evaluate the impacts of foreign direct investment on human rights'. This methodology has been field-tested by five communities, three of which were indigenous communities. Two of the investment projects involved Canadian mining companies. Rights and Democracy calls on the Canadian government to adopt the UN Declaration on the Rights of Indigenous Peoples.

Non-indigenous civil society organisations also participate in broad networks focused on mining, some of which include indigenous members and caucuses, such as the Western Mining Action Network (WMAN)[29] (USA and Canada), the Ontario Mining Action Network (OMAN)[30] (Canada), Mines and Communities[31] (international, UK-based), and the newly formed London Mining Network (UK).[32] In comments on the proposed new Ontario mineral development strategy, OMAN included among its recommendations: 'Free Prior Informed Consent for First Nations prior to the onset of mining activities; First Nations control of resource revenue sharing, land use planning and IBAs [Impact and Benefit Agreements]' (OMAN 2005). The founding document of Mines and Communities, the London Declaration, lists among its demands: 'That surface and sub-surface rights of indigenous peoples and all mining-affected communities be unequivocally respected and enforced, as well as their right to veto unacceptable projects' (Mines and Communities 2001).

Realising solidarity and developing standards in contested terrains

NGOs that prioritise indigenous peoples' rights in their approach to dealing with resource extraction have a better chance of avoiding some of the conflict experienced by some NGOs that are primarily focused on environmental protection or conservation. Some conservation groups have found themselves in conflict with indigenous peoples when they have negotiated with governments and the industry to protect large swaths of land, sometimes in return for agreeing not to oppose mining in another area, without consulting the indigenous peoples affected by these decisions. Environmentalists who aim to protect a particular habitat may find themselves in conflict with indigenous peoples who have determined that they want a mine to go ahead on their land.

28 www.dd-rd.ca, accessed 9 August 2008.

29 www.wman-info.org, accessed 15 August 2008. WMAN has an indigenous caucus.

30 www.miningwatch.ca/index.php?/17/OMAN_OMDS, accessed 15 August 2008. OMAN includes indigenous members.

31 www.minesandcommunities.org, accessed 15 August 2008. Mines and Communities has indigenous participation.

32 Members of the London Mining Network, established on 18 April 2007, are: Colombia Solidarity Campaign, The Corner House, Down to Earth, Forest Peoples Programme, Latin American Mining Monitoring Programme (LAMMP) Partizans (People Against Rio Tinto and its Subsidiaries), PIPLinks, Society of St Columban and Indonesia Human Rights Campaign (TAPOL).

NGOs that prioritise relationships with indigenous peoples also face challenges. They must accept the need for constraint with regard to activism on projects that are not opposed by an indigenous community. In these cases, the relationships provide an opportunity to advise on negotiating economic benefits and better environmental and cultural protection. Similarly, respecting indigenous priorities regarding time and indigenous decision-making processes may cause tensions if these challenge an effective campaign when the company is moving quickly through the regulatory process. Furthermore, a commitment to prioritising indigenous rights is not a guarantee against conflict, particularly when an indigenous community that initially welcomed NGO support to oppose a project becomes divided. More recently, the industry and some government agencies are also starting to play an active role in promoting conflict between indigenous peoples and NGOs that are critical of mining.[33]

Ultimately, effective joint organising, whether to stop a project, limit its damaging effects, or improve economic benefits to indigenous peoples, relies on long-term relationships of trust in which, at the direction of indigenous peoples, NGOs are able to provide useful services, research, legal support, linkages to other key actors and forums and solidarity. For instance, Victoria Tauli-Corpuz, Executive Director and Chairperson of the UN Permanent Forum on Indigenous Issues, gives a behind-the-scenes description of the process and the lobbying that preceded the adoption of the UN Declaration on the Rights of Indigenous Peoples during the First Session of the UN Human Rights Council in June 2006, following decades of negotiation. She notes that 'NGOs who have consistently supported us were also present'.[34]

The following cases describe a range of ways in which members of non-indigenous NGOs have successfully supported indigenous communities engaged in specific struggles with mining companies. These cases illustrate various means by which these struggles drive and shape the development and enforcement of rights-based positions and principles through legal action, international multi-stakeholder dialogue, participatory research, 'direct action', and effective lobbying of home governments of mining companies. In each of these cases, members of NGOs who are engaged in solidarity relationships with mining-affected indigenous communities also participate in national and international CSR standard-setting processes in which they have the opportunity to advocate for the adoption of indigenous rights to lands, territories and resources and of principles, such as free, prior and informed consent, into standards for corporate accountability.

33 For an example of 'manufactured conflict' see the case of some members of the Shuar of Ecuador who accused MiningWatch Canada of genocide for opposing a mine that, in fact, MiningWatch Canada had not been working on. These accusations were published in a national newspaper in Canada and the Canadian Embassy in Ecuador publicly supported the Shuar group that had come out in favour of a Canadian mining project in spite of broad-based Shuar opposition to the mine; www.miningwatch.ca/index.php?/corriente/ecuador_analysis, accessed 15 August 2008.

34 Tauli-Corpuz specifically mentions: 'IWGIA (International Workgroup on Indigenous Affairs), ALMACIGA, Rights and Democracy, the NCIV (Netherlands Centre for Indigenous Peoples), American Friends Service Committee (Quakers)-Canada, DOCIP (Documentation Center for Indigenous Peoples), Amnesty International, International Service for Human Rights, and this time, Cultural Survival was also present' (Tebtebba Foundation 2006).

Western Shoshone: seeking justice through courts and international human rights bodies

The struggle of the Western Shoshone, or Newe, for recognition of their rights is a long and arduous one dating back to the first intrusions on Newe land by prospectors in the 1840s.[35] Their struggle remains intrinsically linked to mineral exploitation as 'Western Shoshone lands are currently the third largest gold producing area in the world' (Bill and Fishel 2007: 5), accounting for 64% of US gold production.[36]

Large-scale mining by numerous multinational corporations and their subsidiaries forms a major threat to Newe land and water through contamination and through depletion of scarce water resources as a result of pit dewatering.[37] Mining also threatens the cultural survival of the Western Shoshone.[38] In 1996, Canada's Oro Nevada Resources started drilling near a sacred hot spring, ignoring a tribal resolution asking them to stay away from that area. The company also made plans to drill in a sacred Newe encampment, violating the American Indian Religious Freedom Act (Mineral Policy Center 1997). These actions prompted the Western Shoshone to campaign at Oro's annual general meeting in 1997, eventually driving Oro out of the area. Currently, Mount Tenabo, a sacred site of local creation stories, is threatened by Barrick Gold's Cortez Joint Venture.[39] In the words of Newe elder Bernice Lalo,

> It's not as if we could just go somewhere, because the land is part of who we are . . . There's no difference between us and the land we walk on. We don't have written language. Our stories belong to that mountain. Our stories belong to that land where our people are buried. Our names are there (MiningWatch Canada and CCISD 2000).

The Treaty of Ruby Valley, signed in 1863 between the US government and the Western Shoshone Nation, provided access to Newe land by settlers and miners. The treaty recognised the Western Shoshone as landowners and entitled them to royalties from extractive activities. But no royalties have ever been paid and the US government, arguing that the Western Shoshone lost their rights to the land through 'gradual encroachment' by non-natives, now acts as landowner, claiming 90% of the land base as 'federal' land, providing huge tracts of land for mining companies and requesting grazing fees from the Western Shoshone.[40] The Western Shoshone have consistently refused to accept a 1979 offer by the US government for $26 million (now worth approximately $140 million) to abolish the treaty, and many refuse to pay grazing fees (Sewall 1999: 29).

35 Newe land runs from southern Idaho, through eastern Nevada, to the Mojave Desert of California.

36 www.nodirtygold.org/western_shoshone_nation_usa.cfm, accessed 15 August 2008.

37 Western Shoshone face mining projects by Newmont, Bravo Venture Group, Nevada Pacific Bold Limited, Barrick Gold Company, Glamis Gold Limited, Gold Corp Inc., Great Basin Gold Limited and US Goldcorp, among others.

38 In 1991 the Western Shoshone National Council set up the Western Shoshone Defense Project whose mission is to protect and preserve Western Shoshone rights and homelands; www.wsdp.org, accessed 29 September 2008.

39 www.sacredland.org/endangered_sites_pages/mt_tenabo.html, accessed 13 October 2008.

40 www.nodirtygold.org/western_shoshone_nation_usa.cfm, accessed 15 August 2008.

Spearheaded by sisters Mary and Carrie Dann, the Western Shoshone have fought for their land rights before the Indian Claims Commission (1962) and in the US court system up to the Supreme Court (1974–1985) with both rulings going against them (Amnesty International USA undated). They have also brought their case before the UN Committee on the Elimination of Racial Discrimination (CERD) (2001, 2005, 2006, 2007), as well as the Inter-American Commission on Human Rights (1993–2002), where they were successful in having their rights recognised and the rulings of the Indian Claims Commission and US Supreme Court challenged (Amnesty International USA undated).

Throughout much of their recent struggle, the Western Shoshone have been supported by national and international NGOs, as well as committed non-indigenous individuals who have worked for the Western Shoshone Defense Project (WSDP) in Crescent Valley. These relationships have helped to document the history of struggle, raise the national and international profile of the issue, provide campaign support, and create international linkages. Project Underground worked with Christopher Sewall at WSDP to document the history and issues facing the Newe and helped them campaign against Oro Nevada (Sewall 1999).[41] Amnesty International USA (undated) took up the case and produced a report. Earthworks[42] features the case on its website. Oxfam America has supported the Western Shoshone since 2003 and provides the WSDP a web-based petition service.[43] MiningWatch Canada hosted Western Shoshone elder Bernice Lalo at an international conference in 2000 and presented a statement from the Western Shoshone at Placer Dome's annual general meeting in Toronto in 2003.

A key strategy of the Western Shoshone is legal action. The Indian Law Resource Center has provided legal advice since 1993, when it helped the Western Shoshone bring their case before the Inter-American Commission on Human Rights, and has documented the case.[44] Great Basin Resource (formerly Mine) Watch[45] provides ongoing technical and campaign support and filed a suit in 2005, together with the Te-Moak Tribe and the Western Shoshone Defense Project, against the US Bureau of Land Management in an attempt to protect Mount Tenabo and Horse Canyon from gold mining.[46] The Indigenous Peoples Law and Policy Program of the University of Arizona now provides legal support on international proceedings at the Organization of American States (OAS) and CERD.

Julie Ann Fishel currently lives in Crescent Valley, provides ongoing strategic and legal assistance to the WSDP, and co-authored the 2007 CERD submission which challenges the Government of Canada to regulate the activities of Canadian mining companies operating on indigenous land abroad (Bill and Fishel 2007). She sums up the importance of the international human rights regime to the struggle of the Western Shoshone: 'Inter-

41 Christopher Sewall, originally from Maine, lived in Crescent Valley and worked with the Western Shoshone for more than ten years.

42 www.earthworksaction.org/take_action.cfm, accessed 13 October 2008.

43 www.oxfamamerica.org, accessed 15 August 2008.

44 www.indianlaw.org, accessed 6 October 2008.

45 www.gbrw.org, accessed 20 August 2008.

46 Mount Tenabo is threatened by the Cortez Gold mine of which Barrick Gold is a majority owner. Barrick Gold is now a party to the law suit.

national human rights laws and bodies are essential to receiving justice. We need international mechanisms to serve as an independent review of the inadequacies of the US judicial system and federal Indian law.'[47]

Earthworks'[48] relationship with the Western Shoshone dates back to 1998, starting with their 'circuit rider' field staff who work with several indigenous communities. These were later joined by campaign staff.[49] A key feature of Earthworks' work with indigenous communities is support for indigenous efforts to preserve sacred sites such as Mount Tenabo. Earthworks has featured the issue in publications (Dahlberg 1999; Earthworks and Oxfam America 2004), has helped Newe leaders attend the annual general meetings (AGMs) of mining companies operating on their land, and facilitated linkages with other US and international indigenous communities affected by mining.[50]

> Our partnerships with indigenous communities and organisations are extremely important because mining activities increasingly take place on indigenous lands. There are the usual environmental concerns, but there is also a broader question of how mining impacts traditional ways of life. Land, water, air—these resources have cultural and spiritual significance to indigenous communities.[51]

Earthworks now features the Western Shoshone case in its 'No Dirty Gold' campaign. The campaign's 'Golden Rules' set standards for mining companies that have been endorsed by 28 jewellery retailers worth roughly US$12 billion in annual sales. 'Free, Prior and Informed Consent of Affected Communities' is one of the standards of the Golden Rules.[52] Earthworks was also instrumental in founding the Initiative for Responsible Mining Assurance (IRMA) in 2006.[53] Multinational mining companies, mining associations, civil society groups and retailers[54] are developing a certification system for metals and minerals based on verified compliance with environmental, human rights and social standards for mining operations. The work of developing standards under

47 www.indianlaw.org, accessed 6 October 2008.

48 Earthworks, together with Oxfam America, heads a downstream market campaign that is to date supported by 28 jewellers and sets criteria for responsible sourcing of metals. Among the criteria are respect for basic human rights outlined in international conventions and law and free, prior and informed consent of affected communities; www.nodirtygold.org/goldenrules.cfm, accessed 15 August 2008.

49 Earthworks' focus on struggle of indigenous peoples with mining grew through its field staff, known as circuit riders, who worked with, among others, the Gros Ventre and Assiniboine Tribes at Fort Belknap, Montana, the Picuris Pueblo in New Mexico, the Dine at Crownpoint, New Mexico, and native Alaskans (Mineral Policy Center 1997).

50 At a Newmont AGM in 2005 the Western Shoshone came together with members of communities from other parts of the world affected by the same company, leading to joint press releases and statements. In the spring of 2005, Earthworks helped coordinate a five-day trip for 17 native Alaskan leaders from communities neighbouring the proposed Pebble Mine to see Nevada mining operations and hear directly from the Western Shoshone and the Yerington Paiute. Members from the Western Shoshone and the Yerington Paiute took the Alaskans out on the land and explained the impacts from water pollution, dewatering, and destruction of cultural sites.

51 Personal communication with Radhika Sarin of Earthworks, 30 March 2007.

52 www.nodirtygold.org/goldenrules.cfm, accessed 15 August 2008.

53 www.responsiblemining.net, accessed 15 August 2008.

54 For a complete list of organisational and individual participants, see www.responsiblemining.net/participants.html, accessed 15 August 2008.

IRMA is under way and two of the 'foundation documents' for IRMA recognise the principle of free prior and informed consent (Miranda *et al.* 2005; CSIRO 2006).[55]

Tintaya and the Mining Ombudsman Project: linking local, national and international NGOs

Oxfam Australia's Mining Ombudsman Project focuses on mining companies that are Australia-based, or listed on the Australian stock exchange. Taking a rights-based approach, the Mining Ombudsman receives complaints from mining-affected communities, assesses the complaints, conducts field investigations and produces reports based on findings. Following a field investigation, the company in question is contacted and made aware of community concerns and the requested remedial action. Depending on a company's response, further action could be taken to resolve the conflict. The Mining Ombudsman Project supports the principle of free, prior and informed consent.[56]

An underlying principle of the work of the Mining Ombudsman is that external reputation pressure on a company 'at home' can improve behaviour overseas:

> Through the Mining Ombudsman project, Oxfam Australia has witnessed the impact that Australian mining companies can have on indigenous peoples. This has been particularly apparent as mining increasingly encroaches on traditional lands in developing countries where local regulations fail to protect the rights of indigenous landholders. Our direct work with indigenous communities and organisations is therefore essential to ensure that their voices are heard here in Australia and is a key step in requiring Australian companies to respect human rights no matter where they operate.[57]

A particularly relevant case in which the Mining Ombudsman provided external reputational pressure and continues to play a mediating role is that of the Tintaya Copper mine in Espinar Province, Peru[58] (see also Anguelovski, Chapter 11 in this volume). Australian-based BHP (now BHP Billiton) took over the operation from USA-based Magma Copper in 1996.[59] Following local community organisation, and documentation of the issues by Peruvian NGOs supported by Oxfam America, the Mining Ombudsman received a request for intervention from five indigenous communities affected by the mine in 2000: 'Their grievances included: forced evictions with little to no compensa-

55 For other documents, see www.responsiblemining.net/documents.html, accessed 15 August 2008.

56 See www.oxfam.org.au/campaigns/mining/ombudsman/consent.html, accessed 15 August 2008; Oxfam Australia 2007. In addition, the Mining Ombudsman is specifically focused on the plight of women in mining-affected communities. The Mining Ombudsman has organised or contributed to conferences for mining-affected women such as the Third International Women and Mining Network Conference in India (2004), the Mining, Women and Communities Workshop in Lae, Papua New Guinea (2004) and the Indigenous Women and Mining Workshop in Karratha, Western Australia (2004). The Mining Ombudsman has also issued publications focused on women's issues (Macdonald and Rowland 2002).

57 Personal communication with Shanta Martin, Mining Ombudsman, 23 March 2007.

58 www.oxfam.org.au/campaigns/mining/ombudsman/cases/tintaya, accessed 6 October 2008.

59 In May 2006 BHP Billiton sold its stake in the Tintaya mine and nearby undeveloped deposits of Antapaccay and Corrocohuayco to Xstrata plc.

tion; loss of livelihoods; pollution; lack of employment opportunities; and increased violence against women' (Oxfam Australia 2004). On 21 December 2004, following almost three years of discussions at a Dialogue Table, a Consolidation Agreement was signed by the communities of Tintaya Marquiri, Alto Huancané, Huancané Bajo, Huano Huano and Alto Huarca, the BHP Billiton Tintaya mining company, Peruvian NGOs CORECAMI[60] and CooperAcción,[61] and Oxfam America, setting out the agreements that had been reached between the communities and the company (Barton 2005; Echave *et al.* 2005).

Importantly, BHP Billiton agreed that new activities on land outside the company's current land holdings will take place only with the free, prior and informed consent of the owners, which may be communities or individuals.[62] The Agreement also provides for land acquisition for displaced community members that is 25–50% greater in area than that which was originally acquired from them by the state and by the company. The company further agreed to contribute a fixed annual amount of US$300,000, later increased to US$330,000, to a development fund in each of the three years following the Agreement. All parties also agreed to ongoing dialogue to resolve human rights and environmental problems.

The Mining Ombudsman has been able to apply its experience in working with directly affected communities to the task of providing comments, in a multi-stakeholder process, on texts being produced by the Australian government for its Leading Practice Sustainable Development Program for the Mining Industry. The Mining Ombudsman was also an invited expert on complaints mechanisms and the ombudsman function at the Canadian government's National Roundtables on Corporate Social Responsibility and the Canadian Extractive Industry in Developing Countries (CSR Roundtables). The final report from this process makes a recommendation that the Canadian government adopt a mining ombudsman (Canada Government 2007).

'Consultation' and the North-South Institute's collaborative research model

A key component of free, prior and informed consent is two-way communication between a project proponent and potentially affected community members who are willing to consider a project. This 'consultation' or 'dialogue' may result in formal negotiation of conditions under which a project may proceed, or, if truly free, may lead to

60 CORECAMI (Regional Coordinator of Communities Affected by Mining) is the regional coordinating group for the Cusco area of CONACAMI (National Coordinating Body of Communities Affected by Mining in Peru); www.conacami.org, accessed 15 August 2008.

61 www.cooperaccion.org.pe, accessed 15 August 2008.

62 Recognition of the principle of free, prior and informed consent by a major mining company is a significant breakthrough. However, even as BHP Billiton was finalising the agreement with the five affected communities, controversy erupted when the company constructed a tailings dam at the headwaters of the Ccañipia River without the prior consent of a local community that was not part of the dialogue table. This raises the question of whether a company will apply a principle universally even if no outside pressure is being brought to bear. BHP Billiton is currently working on a company policy on free, prior and informed consent (personal communication with Shanta Martin, 12 April 2007).

community rejection of the project. The North-South Institute (NSI)[63] recognised that 'there is a dearth of research and literature highlighting the views of Indigenous Peoples themselves' with respect to 'appropriate mechanisms for approaching and involving Indigenous Peoples in decision-making regarding potential mining activities on or near ancestral lands' (Weitzner 2002a: 3). In 2000, the NSI embarked on a collaborative, participatory research project with indigenous organisations in Colombia and Guyana 'to research, document and assess Indigenous experiences of consultations with the mining sector' (Weitzner 2006: 32).

The findings from this research are significant. Commonalities among indigenous participants included the conviction that indigenous peoples are not simply stakeholders to be consulted but 'rights-holders whose identity, autonomy and cultural survival is inextricably linked to their relationship with the land'; that Western and indigenous views of what constitutes development differ; and that there is conflict between state and indigenous views regarding ownership of sub-surface resources on indigenous land (Weitzner 2002a: 4).

Equally important is the finding that in many cases consultation processes have been 'destructive in and of themselves' (Weitzner 2002a: 5). Reasons for this include that industry participants too often assume a project will go ahead, leading to an attitude towards consultation that sees it as either mere formality, or simply a means to determine which negative impacts will need to be mitigated; or even misrepresents the process itself as a validation of the project, as opposed to recognising consultation as an exercise in self-determination. In the worst cases, the consultation process includes serious interference in traditional decision-making structures, weakening or replacing 'traditional authority structures by imposing other forms of decision-making and conflict management' (Weitzner 2002a: 5). Too often, consultation processes lead to, or exacerbate, internal conflicts and contribute to increased substance abuse and violence (Weitzner 2002a: 5).

Participants recognised that many of these negative impacts are the result of unequal power relations between the consultation participants. Indigenous participants in the research project recognised that one way to adjust the power balance would be for governments and mining companies to accept the principal of free, prior and informed consent (FPIC) (Weitzner 2002a: 5-6). Fundamentally, according to the NSI's senior researcher, Viviane Weitzner, there is a need to

> de-colonize the language. The term 'consultation' tells us a lot about the power imbalance inherent in the process. For a consultation to take place, someone is doing the consulting (active), and someone is being consulted (passive). But with FPIC, communities are giving or withholding consent (active).[64]

In 2002, the NSI published a synthesis report (Weitzner 2002b) and initiated a second phase of the project aimed at capacity-building and multi-party dialogue 'with a view to influencing policy and practice at the national level so it is more aligned with Indigenous perspectives, aspirations and rights' (Weitzner 2006).

63 The NSI's mandate is policy-relevant research on relations between industrialised and developing countries with respect to development questions. The North-South Institute is based in Ottawa, Ontario, Canada: www.nsi-ins.ca, accessed 6 October 2008.

64 Personal communication with Viviane Weitzner, 15 April 2007.

As part of this second phase, the NSI has expanded its Southern participation to indigenous communities in West Suriname and also responded to requests from South American indigenous organisations to learn from the experience of Canadian indigenous peoples in negotiating with mining companies.[65] As a result, Florence Catholique and Delphine Enzoe of the Lutsel K'e Dene First Nation travelled to Suriname in 2005 to share their knowledge of negotiating an Environment Agreement, a Socio-Economic Agreement and an Impact and Benefit Agreement (IBA) for BHP Billiton's Ekati diamond mine.[66] A lesson shared by the Dene is the need for complete information and adequate time for a community to absorb and assess information:

> We didn't know anything about mining companies, their influence and powers. When they first came and found diamonds, the federal government proposed an IBA. We didn't know what an IBA was. We had 60 days to come up with an IBA. This wasn't sufficient time (Lutsel K'e Dene community negotiator, quoted in Weitzner 2006: 10).

The NSI's senior researcher participated in a standards working group and contributed a paper on the need to adopt the principle of free, prior and informed consent in standards for the Canadian extractive sector to the Canadian Government's CSR Roundtables. She was also an active participant in the Canadian Network on Corporate Accountability (CNCA) that provided guidance to Civil Society Advisory Group members of the CSR Roundtables. The CNCA has adopted a firm position in support of free, prior and informed consent.[67]

Kanak leadership: building international solidarity

New Caledonia, called Kanaky by its indigenous inhabitants, is a Pacific island territory, annexed by France in 1853, which contains an estimated 25% of the world's known nickel reserves. Kanaks, now about 45% of the population, were first granted citizenship in 1946 and given the right to vote in 1957. A Kanak uprising for independence in the early 1980s led to the Matignon Accord in 1988 and a promised independence referendum in 1998. In 1998 the referendum became a vote for another accord, the Noumea Accord, the best features of which are formal recognition of Kanak 'civilisation' and the establishment of a Kanak Customary Senate (Sénat Coutumier) having consultative powers over all matters pertaining to Kanak traditional practices and values, as well as relationships to the land, fresh water, sea and air. (For more on the history and background to this issue, see Coumans 2002a, 2002b, 2003: 13-19, 2005; Coumans and Beaumont 2005; Ali and Grewal 2006.) The newly recognised authority of the Kanak people, expressed in the Sénat Coutumier, was immediately tested in the intense struggle against Cana-

65 The NSI now has a project in place on indigenous-to-indigenous training (Canada–South America and vice versa) to fill in some of the capacity-building gaps identified.

66 This exchange followed a previous exchange facilitated by MiningWatch Canada between Lutsel K'e Dene from the Northwest Territories of Canada and members of the Attiwapiskat First Nation of Northern Ontario, Canada who face diamond mining on their land by de Beers.

67 The CNCA was established to participate in, and provide guidance and direction to the Civil Society Advisory Group members of the Canadian government's CSR Roundtables. Its office and website is currently housed at the Halifax Initiative. For the CNCA's response to the Final Report of the CSR Roundtables, see www.halifaxinitiative.org, accessed 15 August 2008.

dian mining company Inco's (now Vale's) Goro Nickel Project in the Southern Province. The Sénat Coutumier rejected Inco's environmental impact assessment (EIA) in 2002. This rejection was ignored and the project was granted a mining permit in 2004.

MiningWatch Canada received a request to engage on this issue in 2000, one year into its organisational existence.[68] The visit to Canada of a first delegation from New Caledonia was facilitated by MiningWatch Canada's Asia-Pacific Coordinator in 2001.[69] The delegation included Georges Mandaoue, newly elected President of the Sénat Coutumier, and Regis Vendegou, Secretary General of the Sénat Coutumier.[70] This delegation met Inco executives in Toronto, visited Sudbury, home of Inco's largest and oldest operations, and met members of the indigenous Innu Nation in Labrador. The Innu have a history of struggle against Inco's Voisey's Bay Nickel Project and have gained a lot of experience in negotiating a legally binding IBA with Inco. Also in 2001, the Kanak Comité Rhéébù Nùù[71] was established to protect fledgling Kanak rights and authority in the face of the threat posed by the Goro mine.

The first visit to Canada led to a second trip by a larger delegation in 2003. By this time, a small international committee had been formed to work with Kanak leadership on the mining-related issues in New Caledonia, as well as to seek World Heritage listing for the territory's extraordinary reef system, parts of which are threatened by mining.[72] The second high-level delegation included leaders from the Sénat Coutumier, as well as

68 MiningWatch Canada is supported by environmental, social justice, Aboriginal and labour organisations across Canada and responds to requests for assistance from mining-affected communities in Canada and from communities affected by Canadian mining companies around the world. MiningWatch Canada has always maintained a focus on indigenous rights, which is reflected in indigenous representation on its board of directors. The first conference held by MiningWatch Canada, as a new organisation, was with indigenous communities from across Canada. See Innu Nation and MiningWatch Canada 1999.

69 MiningWatch Canada's Asia-Pacific Programme is coordinated by Catherine Coumans.

70 The two other members of this first delegation were not Kanak. They were Jacky Mermoud, then Secretary General of Action Biosphere, and Rick Anex, then member of Action Biosphere, Treasurer of Corail Vivant and Secretary General of the Pacific Green Party.

71 Comité Rhéébù Nùù was established in 2001. *Rhéébù nùù* means 'eye of the land'. While often described as a local group opposing the Goro mine, Rhéébù Nùù in fact has membership by Kanaks from all over New Caledonia and its concerns are understood by Kanaks to be much further-reaching than Goro, or even mining, alone. Rhéébù Nùù is understood to represent Kanak aspirations for recognition of their indigenous rights in all arenas of life in Kanaky. See www.rheebunuu.com, accessed 15 August 2008.

72 The committee includes Techa Beaumont of the Australian Mineral Policy Institute, Senator Christine Milne in Australia and Catherine Coumans from MiningWatch Canada. The Mineral Policy Institute works on indigenous rights issues with respect to mining in Australia, e.g. the Yakabindie/Jones Creek proposed mine, the Windarling Mine and the Ranger Uranium Mine, as well as internationally, e.g., the El Cerrejon Mine in Columbia, the Yanacocha Mine in Peru and the Ramu Nickel Mine in Papua New Guinea. Senator Milne, of the Green Party from Tasmania, is also Global Programme Vice President of the International Union for Conservation of Nature. This small committee benefits from the dedicated assistance of Saramin Jacques Boengkih in Kanaky. Mr Boengkih is Executive Director of Agence Kanak de Développement. In 2008 New Caledonia reefs and ecosystems were accepted for World Heritage status.

from Rhéébù Nùù.[73] The ten-day trip was organised by MiningWatch Canada's Asia-Pacific Coordinator as a mobile workshop on IBAs and on negotiating with corporations. The delegation met, among others, a lawyer for the Innu who was involved in negotiating the IBA between Inco and the Innu; seven executives from Falconbridge (now Xstrata plc) about the Raglan IBA between the company and four Inuit communities in Northern Quebec; Makavik Corporation, the organisation set up by the communities covered by the Raglan Agreement to represent their interests; Inco executives; the then Executive Director of the Canadian Arctic Resources Committee with expertise on IBAs for indigenous communities affected by diamond mining in the Northwest Territories; the then National Chief Matthew Coon Come and members of the Assembly of First Nations; and members of the Grand Council of the Crees.[74]

The delegation held a press conference in Canada's parliamentary press gallery in which Raphael Mapou of Rhéébù Nùù noted: 'The similarities between our situation and those of Canadian indigenous peoples, which INCO knows well, should have led INCO to seek a better relationship with us than has been the case.'[75] The Kanaks also issued a joint press release with the Grand Council of the Crees. Philip Awashish, spokesperson for the Grand Council, noted that 'the experience of our two nations, the denial of our rights, of our culture and traditions by governments and corporate interests, parallel each other. And so our mutual support is only natural.'[76]

The Kanak delegation heard that indigenous peoples can declare and assert their rights, even in a political context that does not formally recognise those rights; that international lobbying, public relations and litigation are key aspects of a successful struggle;[77] and that Canadian companies cannot credibly deny knowledge of the consultation process that is required by indigenous peoples, and that may lead to a negotiated and legally binding settlement of all issues related to the impacts and benefits of a mine. They also heard that social, environmental and economic issues of concern to indigenous peoples should be included in a negotiated settlement; environmental review processes should facilitate indigenous communities in defining the issues to be addressed and providing them with independent expert assistance to facilitate their participation; and 'direct action' may be needed in order to bring governments and companies to the table for negotiations.

73 Raphaël Mapou, then Chairman of the *Rhéébù Nùù* Committee (RNC), Sylvestre Newedou, member of the Southern Province Assembly and member of the RNC, François Vouty, deputy-mayor of Yate and member of the RNC, Roch Wamytan, High Chief of St Louis tribe, member of the RNC, Minister for Aboriginal Affairs in the New Caledonian Government, Georges Mandaoue, senator, past president, Vincent Akaro, senator, Sarimin Jacques Boengkih, cultural and environmental consultant and interpreter.

74 The delegation also met members of the Socially Responsible Investment Community; two Parliamentary Committees; the Secretary of State for Asia-Pacific; officials of the Canadian Environmental Assessment Agency; an expert on the Voisey's Bay Environmental Panel Review; the office of the Federal Minister for Indigenous and Northern Affairs; and the Export Development Canada (Canada's Export Credit Agency, which was considering supporting the mine financially).

75 Press release, www.miningwatch.ca/index.php?/New_Caledonia/INCO_closes_Goro_Nic, accessed 15 August 2008.

76 Press release, www.miningwatch.ca/index.php?/New_Caledonia/Grand_Council_of_the, accessed 15 August 2008.

77 *Ibid.*

The Kanak delegation incorporated what it heard in Canada in an audio-visual presentation and a booklet that was presented in Kanak villages. Rhéébù Nùù and the Sénat Coutumier called on Inco to halt construction at the Goro site and to enter into comprehensive negotiations with the Kanak people about all social and environmental aspects of the proposed project. Rhéébù Nùù established a website and launched legal challenges against Inco in New Caledonia and Paris, which led to the revocation of Inco's 2004 mining permit[78] and to an order for Inco to halt construction of a waste dump in a biologically sensitive area.[79] Rhéébù Nùù convinced the authorities of the Southern Province to commission an independent review of Inco's EIA, which was highly critical of the document and is leading Vale to make major adjustments to the project. Previously unknown, Rhéébù Nùù's positions and actions are now frequently referenced in the global and financial press.

Construction of the Goro project was not halted. There is a mediated dialogue taking place between Goro Nickel, Kanak authorities and the Southern Province, but at the same time there is broad Kanak-led public opposition to the project's planned effluent pipe into the sea. Between 2004 and 2006, Kanaks blockaded the Goro site three times. In 2006, construction was halted for about three weeks. A striking symbol at the blockade was the sight of two flags flying side by side. One was the Kanak flag, and the other was the UN flag. When the French military police finally moved in to forcefully remove the blockades, they started by destroying a small religious shrine and by burning the two flags in the road.[80]

Based on his experience and understanding of the issues related to indigenous peoples and mining struggles and facilitated by his relationship with MiningWatch Canada, Sarimin Jacques Boengkih from New Caledonia was invited as an indigenous expert witness in the Canadian government's CSR Roundtables. Boengkih was effective in demonstrating the need for the Canadian government to provide leadership in regulating the behaviour of its companies operating overseas in such a way that indigenous rights will be protected even if Canadian mining companies are operating in states, or colonies, that do not fully recognise these rights. He noted the importance of Canadian government involvement in setting standards for its companies' operations in developing countries, both as a means of protecting universally recognised indigenous rights, and as a means of averting potentially destabilising and costly conflict.

Canatuan Subanon: taking the struggle to the Government of Canada

The indigenous Canatuan Subanon of Zamboanga del Norte, Mindanao, in the Philippines are no strangers to international campaigning in their struggle to protect their

78 www.miningwatch.ca/index.php?/New_Caledonia/Goro_Licence_Revoked, accessed 15 August 2008. While the mining permit has remained revoked, it has not stopped the construction, which is proceeding on a construction permit.

79 This court order was challenged by Companhia Vale do Rio Doce (CVRD, now Vale) and overturned by the court on a technicality.

80 Personal communication with Sarimin Jacques Boengkih, November 2006.

rights and ancestral land from Canadian-based TVI Pacific.[81] Supported by Philippine Indigenous Peoples Links (PIPLinks) and CAFOD in the UK since the mid-1990s, and MiningWatch Canada and Christian Aid, UK, since 2000, the Canatuan Subanon are also supported in the Philippines by local organisations, such as the church-based DIOPIM Committee on Mining Issues (DCMI)[82] and nationally based organisations such as the Legal Rights and Natural Resources Center-Kasama sa Kalikasan/Friends of the Earth Philippines (LRC-KsK/FOE)[83] and Tebtebba Foundation.[84]

The Canatuan Subanon, and other members of the municipality of Siocon, struggle with a wide range of issues as a result of TVI Pacific's operations on the ancestral land of the Subanon. These include intrusion on the ancestral domain of the Canatuan Subanon without their prior consent; forced evictions of both Subanon and settlers; and violence, intimidation and denial of access to their lands by TVI-financed paramilitary forces. They also include defamation on TVI Pacific's website of specific Subanon who oppose the company; desecration of a sacred mountain; contamination of rivers originating at the mine site; and interference in the internal leadership structures and organisations of the Canatuan Subanon, leading to divisions and to the recognition as hereditary leader (*timuay*), by TVI Pacific, of a person who has no claim to this title and is supplanting a hereditary leader who opposes the company's operations.[85] These issues have been fully documented, most recently by a high-level fact-finding delegation (Columban Fathers 2007; see also Christian Aid and PIPLinks 2004: 29-41)[86] and in a Human Rights Impact Assessment.[87]

With the assistance of PIPLinks, Subanon leaders have testified about their situation in Geneva before the UN Working Group on Indigenous Populations in 2001, 2002, 2004 and 2005 and before the UN Committee on the Elimination of all Forms of Racial Discrimination in 2008. Their case has also attracted the attention of Rodolfo Stavenhagen, UN Special Rapporteur on the situation of human rights and fundamental freedoms of indigenous peoples, who tried to visit the area in 2002, but was not granted permission by the Philippine government.

81 The Canatuan Subanon are represented by their own organisation, the Siocon Subanon Association Apu' Manglang Pusaka. They also participate in an organisation called the Save Siocon Paradise Movement which represents the Canatuan Subanon of the municipality of Siocon, Christianised townspeople of the municipality of Siocon and coastal Muslims of Siocon. They are also supported by a broader Subanon organisation called Pigsalabukan Bansa Subanon.

82 www.dcmiphil.org, accessed 15 August 2008. DCMI is located in Dipolog, the capital of Zamboanga del Norte. The DIOPIM area covers Dipolog, Iligan, Ozamis, Pagadian, Ipil and Marawi.

83 www.lrcksk.org, accessed 9 August 2008.

84 www.tebtebba.org, accessed 15 August 2008.

85 After many years in which one traditional leader who opposed TVI Pacific's operations found himself the subject of public criticism by the company and replaced by a *timuay* favourable to TVI Pacific but with no hereditary legitimacy, the National Commission on Indigenous Peoples finally, in 2007, confirmed the rightful *timuay* status of Jose (Boy) Anoy.

86 One of the members of the fact-finding delegation was Clare Short, Member of Parliament in the British House of Commons, who declared herself 'enormously shocked' by mining operations in the Philippines.

87 The Human Rights Impact Assessment was conducted in a participatory fashion between Montreal-based Rights and Democracy and local and international partners. See www.dd-rd.ca/site/media/index.php?lang=en&subsection=news&id=672, accessed 15 August 2008.

Following a visit to Canatuan by staff from MiningWatch Canada and Kairos in 2004,[88] and a return visit by a delegation from the town of Siocon to Canada later that year, a formal invitation was issued to present the case before the Canadian parliamentary Sub-committee on Human Rights and Democratic Development. In March of 2005, Onsino Mato and Godofredo Galos from Siocon testified *in camera*[89] about the problems they suffered from TVI Pacific's operations.[90] This testimony led to three further hearings on mining by the sub-committee and to a report that received all-party unanimous support from both the Sub-committee on Human Rights and Democratic Development and the Standing Committee on Foreign Affairs and International Trade. This report calls on the Government of Canada to do more to regulate the behaviour of Canadian mining companies operating in developing countries.[91] With respect to indigenous peoples the report notes that 'particular attention should be paid to the rights of indigenous peoples as currently specified in the United Nations Draft Declaration on the Rights of Indigenous Peoples' (Canada, House of Commons 2005). The report also calls on the government of Canada to 'conduct an investigation' into the impacts on indigenous and human rights of TVI Pacific's Canatuan project and to '[e]nsure that it does not promote TVI Pacific Inc. pending the outcome of this investigation' (Canada, House of Commons 2005).

The Government of Canada declined to investigate TVI Pacific's activities, but did decide to implement the CSR Roundtables in order to explore possibilities leading to 'strengthening of existing programmes and policies in this area, and, where necessary, to the establishment of new ones' (Canada, House of Commons 2005). MiningWatch

88 Catherine Coumans from MiningWatch Canada and Nancy Slamet and Julie Graham from Kairos visited the village of Canatuan, Siocon, Zamboanga del Norte and the TVI Pacific mine site in October 2004. Following this site visit they also spent five hours in the Canadian Embassy in Manila with Timuay (traditional leader) Jose Anoy of the Cantuan Subanon and Godofredo Galos of the townspeople detailing the indigenous rights and environmental concerns of local people about TVI Pacific's development.

89 Days before the hearing both MiningWatch Canada and the parliamentary sub-committee received letters from TVI Pacific's lawyers warning of repercussions should MiningWatch Canada or the Filipinos make statements that TVI Pacific would consider damaging to its interests. Although witnesses before a parliamentary committee have immunity so that they can speak freely, the chair of the sub-committee was sufficiently concerned about potential repercussions for the Philippine witnesses once they returned to the Philippines that he offered Onsino Mato and Godofredo Galos the option of testifying *in camera*. For more on the legal exchange, see www.miningwatch.ca/index.php?/Canatuan_Project_TVI/Stapon_letter_to_MWC, accessed 15 August 2008; www.miningwatch.ca/index.php?/Canatuan_Project_TVI/Response_to_Stapon, accessed 15 August 2008.

90 Catherine Coumans of MiningWatch Canada and Diana Bronson of Rights and Democracy also testified. Their testimony can be read at www.miningwatch.ca/index.php?/Canatuan_Project_TVI/Hansard_050323, accessed 15 August 2008. In addition to testimonies, submissions were made to the Committee from Victoria Tauli-Corpuz; Ofelia 'Inday' Dafi (from DCMI); Bishop Jose Manguiran; Siocon Peace and Development Advocates League; Timuay Jose (Boy) Anoy; Foreign Affairs Canada; International Trade Canada; Organisation for Economic Cooperation and Development; DCMI; Kairos; Alan Laird; Mennonite Central Committee Canada; National Commission on Indigenous Peoples; National Council of Churches in the Philippines.

91 The Standing Committee on Foreign Affairs and International Trade's Fourteenth Report (Canada, House of Commons 2005) called on the Government of Canada to, among other things, 'establish clear legal norms in Canada to ensure that Canadian companies and residents are held accountable when there is evidence of environmental and/or human rights violations associated with activities of Canadian mining companies'.

Canada's Asia-Pacific Coordinator participated in the CSR Roundtables as a member of the Government's Advisory Group, focusing particularly on the development of standards for extractive companies operating overseas. As an Advisory Group member she was able to advocate for the participation, as experts, of indigenous leaders Sarimin Jacques Boengkih from New Caledonia and Victoria Tauli-Corpuz from the Philippines.

Indigenous rights in the context of global CSR standard-setting: government and industry responses

The cases in this chapter illustrate a range of ways in which indigenous peoples who are coping with resource extraction are participating in solidarity relationships with rights-based NGOs and international networks. These cases also show how non-indigenous members of NGOs who partner with indigenous peoples in mining struggles may become committed advocates for key indigenous rights-based principles in national and international multi-stakeholder standard-setting processes.

Significantly, these solidarity relationships are taking place in a context of rapidly evolving responses to indigenous rights struggles by governments and by extractive industry players. The Canadian government's multi-stakeholder CSR Roundtables process of 2006–2007 provided a window on the aspirations and strategies of Canadian government departments seeking ways to better support Canadian mining companies facing opposition overseas, as well as on industry responses to local-level indigenous opposition.[92] For example, an existing Canadian International Development Agency (CIDA) programme known as Indigenous Peoples Partnership Program (IPPP) was promoted by both industry and government officials in the CSR Roundtables as a likely vehicle to increase CIDA's involvement in discussions about the economic benefits of

92 For official documents related to the CSR Roundtable process, see Canada Government 2006; and for civil society perspectives, see www.halifaxinitiative.org, accessed 15 August 2008. The government Steering Committee for the CSR Roundtable process was chaired by the Department of Foreign Affairs and International Trade and included officials from Natural Resources Canada, Industry Canada, Environment Canada, the Canadian International Development Agency, Indian and Northern Affairs Canada, the Department of Justice, Export Development Canada and the Privy Council Office. The Advisory Group, of which the author of this chapter was a member, was made up of representatives from industry associations, extractive companies, civil society organisations, labour, academics and the socially responsible investment sector. For more on this process and the Final Report from the Advisory Group, see geo.international.gc.ca/cip-pic/library/Advisory%20Group%20 Report%20-%20March%202007.pdf, accessed 15 August 2008.

resource extraction with indigenous peoples in countries hosting Canadian mining companies.[93]

The Canadian CSR Roundtables also highlighted the normative roles being played in standard-setting processes by the rights-based principles of the UN[94] and the risk-based policies of the World Bank Group and major financial institutions.[95] While the World Bank and the International Finance Corporation (IFC) have developed a number of 'risk-based' guides specifically for projects that impact on indigenous peoples (see, for example, IFC 2006, 2007, undated a, undated b; World Bank 2005), the World Bank Group recognises free, prior and informed *consultation*, as opposed to *consent*, and standards such as the IFC Performance Standards fall short of human rights principles in a number of respects.[96]

In attempting to set standards for Canadian extractive companies operating in developing countries the Advisory Group for the CSR Roundtables was divided. Industry representatives argued for adoption of the IFC Performance Standards. Civil society representatives of the Advisory Group argued for the development of standards rooted in internationally accepted human rights principles. In this case, a compromise was reached, but one in which the discussion about whether free, prior and informed consent will be included in the Canadian Standards was deferred. (For a detailed discussion, see Canada Government 2007: 10-13.)

While the global mining industry, through its various industry associations, is focusing attention on developing beneficial relations with indigenous peoples, key principles such as free, prior and informed consent have not been adopted in any industry associations' code of practice. In March 2006, the International Council on Mining and Metals (ICMM) released a draft Position Statement on Mining and Indigenous Peoples' Issues (ICMM 2006). On 17 April 2007, ICMM provided the results of a poll on its Posi-

93 CIDA has already been using its programme IPPP, partnering with Indian and Northern Affairs Canada and with Canadian embassies, to facilitate trips by Canadian indigenous peoples to countries where Canadian mining companies are experiencing conflict with local indigenous peoples. Additionally, the Canadian Embassy in Manila, which has been very active in protecting the interests of TVI Pacific both before and after the company became the focus of a Parliamentary Committee hearing (see section on the Canatuan Subanon in this chapter), has stepped up its efforts to address the broader issue of indigenous opposition in the Philippines to the presence of Canadian mining companies. From 20 November to 1 December 2006, the Canadian Embassy, in association with the embassies of Australia, USA and UK, several Canadian, US, Australian and Philippine private sector interests, the World Bank and invited indigenous representatives from Canada, conducted a 'Train the Trainers' session focused on indigenous peoples in Manila where the possible advantages of mineral development on indigenous lands were highlighted. The Embassy, in an internal report on the meeting, noted as a 'memorable moment' 'when a non-declared up until that point anti-mining NCIP [National Commission on Indigenous Peoples] Provincial Officer stood up to state that: "Until today, I was anti-mining. Now, I am prepared to listen to what the companies have to say" ' (personal communication, Canadian Embassy, Manila, 8 February 2007).

94 In addition to general human rights guidance rooted in internationally recognised UN declarations, covenants and treaties, captured in the United Nations Norms on the Responsibilities of Transnational Corporations and other Business Enterprises with Regard to Human Rights, there are also codes of conduct specifically for businesses that are rooted in the UN such as the United Nations Global Compact and the Voluntary Principles on Security and Human Rights.

95 Such as the International Finance Corporation's Safeguard Policies and related guidance documents and the Equator Principles.

96 The role of the World Bank in setting international standards and rules for engagements between mining companies and local communities is significant and growing. See, e.g., Szablowski 2007.

tion Statement that found it reflected: 'Inadequate recognition of historic and current disadvantage experienced by IPs [Indigenous Peoples]; Inadequate recognition of Indigenous Peoples' rights, particularly to Free Prior and Informed Consent.'[97]

In the Canadian CSR Roundtable process, industry representatives on the Advisory Group and some government representatives made a determined effort to shift focus away from a 'rights-based' approach to the issues, which would include the right of indigenous peoples to say 'no' to a mine, to an approach focused on assuring better 'development' outcomes. CIDA representatives on the CSR Roundtables attempted to get 'policy cover' out of the round-table process for directing more aid dollars to development projects around Canadian mine sites overseas. A presentation by John Ruggie, UN Special Representative for Business and Human Rights, during one of the CSR Roundtables appeared to provide support for the notion of directing government aid dollars to communities affected by a country's overseas mining operations (Ruggie 2006).

Both industry representatives and Canadian government participants repeatedly noted that 'responsible' NGOs need to do more to assist in reducing conflict and local-level opposition to Canadian mining projects overseas by assuring good development outcomes around Canadian mines. The implied analysis is that issues such as indigenous peoples' right to self-determination and to control over development on their lands and territories can be addressed by making sure that there are greater local-level economic development benefits derived from mining. The diverse civil society organisations featured in this chapter have in common that they can best be characterised as rights-based advocacy NGOs. Increasingly, NGOs that support a community's right to reject a project are being sharply contrasted by mining industry and government representatives with 'development' NGOs, which are described as desirable 'partners' for industry, governments and indigenous peoples.[98]

This 'development' offensive against advocacy NGOs engaged in rights-based solidarity relationships with indigenous peoples affected by mining presents a new opportunity for these NGOs to advance alternative development models rooted in their solidarity relationships with indigenous peoples.

97 The full survey report summarises the feedback provided by the respondents through the consultation and is being used to define the next steps in ICMM's ongoing work on indigenous peoples issues; www.icmm.com/page/2075/icmm-releases-findings-of-survey-on-indigenous-peoples-issues, accessed 10 August 2008.

98 High-profile and substantial financial deals between mining companies and development NGOs are becoming increasingly common. Barrick Gold recently closed a deal with World Vision for US$1.3 million over five years for development work at Barrick's controversial Lagunos Norte mine in Peru (*Daily News* 2007). On the flipside of this development is an increasingly vitriolic backlash against NGOs who assist communities in opposing mine development. See, e.g., the recent video *Mined Your Own Business* about the opposition to Gabriel Resources proposed mine in Romania and public statements made by Patrick Moore (Nones 2007).

References

Ali, S.H., and S.W. Grewal (2006) 'The Ecology and Economy of Indigenous Resistance: Divergent Perspectives on Mining in New Caledonia', *The Contemporary Pacific* 18.2: 361-92.

Amnesty International USA (undated) 'Western Shoshone', *Just Earth Report*, www.amnestyusa.org/justearth/indigenous_people/western_shoshone.html, accessed 15 August 2008.

Assembly of First Nations (2006) *Patterns of Deception: Canada's Failure to Uphold the Honour of the Crown. A Commentary on the Government of Canada's Paper 'Canada's Position: United Nations Draft Declaration on the Rights of Indigenous Peoples June 29, 2006'* (Ottawa: Assembly of First Nations): 21-23.

Barton, B. (2005) 'A Global/Local Approach to Conflict Resolution in the Mining Sector: The Case of the Tintaya Dialogue Table' (MA thesis, Fletcher School of Law and Diplomacy, Tufts University, Medford, MA; fletcher.tufts.edu).

Bill, L.R., and J.A. Fishel (2007) *Report on the Effects of Canadian Transnational Corporate Activities on the Western Shoshone Peoples of the Western Shoshone Nation: Submitted to the Committee on the Elimination of Racial Discrimination (CERD) 70th Session by the Western Shoshone Defense Project in Relation to Canada's 17th and 18th Periodic Reports to CERD February 2007*: 5; www.wsdp.org/Shoshone_Shadow_Repost.doc, accessed June 2008.

Canada, Government of, Department of Foreign Affairs and International Trade (2007) *National Roundtables on Corporate Social Responsibility (CSR) and the Canadian Extractive Industry in Developing Countries Advisory Group Report 2007*; geo.international.gc.ca/cip-pic/current_discussions/csr-roundtables-en.asp, accessed 9 August 2008.

Canada, Government of, Prospectors and Developers Association of Canada, Mining Association of Canada and the Canadian Aboriginal Mining Association (2006) *Mining Information Kit for Aboriginal Communities* (Ottawa: Government of Canada; www.mining.ca/www/media_lib/MAC_Documents/Publications/English/Mining_Toolkit2006E.pdf, accessed 15 August 2008).

Canada, House of Commons, Standing Committee on Foreign Affairs and International Trade (2005) *Mining in Developing Countries: Corporate Social Responsibility*, 14th Report, 1st Session, 38th Parliament, June 2005; cmte.parl.gc.ca/cmte/CommitteePublication.aspx?COM=8979&Lang=1&SourceId=178650, accessed 20 August 2008.

Christian Aid and PIPLinks (Philippines Indigenous Peoples' Links) (2004) *Breaking Promises, Making Profits: Mining in the Philippines* (London: Christian Aid/PIPLinks; www.piplinks.org/development_issues/ philippines_report.pdf, accessed 20 August 2008): 29-41.

Colchester, M., and F. MacKay (2004) *In Search of Middle Ground: Indigenous Peoples, Collective Representation and the Right to Free, Prior and Informed Consent* (Moreton-in-Marsh, UK: Forest Peoples Programme).

——, A.L. Tomayo, R. Ravillos and E. Caruso (eds.) (2003) *Extracting Promises: Indigenous Peoples, Extractive Industries and the World Bank* (Moreton-in-Marsh, UK: Forest Peoples Programme; Baguio Cuty, Philippines: Tebtebba Foundation).

Columban Fathers (2007) *Mining in the Philippines: Concerns and Conflicts* (Solihull, UK: Columban Fathers).

Coumans, C. (2002a) 'Backgrounder: What's INCO doing in New Caledonia?', 6 March 2002; www.miningwatch.ca/index.php?/New_Caledonia/Backgrounder_Whats_I, accessed 20 August 2008.

—— (2002b) 'Inco in New Caledonia: Protests Erupt Over Shut Down of Prony and Goro', 16 September 2002; www.miningwatch.ca/index.php?/Inco_in_New_Caledoni/Protests_Erupt_over_, accessed 20 August 2008.

—— (2003) 'Kanaky/New Caledonia: Goro Nickel Mine', in NGO Working Group on the EDC (ed.), *Seven Deadly Secrets: What the Export Development Canada Does Not Want you to Know* (Ottawa: Halifax Initiative): 13-19.

—— (2005) 'Background: Recent History of the Ongoing Struggle by Kanak People to Make Inco Respect Their Rights', 10 February 2005; www.miningwatch.ca/index.php?/Inco_in_New_Caledoni/Recent_History_Goro, accessed 20 August 2008.

—— and T. Beaumont (2005) 'Nomination for Rhéébù Nùù Committee John Humphrey Freedom Award, 2005, prepared by Catherine Coumans of MiningWatch Canada and Techa Beaumont of Mineral Policy Institute', unpublished.

CSIRO (Commonwealth Scientific and Industrial Research Organisation Australia) (2006) 'Mining Certification Evaluation Project'; www.minerals.csiro.au/sd/SD_MCEP.htm, accessed 9 August 2008.

Dahlberg, K. (1999) *14 Steps to Sustainability* (Washington, DC: Earthworks; www.earthworksaction.org/publications.cfm?pubID=17, accessed 9 August 2008).

Daily News (2007) 'Barrick Teams with World Vision', *Daily News*, 1 April 2007; www.canadianminingjournal.com/issues/ISArticle.asp?id=67141&issue=04012007, accessed 15 August 2008.

Earthworks and Oxfam America (2004) 'Dirty Metals: Mining, Communities and the Environment'; www.nodirtygold.org/dirty_metals_report.cfm, accessed 9 August 2008.

Echave, J., K. Keenan, M.K. Romero and Á. Tapia (2005) *Los procesos de diálogo y la administración de conflictos en territorios de comunidades: El caso de la Mina de Tintaya en el Perú* (Lima, Peru: CooperAcción).

Forest Peoples Programme and Tebtebba Foundation (2006) *Indigenous Peoples' Rights, Extractive Industries and Transnational and Other Business Enterprises* (Moreton-in-Marsh, UK: Forest Peoples Programme; Baguio City: Tebtebba Foundation).

ICMM (International Council on Mining and Metals) (2003) 'Ten Principles for Sustainable Development Performance'; www.icmm.com/our-work/sustainable-development-framework/10-principles, accessed 9 August 2008.

—— (2006) 'Draft Position Statement on Mining and Indigenous Peoples Issues'; www.icmm.com/page/2075/ICMM-releases-findings-of-survey-on-indigenous-peoples-issues, accessed 6 October 2008.

IFC (International Finance Corporation) (2006) 'International Finance Corporation Performance Standard 7: Indigenous Peoples'; www.ifc.org/ifcext/enviro.nsf/Content/PerformanceStandards, accessed 10 August 2008.

—— (2007) 'ILO Convention 169 and the Private Sector: Questions and Answers for IFC Clients'; www.ifc.org/ifcext/sustainability.nsf/Content/Publications_GoodPractice_Labor, accessed 10 August 2008.

—— (undated a) 'Indigenous Peoples Fact Sheet', www.ifc.org/.../eir.nsf/AttachmentsByTitle/IndigenousPeople1/$FILE/INDIGENOUS+PEOPLES+FACT+SHEET.pdf.

—— (undated b) 'Stakeholder Engagement: A Good Practice Handbook for Companies Doing Business in Emerging Markets', www.ifc.org/ifcext/sustainability.nsf/Content/Publications_GoodPractice_StakeholderEngagement, accessed 20 August 2008.

Macdonald, I., and C. Rowland (eds.) (2002) *Tunnel Vision: Women, Mining and Communities* (Melbourne: Oxfam Community Aid Abroad).

Mineral Policy Center (1997) *Mining and Sacred Sites: Summaries of Native American Mining Issues* (Washington, DC: Mineral Policy Center).

Mines and Communities (2001) 'London Declaration', 20 September 2001; www.minesandcommunities.org//article.php?a=8470, accessed 6 October 2008

Mining Association of Canada (2006) 'Towards Sustainable Mining'; www.mining.ca/www/Towards_Sustaining_Mining/index.php, accessed 15 August 2008.

MiningWatch Canada and CCISD (Canadian Consortium for International Social Development) (2000) *On the Ground Research: A Workshop to Identify the Research Needs of Communities Affected by Large-Scale Mining, 14–16 April 2000* (Ottawa: MiningWatch Canada/CCISD).

Miranda, M., D. Chambers and C. Coumans (2005) 'Framework for Responsible Mining: A Guide to Evolving Standards'; www.frameworkforresponsiblemining.org, accessed 20 August 2008.

Nones, J.A. (2007) 'Newmont Leads Industry in Mining Sustainability', *ResourceInvestor*, 14 May 2007; www.resourceinvestor.com/pebble.asp?relid=31783, accessed 10 August 2008.

OMAN (Ontario Mining Action Network) (2005) 'Comments on Proposed New Ontario Mineral Development Strategy'; www.miningwatch.ca/index.php?/17/OMAN_OMDS, accessed 9 August 2008.

Oxfam Australia (2004) 'Mining Ombudsman Report 2004'; www.oxfam.org.au/campaigns/mining/reports, accessed 9 August 2008.

—— (2007) 'Free, Prior and Informed Consent'; www.oxfam.org.au/campaigns/mining/docs/FPIC_statement.pdf, accessed 20 August 2008.

PDAC (Prospectors and Developers Association of Canada) (2008) 'Sustainable Development Framework for Exploration'; www.pdac.ca/pdac/advocacy/csr/index.html, accessed 15 August 2008.

Ruggie, J.G. (2006) 'Remarks at Public Session', *National Roundtable on Corporate Social Responsibility and the Canadian Extractive Industry in Developing Countries*, Montreal, 14 November 2006; geo.international.gc.ca/cip-pic/library/CSR_Montreal_Presentation_J_Ruggie.pdf, accessed 15 August 2008.

Sewall, C. (1999) *Digging Holes in the Spirit: Gold Mining and Survival of the Western Shoshone Nation* (Oakland, CA: Western Shoshone Defense Project and Project Underground).

Tauli-Corpuz, V. (2005) 'Visions and Movements of Indigenous Peoples for a New Community', *The Ecumenical Review*, April 2005; findarticles.com/p/articles/mi_m2065/is_2_57/ai_n15954375, accessed 20 August 2008.

—— and J. Carino (eds.) (2004) *Reclaiming Balance: Indigenous Peoples, Conflict: Resolution and Sustainable Development* (Baguio City, Philippines: Tebtebba).

Tebtebba Foundation (2006) *UN Declaration on the Rights of Indigenous Peoples and Programme of the Second International Decade of the World's Indigenous People* (Baguio City, Philippines: Tebtebba Foundation).

Weitzner, V. (2002a) *Cutting-Edge Policies on Indigenous Peoples and Mining: Key Lessons for the World Summit and Beyond* (Ottawa: North-South Institute).

—— (2002b) *Through Indigenous Eyes: Towards Appropriate Decision-Making Processes Regarding Mining on or near Ancestral Lands* (Ottawa: North-South Institute).

—— (2006) *'Dealing Full Force': Lutsel K'e Dene First Nation's Experience Negotiating with Mining Companies* (Ottawa: North-South Institute).

World Bank (2005) 'Revised Operational Policy and Bank Procedure on Indigenous Peoples (OP/BP 4.10)'; web.worldbank.org/WBSITE/EXTERNAL/TOPICS/EXTSOCIALDEVELOPMENT/EXTINDPEOPLE/0,,menuPK:407808~pagePK:149018~piPK:149093~theSitePK:407802,00.html, accessed 20 August 2008.

4

Understanding corporate–Aboriginal agreements on mineral development

A conceptual framework

Ciaran O'Faircheallaigh

Department of Politics and Public Policy, Griffith University, Australia

Negotiation of legally binding agreements between mining companies and Aboriginal groups represents a critical aspect of CSR in resource-rich industrialised countries such as Australia and Canada and, increasingly, in developing countries (Banks and Ballard 1997: Appendix 1; Brew 1998; ICME 1999; IIED and WBCSD 2002; Langton *et al.* 2004; O'Faircheallaigh 2006a; O'Reilly and Eacott 1999; Sosa and Keenan 2001). The negotiation of agreements is generally regarded as a positive sign of a new willingness by mining companies to engage with Aboriginal groups in a serious and sustained manner, to share with them the wealth generated by mining on Aboriginal lands, and to allow them a say in the manner in which mines are developed and operated (Environmental Law Institute 2004: 11, 13-14; ICME 1999: vii; Keon-Cohen 2001; Meyers 1996; Miranda *et al.* 2005: 69-70; Senior 1998). Typical of this view is the statement by the CEO of Falconbridge Ltd on the signing of an agreement between his company and the Nunavik Inuit for the development of the Raglan nickel mine in Northern Quebec: 'With its commitment to the people of Nunavik, the Raglan Agreement stands as a landmark in Canadian mining history and in the development of the Canadian North. Raglan marks the beginning of a new era in mining' (cited in ICME 1999: 30).

But what do we actually know about the consequences for Aboriginal peoples of entering into agreements with mining companies? A growing literature on such agreements provides information on the content of agreements (Kennett 1999; O'Faircheallaigh and Corbett 2005; O'Faircheallaigh 2006b; Sosa and Keenan 2001) and discusses the processes and strategies involved in negotiating agreements and in particular the requirements for 'successful' negotiations by Aboriginal groups (Barsh and Bastien 1997; O'Faircheallaigh 1996, 2000; Weitzner 2006). There is a more limited literature focusing on criteria for evaluating agreements (O'Faircheallaigh 2004b), and dealing with some more general issues that arise across agreements, such as their enforceability and the question of whether Aboriginal–company agreements should provide consent for mining (Keeping 1998; Kennett 1999; O'Reilly and Eacott 1999).

Not all of this literature is uncritical of agreements and some of it highlights, for instance, a lack of Aboriginal community involvement in negotiation processes and the limited benefits gained from agreements by some Aboriginal groups (O'Faircheallaigh 2008; Weitzner 2006). However, even the more critical analysis tends to focus heavily on the agreements themselves, on the processes that give rise to them, their content, the immediate impact of their provisions, and how more positive outcomes could be achieved from the agreement making process (O'Faircheallaigh 2006a, 2008; O'Reilly and Eacott 1999; Weitzner 2006). There is very little in the literature that attempts to provide an overall conceptual understanding of the wider legal, political and institutional ramifications of agreement-making between Aboriginal groups and mining companies. In particular, there is little analysis of the way in which agreement-making influences other options and strategies available to Aboriginal groups faced with mineral development on their ancestral land; or of the implications of Aboriginal agreements with private corporations for relations between Aboriginal peoples, the state, other political interests and civil society more generally.

This chapter makes a preliminary and exploratory contribution to such a conceptual understanding. It does so by seeking to locate Aboriginal–mining company agreements within a broader set of relationships between Aboriginal peoples and other elements of a liberal democratic political system within which mining projects are approved and regulated. The specific context for the analysis is therefore settler states that have significant Aboriginal populations, in particular Australia, Canada, New Zealand and the USA.

The chapter first considers the 'counterfactual': that is, a situation in which no contractual arrangements exist between affected Aboriginal groups and mining companies wishing to develop resources on Aboriginal land. It then examines how the creation of a contractual relationship through negotiation of project-based agreements between Aboriginal groups and mining companies (and in certain cases government also) affects the legal and political status of Aboriginal groups and the nature of their relationship with other elements of the political system. The comparison between the 'counterfactual' and the 'agreement' scenarios highlights some major conceptual and practical issues raised by Aboriginal participation in contractual arrangements. It also facilitates analysis of a fundamental issue that rarely features in the literature on Aboriginal–mining company agreements: whether, and under what conditions, entering agreements is or is not likely to create net benefits for Aboriginal groups.

The counterfactual: absence of an agreement

Figure 4.1 illustrates the counterfactual, where no contractual arrangement exists between Aboriginal peoples, mining companies and governments in relation to a mining project. The situation this creates for Aboriginal groups can be summarised as follows.

Aboriginal people are unconstrained in pursuing political strategies designed to halt project development or change the nature or timing of development that does occur. Reflecting this fact, the circle representing Aboriginal people lies entirely in the 'Political' zone of Figure 4.1. Aboriginal groups can, for instance, seek public support through the media, build political alliances with NGOs such as environmental and church groups, lobby government and mobilise pressure on corporations and their shareholders (activities represented by the black lines emerging from the 'Aboriginal' circle in Fig. 4.1). For example, Innu and Inuit landowners in Labrador used a number of these strategies to delay the development of the proposed Voisey's Bay nickel project in the late 1990s. The Mirrar, Aboriginal traditional owners of the land on which the proposed Jabiluka uranium project in Australia's Northern Territory is located, used a combination of all of them to oppose development of the deposit. They were ultimately successful, with Rio Tinto agreeing to refill a portal that had been constructed to start mine development and committing not to recommence development without the consent of the Mirrar (Gibson 2006; Katona 2002).

Aboriginal access to components of the judicial and regulatory system that are relevant to project approval and management (represented by the line at the top of Fig. 4.1 and marked 'Litigate, Object, etc.') is unconstrained by any contractual obligations. Aboriginal people can exercise rights available to citizens generally or rights arising from any specific property or other Aboriginal interests they hold. Those rights may allow them, for instance, to challenge the level of environmental assessment proposed for a project; to take legal action to prevent damage to Aboriginal cultural heritage or the environment; or to sue for compensation if such damage occurs. Using legal and procedural rights and political strategies, Aboriginal groups may be able to influence the terms of contractual and regulatory instruments negotiated between the state and the developer: for instance, by helping to shape the conditions attached to environmental approvals and mining leases.

While Aboriginal groups are in principle free to access all available legal and political strategies to influence decisions regarding development on their ancestral lands, in practice a number of factors may constrain their capacity to do so. The first involves their often limited access to financial resources and relevant expertise. It is costly to bring people together to discuss and agree on proposed strategies, to engage in legal action, or to make representations to governments or corporations in distant capital cities. The cost of acquiring access to the technical and political expertise required to effectively pursue legal and political strategies is also high.

The second constraint involves the fact that in many contexts legal recognition of Aboriginal rights and interests may be limited, and so in practice few judicial or regulatory avenues may be open to them. This is particularly so where Aboriginal peoples have been heavily impacted by European settlement. Thus a number of Aboriginal groups in regions of eastern and southern Australia settled in the early 19th century; for instance,

FIGURE 4.1 Mineral development process without aboriginal–mining company agreement

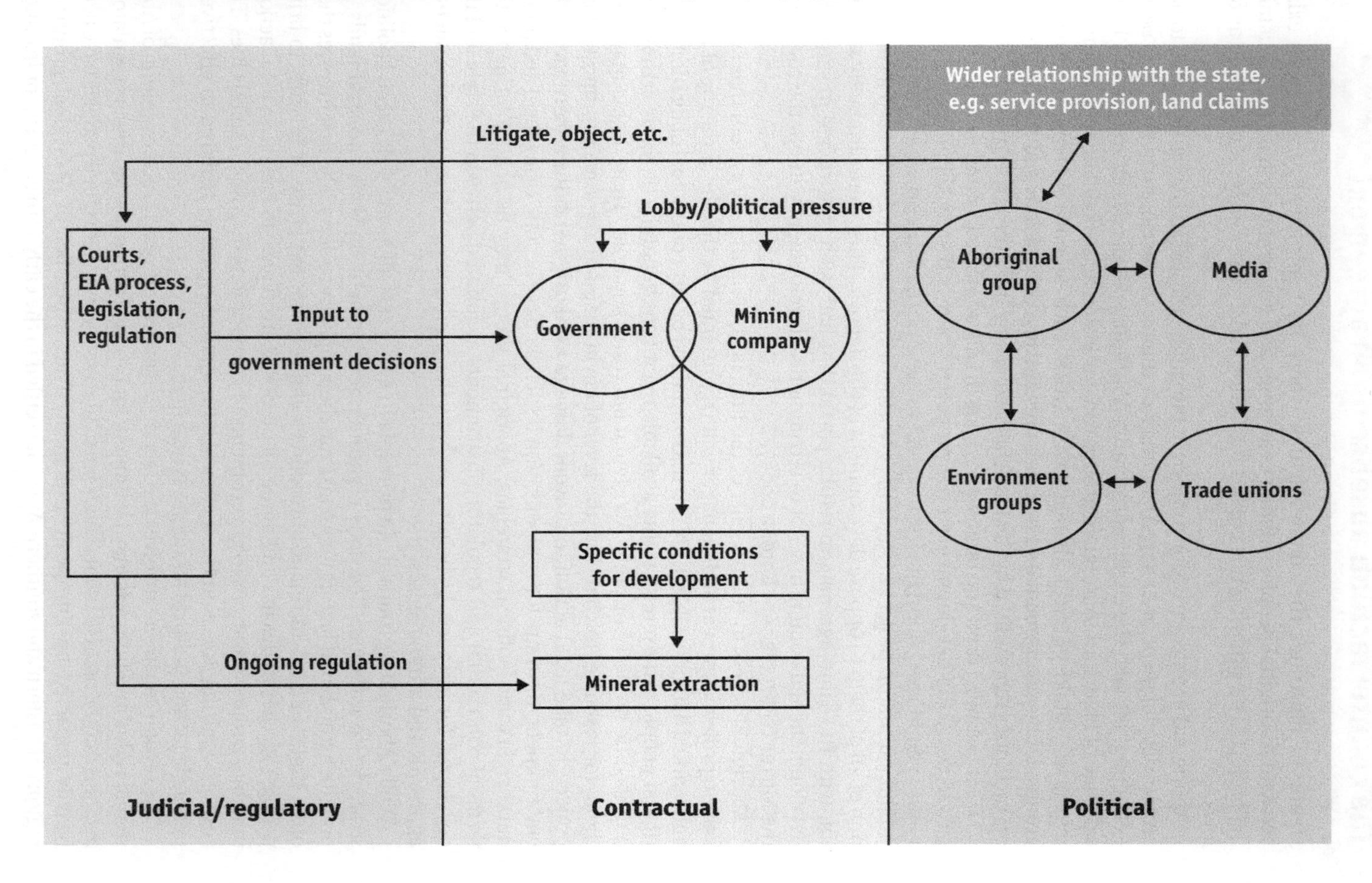

the Yorta Yorta traditional owners of the Murray River basin in Victoria have been unable to secure recognition of their native title (Strelein 2006), and so may face problems in gaining standing in relevant legal processes.

Political strategies and use of legal rights are generally more effective at minimising the potentially negative environmental and cultural effects of a project than at maximising their potential economic and social benefits for Aboriginal groups, for a number of reasons. First, legal processes are generally reactive and respond to actions by individuals or corporations that wish to prevent damage to property or other interests or seek compensation for damage that has occurred. They are not oriented towards creating economic opportunities for specific groups in society. Second, project approval processes such as environmental impact assessments (EIAs) tend to be biased towards minimising the biophysical effects of a project rather than towards addressing either its negative or potentially positive economic and social aspects. Thus the Final Report of the Mining, Minerals and Sustainable Development (MMSD) initiative, for instance, noted that social factors have only recently started to 'creep' into environmental impact assessment (IIED and WBCSD 2002: xxi). Third, political strategies such as the creation of alliances with environmental NGOs or the use of media campaigns are generally easier to pursue when they focus on prevention of damage to ecosystems or to 'traditional' Aboriginal culture than when they focus on ensuring that Aboriginal people share in the economic benefits of mining. For example, Aboriginal groups in Western Australia are finding it more problematic to recruit the support of environmentalists for their demands to obtain an equitable share of benefits from current resource development projects than they did to gain support for opposing resource projects that threatened Aboriginal sacred sites in the early 1980s (Hawke and Gallagher 1989; Kimberley Land Council 2008).

Another drawback for Aboriginal people is that they have little opportunity to be directly involved in managing the environmental and other impacts of projects on an ongoing basis. They may be able to utilise EIA processes to influence the nature of the environmental protection regimes that are initially established. However, the decisions that emerge from EIA processes tend to be 'once and for all' and not to be revisited during the life of a project (Howitt 2001: 337-38; Joyce and MacFarlane 2001: 3, 12; O'Faircheallaigh 2007: 322-33), excluding any opportunity for an ongoing role in environmental decision-making and management for Aboriginal people. The environmental management regimes established pursuant to legislation and EIA decisions are applied to projects by companies and governments. This can be a major problem for Aboriginal people who believe they have responsibilities for looking after their ancestral land that cannot be delegated to others.

Turning to responsibility for enforcing, or the availability of legal mechanisms or institutional capacity to enforce, conditions imposed on resource extraction, in principle there is no ambiguity. Government bears the sole responsibility and has available to it a range of bureaucratic and legal resources to ensure corporate compliance. This is the case whether these conditions derive from contractual arrangements between government and developers or from general legislation and regulation (see Fig. 4.1). This does not of course mean that government will in practice act to ensure such compliance (see, for instance, Anguelovski, Chapter 11 this volume; Crate and Yakovleva, Chapter 12 this volume; M. Smith 2007).

The impact of Aboriginal–mining company agreements

When Aboriginal groups enter contractually binding agreements with mining companies, some of the processes and relationships represented in Figure 4.1 and discussed above change significantly. An analysis of the changes that occur is critical in achieving a broad, conceptual understanding of the impact and implications of agreement-making. To date, analysis of this sort has been very limited. Figure 4.2 represents a starting point for the discussion. This portrays the new situation, with the sphere representing Aboriginal people now overlapping the 'Political' and 'Contractual' zones of the diagram, and areas, processes or relationships that may change as a result of negotiated agreements highlighted as thick (and in most cases broken) lines.

One of the most significant changes involves the access of Aboriginal groups to the judicial and regulatory systems (represented by the broken line at the top of Fig. 4.2). At least three features of negotiated agreements are relevant here. First, recent agreements in the case study countries almost always involve Aboriginal support for the project concerned, and/or for the grant of specific titles or approvals required for the project to proceed. For example, Kennett (1999) notes that many agreements in Canada contain specific provisions that commit the Aboriginal party either to support the project involved or to refrain from opposing it in environmental assessment or regulatory proceedings. One agreement states that in consideration for the mining company entering into the agreement, the Aboriginal signatories 'will not object to the issuance of any licences, permits, authorizations or approvals to construct or operate the Project' (Kennett 1999: 45-46). Another agreement involves broader commitments, requiring the Aboriginal party not to 'institute any legal proceedings or engage in or undertake any other actions or activities to prevent or delay authorization of the [mining project]'. A number of agreements commit the Aboriginal parties to not opposing projects in the event that they become subject to an environmental assessment as a result of actions taken by non-signatories to the agreements (Kennett 1999: 45-46).

It follows that Aboriginal groups may be contractually constrained in their ability, for instance, to object to government approval of a project either in principle or in its current form. Thus, for example, the operator of one project in Canada utilised the existence of such clauses to argue that an Aboriginal signatory to the agreement was prohibited from objecting to the grant of a water licence required to allow expansion of the project.

Second, some agreements contain provisions preventing Aboriginal groups from utilising specific legal or regulatory avenues that would otherwise be available to them. For example, under one recent Australian agreement the Aboriginal parties undertake not to 'lodge any objections, claims or appeals to any Government authority . . . under any [state] or Commonwealth legislation, including any Environmental Legislation' (O'Faircheallaigh and Corbett 2005: 637). Third, agreements may contain dispute resolution processes that preclude the parties from initiating legal proceedings to resolve disputes, or require all other potential avenues for resolving disputes to be exhausted before they do so. In combination, such provisions can create a fundamental shift in the ability of Aboriginal groups to exercise legal rights they would otherwise have available and more generally to access legal and regulatory regimes relevant to resource extraction.

A second important area involves the ability of Aboriginal groups to build relationships with and to mobilise other political actors. The common requirement for Aborig-

FIGURE 4.2 Mineral development process with aboriginal–mining company agreement

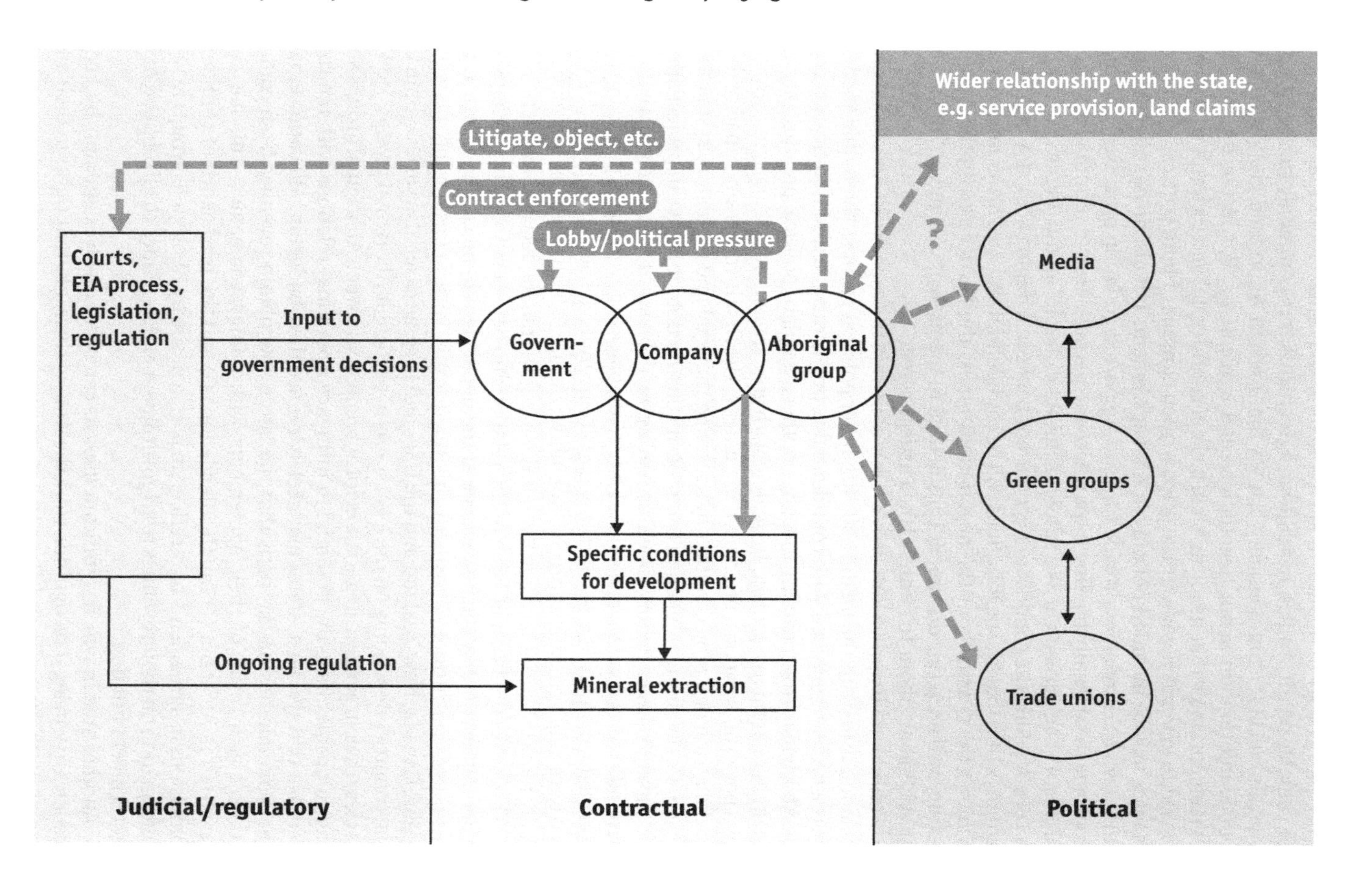

inal groups to support a project immediately limits their capacity to manoeuvre politically, particularly in relation to environmental and other groups that might otherwise be valuable political allies. In addition, agreements very commonly (indeed almost universally) include confidentiality clauses that prevent Aboriginal groups from making public information about negotiations and agreements. For instance, one Canadian agreement states that its terms 'shall be kept confidential and its terms shall not be disclosed to any party, without the prior written consent of both parties'. Confidentiality clauses may be included not only in final agreements but also in negotiation protocols under which companies provide funds to support negotiation processes; and they may continue to be legally binding even where the parties agree to terminate a negotiation protocol or an agreement as a whole. Thus, for example, an Australian negotiation protocol governing the conduct of negotiations between a mining company and Aboriginal groups states that 'the negotiations will be confidential and the Parties must not say anything to the public about the negotiations unless the Parties agree otherwise . . . the Parties will keep this document confidential to themselves . . . and not make any public statements about [it] unless [the Parties] agree.'

Confidentiality provisions can severely constrain the capacity of Aboriginal groups to communicate with the media and with other political groups. The requirement to support a project combined with confidentiality provisions can also significantly constrain an Aboriginal group's ability to lobby or otherwise place political pressure on a government in relation to a project. In dealing with government, most Aboriginal groups have two powerful weapons, often used in tandem. The first involves any capacity they have to delay or halt a project, either by accessing the legal and regulatory systems and, for example, obtaining injunctions on project construction or delays in project approvals; or through direct action aimed at halting or delaying development activity on the ground. The second involves the ability to embarrass government politically by using the media to appeal to its constituents (Gibson 2006; Trebeck, Chapter 1 this volume). If contractual agreements preclude or inhibit the use of both weapons, this may substantially reduce Aboriginal capacity to influence government decision-makers.

This raises the broader issue of the relationship between Aboriginal groups and the state, represented by the broken line at the upper right hand corner of Figure 4.2. The legal and constitutional basis for this relationship varies considerably between Australia, Canada, New Zealand and the USA, and in some cases also varies within individual countries depending on the legal status of particular Aboriginal groups. However, it is clear that, in general, negotiations of agreements between Aboriginal groups and mining companies have the potential to influence Aboriginal relations with the state in a number of ways. First, states may seek to reduce their budgetary allocations to Aboriginal communities on the basis that the latter now obtain revenues from commercial sources as a result of their agreements with mining companies. This has certainly occurred in Australia (O'Faircheallaigh 2004a), and the prevalence of confidentiality provisions in agreements may reflect, in part, a desire by Aboriginal groups to withhold information on their revenues from government and so reduce the likelihood of a cut in government funding.

Another area in which significant impacts can occur involves attempts by Aboriginal peoples to win legal recognition from the state of their inherent rights to their ancestral states. Both Australia and Canada, for instance, have been and continue to be extensively involved in negotiations and/or litigation with Aboriginal groups regarding either recognition of their 'native title' for the first time through negotiation of comprehensive land claim settlements (Canada) or determinations of native title (Australia), or regarding implementation of treaty obligations that the state has historically ignored. The discovery of a major mineral deposit on an Aboriginal group's land often focuses state attention on land tenure issues, in many cases in response to corporate pressure on state agencies and on political leaders to have these issues resolved as a precondition for undertaking major capital investments.

The implications of a stronger state focus on resolving land tenure issues as a result of major mineral discoveries are unclear and require further research. On the one hand, Aboriginal groups, frustrated over many years by their inability to gain the attention of government decision-makers and by the unwillingness of government to properly resource land claim processes, may suddenly find that the purse strings are loosened and that their access to decision-makers is enhanced. They may be able to use the leverage provided by the government's desire to secure project approval in order to achieve their goals in relation to much wider areas of land than that involved in the project concerned. For example, Innu and Inuit groups in Labrador had made limited progress on settling their claims with Canada prior to the discovery of the Voisey's Bay nickel deposit. The prospect of the project's development resulted in a much more active stance by Canada in relation to the negotiations, and led to the conclusion of Interim Land Claim Agreements with both groups in 2002.

On the other hand, the state is usually driven by a desire to resolve land tenure issues both speedily and in a way that facilitates project development, and as a result Aboriginal groups may find themselves under enormous pressure to accept land claim settlements that do not satisfy their wider, long-term aspirations. Significant issues also arise in relation to implementation of settlements. If government is driven primarily by a desire to have projects approved, then once this has occurred its commitment and willingness to provide resources may quickly evaporate. The 2001 Western Cape Communities Co-existence Agreement (WCCCA) between the State of Queensland, Comalco Ltd and Aboriginal traditional owners highlights this issue. To facilitate the agreement Queensland undertook to transfer substantial areas of land not required by Comalco for mining to its traditional Aboriginal owners, but seven years later not a single hectare has actually been transferred.

Agreement provisions regarding Aboriginal support and confidentiality can also result in fundamental changes in the way in which Aboriginal groups relate to mining companies. As Trebeck explains in Chapter 1, the willingness of corporations to undertake CSR initiatives in relation to any social group depends, in large measure, on the capacity of that group to inflict damage on the corporation by threatening its social licence to operate. In her words, groups must apply 'an ever-present threat of the loss of social licence to operate in order to ensure that companies recognise and address [their] demands . . . civil society organisations need to maintain surveillance and pressure to ensure it is always in the corporate interest to respond to community demands' (Trebeck, Chapter 1 this volume, page 20). She notes in particular that the capacity of groups to threaten the reputation of corporations is a 'crucial lever'. Where agreements bind Aboriginal groups to support corporate activities and silence them through confidentiality provisions, they have substantially surrendered their ability to threaten a company's licence to operate. It may of course be the case that this threat is no longer needed, because agreements contain legally enforceable provisions that ensure the ongoing performance by a company of certain CSR obligations. Two points remain. First, the nature of the relationship between Aboriginal groups and companies has profoundly changed. Second, the question of whether obligations taken on by corporations through agreements with Aboriginal groups are both substantial and enforceable and so represent a 'fair trade' for the forbearance promised by those groups cannot be resolved *a priori*, but only through an examination of the provisions of individual agreements. Another important issue here involves the length of time over which agreements apply, which is typically for the whole of project life and for major projects is often measured in decades rather than years. If Aboriginal groups discover after the event that the trade-off they have made is not to their advantage, it may be a very long time before they have an opportunity to change the situation.

This brings us to the new line in Figure 4.2 linking the interface between the 'company' and 'Aboriginal group' circles directly to the setting of specific conditions for mineral extraction. This represents a further and very significant change from the 'counterfactual', and indeed provides the key rationale for Aboriginal groups to enter binding agreements. It provides a mechanism through which they can simultaneously pursue two key sets of objectives.

First, binding agreements offer opportunities to share in the economic benefits generated by resource extraction. For instance, they can offer Aboriginal groups access to an income stream in the form of royalty or other payments. This can assist in meeting a community's short-term and often urgent need to fund services such as housing, health and education and to augment Aboriginal incomes that are usually a fraction of the national average. As noted above there is a danger that government expenditures will fall, negating some of this impact, but on the other hand if used judiciously mining revenues can be used to leverage additional public expenditure. For example, the Gagudju traditional owners of the Ranger uranium mine used their royalties to start new education and health initiatives that government had refused to fund, but, once these were operational, they negotiated for government to take over their funding. The traditional owners of the Argyle diamond mine have used part of their income stream to create a 'partnership fund', from which money can only be committed if government is willing to provide matching funds. On this basis they have established projects in areas including kidney health, health education, sports, law, culture and youth suicide prevention (Gelganyem Trust and Kilkayi Trust 2006, 2007). At a broader level, access to mining income can provide Aboriginal groups a degree of autonomy from the state, allowing them to establish their own priorities rather than having to accept the state's priorities as a condition for access to public funding; and adding to their negotiating power in dealing with the state in relation to, for instance, service delivery, land title and management, and governance.

In the longer term, income streams from mining create the potential for Aboriginal groups to establish capital funds that will generate income into the future and indeed long after mining has ceased. For instance, the Aboriginal signatories to the WCCCA decided to invest in excess of 50% of their revenues in a long-term investment fund. Income is reinvested for 20 years, after which the capital base is preserved and interest is available to fund current spending. The capital fund already sits at more than A$30 million, and by 2021 it will have the capacity to generate ongoing and substantial income.

Agreements can also offer preferential access to employment and training and business development opportunities for members of Aboriginal groups or for Aboriginal corporations. Income levels tend to be considerably lower, and unemployment levels considerably higher, in Aboriginal communities than in mainstream ones, and access to such opportunities can be critical in helping to overcome Aboriginal economic disadvantage. To take one example, Aboriginal people accounted for less than 5% of the workforce of Argyle Diamonds Ltd when negotiations for an agreement with traditional owners commenced in 2001. Today they account for over 20% of the workforce, generating employment for some 100 additional Aboriginal employees. At average 2007 mining industry earnings of A$90,000, this represents additional income of some A$9,000,000 annually for Aboriginal workers, families and communities.

Second, agreements can provide opportunities to be involved proactively, and on an ongoing basis, in managing the cultural, social and environmental impacts of resource extraction. These might include a key and possibly leading role in defining, identifying and protecting Aboriginal cultural heritage; participation in joint company–Aboriginal environmental management regimes; and involvement in decisions regarding project expansions and project closure. For instance, under the WCCCA, traditional owners are funded to operate a cultural heritage protection system intended to avoid damage to sites of significance; receive annual payments to help support a ranger programme designed to control impacts associated with the activities of mining town residents and tourists; have an opportunity to comment on all applications for environmental permits; and must be consulted by the mine operator regarding implementation of its environmental management system and any major project changes that may have a significant impact on the environment. As discussed earlier, a mineral development system missing an Aboriginal–corporate contractual arrangement, as represented in Figure 4.1, offers limited opportunities for pursuing this sort of ongoing involvement in managing environmental and cultural impacts.

Turning to the issue of enforcing conditions attached to mining projects, as indicated above in the 'no agreement' model in Figure 4.1 responsibility lies with government, which has at its disposal appropriate enforcement mechanisms and the resources to apply them. With the introduction of a company–Aboriginal agreement, part of the responsibility for enforcement falls on Aboriginal peoples, as indicated by the insertion of 'Contract enforcement' alongside the line linking the Aboriginal group to the Courts (Fig. 4.2). To date, major problems have arisen in relation to enforcement of conditions negotiated between mining companies and Aboriginal groups because many agreements do not include the sorts of provision that are required to maximise the chances that they will be fully and effectively enforced and implemented (O'Faircheallaigh 2002, 2003). Critical requirements include:

1. Allocation of the financial and human resources required for enforcement and implementation. This does not simply involve funding for the direct costs of specific initiatives (for instance, for apprentices' wages, or for the cost of delivering cultural awareness training). It also includes resources to support the general implementation of agreements, to ensure that all of the initiatives contemplated by an agreement actually happen when and how they are intended to happen

2. Identification of goals that are clear and precise and linked to specific timeframes. For instance, while there is debate within some mining companies regarding the merits of targets for Aboriginal employment, international experience shows that absence of specific, numerical goals makes it extremely difficult to generate the organisational momentum required to significantly change the composition of the workforce
3. Clear specification of who is responsible for delivering on agreement commitments, and allocation of responsibility to individuals who have the authority and capacity to deliver
4. Inclusion of incentives and penalties designed to ensure that commitments are adhered to, especially over the long term. Turnover in mining companies and among non-Aboriginal staff in Aboriginal organisations is high, and as a result those responsible for implementation may not have been involved in negotiating agreements and may have little personal stake in their successful implementation. In this situation especially, substantial, and if possible automatic, penalties and incentives are required
5. Systematic, ongoing information-gathering and monitoring is essential to establish progress towards goals, attract attention to implementation failures, and provide a basis for developing alternative or additional initiatives, if necessary. Information must be reported to the affected parties on a regular basis and in a form that allows its significance to be easily understood. While the need for regular monitoring of activities and of goals against objectives is very firmly established in relation to the environmental regimes applied to mining companies, it is still the case that few mining company–Aboriginal agreements provide for any systematic monitoring of progress or achievements against agreement goals
6. Creation of structures such as joint Aboriginal community–mining company coordination committees whose *primary* purpose is to ensure that implementation occurs. These structures need to be appropriately resourced and to include the key individuals from mining companies and Aboriginal organisations with authority in relation to the issues involved. The regular involvement of senior personnel with decision-making authority is an essential prerequisite

In some cases implementation problems will be of a sort or on a scale that cannot be addressed within the existing terms of an agreement. Indeed, given that many mining agreements have long terms, frequently in excess of 20 years, there is a strong likelihood that during their lives the broader social and economic environment will change substantially, raising issues regarding the appropriateness of agreement goals and of the mechanisms provided for their implementation. Thus there is also a need for effective mechanisms that, while protecting the underlying interests of the parties, facilitate regular review of agreements and ensure that they can be amended in a timely manner and without undue cost.

Achieving the requirements for effective enforcement and implementation places a substantial burden on Aboriginal groups, but one they must meet successfully if the potential benefits of contractual agreements are to be realised in practice.

Implications for research and for Aboriginal negotiation strategies

This discussion raises a series of significant questions, each of which could become the focus of a substantial research effort. What exactly is the impact of contractual agreements with mining companies on the access of Aboriginal people to the judicial and regulatory system, and how does this affect their ability to minimise negative impacts of mining projects on their ancestral lands? In the absence of agreements, do Aboriginal groups actually have the capacity to utilise the judicial and regulatory system? In other words, *in practice* what do they lose if they accept constraints on their rights to access that system? What is the effect of the common requirement in agreements for Aboriginal support of projects on the political flexibility of Aboriginal groups? Where does the impetus for confidentiality provisions originate, and to what extent do they restrict the freedom of Aboriginal groups in lobbying government and in building and maintaining political alliances and public support? How do contractual arrangements affect wider relationships between Aboriginal groups and the state, especially in relation to the state's obligations to provide services to its (Aboriginal) citizens, and to resolution of wider issues regarding land tenure and treaty rights? Can Aboriginal groups secure a share of economic benefits from mining projects and become directly involved in managing their environmental and cultural impacts in the absence of contractual agreements? Can contractual arrangements be negotiated with mining companies that allow a sharing of benefits and direct Aboriginal participation in project management, yet do not close off judicial, regulatory and political options that might strengthen the capacity of Aboriginal groups to control and shape the impact of mineral development on their ancestral lands?

Even in the absence of more systematic research on these issues, the earlier discussion seems to have some clear implications for Aboriginal negotiation strategies. First, it is important for Aboriginal groups to undertake, at an early stage in project development, a mapping exercise of the sort attempted in Figures 4.1 and 4.2 so that they can consider how negotiation of a contractual relationship with a mining company may affect their engagement with the political and judicial/regulatory system as a whole, including their existing interaction with government in areas such as service provision and land claim negotiations. In my experience of negotiations in Australia and Canada, this is in fact rarely done. Such an exercise can reveal both threats (for instance, a group's inability to maintain valued political alliances, or a decline in government service provision), opportunities (for example, an increased capacity to engage with government decision-makers) and challenges (for instance, the need to develop a capacity to enforce contractual obligations and ensure effective implementation). In the absence of such an exercise and of preparatory work following on from it, Aboriginal groups may be poorly prepared to deal with threats and poorly placed to grasp opportunities and meet challenges.

Second, it is important to question the assumption (prevalent in much writing on CSR) that negotiation of contractual arrangements with mining companies represents the preferred (indeed the only feasible) method of resolving issues raised by mineral development on Aboriginal land. There are in fact legal and political strategies available to Aboriginal groups, and the potential costs and benefits of these strategies need to be assessed in comparison to the costs and benefits offered by agreement-making.

Pursuit of negotiated agreements with mining companies and use of legal and political strategies are not of course mutually exclusive. Indeed, it can be argued that Aboriginal groups maximise their gains from negotiated agreements where negotiations are *accompanied by* the strategic use of litigation, of direct action designed to attract media attention and political support, and building of alliances with other political interests such as environmental groups or trade unions. This raises a third point. It is clearly advisable for Aboriginal groups to maintain their freedom of movement for as long as possible and to the greatest degree possible. It may be inadvisable, for instance, to accept confidentiality provisions in a negotiation protocol, as this may prevent mobilisation of the media and of political allies during the negotiation process. More generally, all proposals to enforce confidentiality (beyond matters that are commercially sensitive for the developer) should be carefully scrutinised, as should any proposed agreement provisions restricting Aboriginal access to the judicial and regulatory system or requiring broad Aboriginal support for a project (as opposed to specific Aboriginal approvals that are an essential precondition for a project to proceed).

This does not imply that Aboriginal groups can both gain the benefits associated with contractual agreements and at the same time retain all of the freedom and options available to them in the absence of such agreements. It does imply that entering contracts with corporate interests has wider and important implications for relationships between Aboriginal groups, the state and civil society. These implications need to be carefully considered both in shaping negotiation strategies and, ultimately, in determining whether contractual relationships represent the best way of pursuing Aboriginal interests.

References

Banks, G., and C. Ballard (eds.) (1997) *The Ok Tedi Settlement: Issues, Outcomes and Implications* (Pacific Policy Paper 27; Canberra: Australian National University, National Centre for Development Studies and Resource Management in the Asia Pacific).

Barsh, R.L., and K. Bastien (1997) *Effective Negotiation by Aboriginal Peoples: An Action Guide with Special Reference to North America* (Geneva: International Labour Office).

Brew, R. (1998) 'The Lihir Experience: Project Development Issues in Papua New Guinea', paper presented at *PNG Mining and the Community Conference*, Madang, Papua New Guinea, 26–29 July 1998.

Environmental Law Institute (2004) *Prior Informed Consent and Mining: Promoting the Sustainable Development of Local Communities* (Washington, DC: Environmental Law Institute).

Gelganyem Trust and Kilkayi Trust (2006) *Annual Report April 2005–June 2006* (Kununurra, WA: Gelganyem Trust and Kilkayi Trust).

—— (2007) *Future, Country: Gelganyem, Kilkayi* (Kununurra, WA: Gelganyem Trust and Kilkayi Trust).

Gibson, R. (2006) 'Sustainability Assessment and Conflict Resolution: Reaching Agreement to Proceed with the Voisey's Bay Nickel Mine', *Journal of Cleaner Production* 14.3–4: 334-48.

Hawke, S., and M. Gallagher (1989) *Noonkanbah, Whose Land, Whose Law?* (Fremantle, WA: Fremantle Arts Center).

Howitt, R. (2001) *Rethinking Resource Management: Justice, Sustainability and Indigenous Peoples* (London/New York: Routledge).

ICME (International Council on Metals and the Environment) (1999) *Aboriginal Peoples and Mining: Case Studies* (Ottawa: ICME).

IIED (International Institute for Environment and Development) and WBCSD (World Business Council for Sustainable Development) (2002) *Breaking New Ground: Mining, Minerals and Sustainable Development. Final Report of the MMSD Project* (London/Sterling, VA: Earthscan; www.iied.org/mmsd/finalreport/index.html, accessed 20 August 2008).

Joyce, S.A., and M. MacFarlane (2001) *Social Impact Assessment in the Mining Industry: Current Situation and Future Directions* (London: International Institute for Environment and Development Mining, Minerals and Sustainable Development Project).

Katona, J. (2002) 'Mining Uranium and Aboriginal Australians: The Fight for Jabiluka', in G. Evans, J. Goodman and N. Lansbury (eds.), *Moving Mountains: Communities Conflict Mining and Globalisation* (London: Zed Books): 195-206.

Keeping, J. (1998) *Thinking about Benefits Agreements: An Analytical Framework* (Yellowknife, NT: Canadian Arctic Resources Committee).

Kennett, S.A. (1999) *A Guide to Impact and Benefits Agreements* (Calgary, AB: Canadian Institute of Resources Law, University of Calgary).

Keon-Cohen, B. (2001) (ed.) *Native Title in the New Millennium* (Canberra: Australian Institute of Aboriginal and Torres Strait Islander Studies).

Kimberley Land Council (2008) 'Kimberley Land Council Criticises Misrepresentation of Traditional Owners', Media Release, 5 March 2008; www.klc.org.au/media/080305mr_INPEX_Protest.pdf, accessed 10 August 2008.

Langton, M., M. Tehan, L. Palmer and K. Shain (eds.) (2004) *Honour among Nations? Treaties and Agreements with Aboriginal People* (Carlton, VIC: Melbourne University Press).

Meyers, G. (ed.) (1996) *The Way Forward: Collaboration 'In Country'. Proceedings of the Aboriginal Land Use Agreements Conference, Darwin, NT, 26–29 September 1995* (Perth, WA: National Native Title Tribunal).

Miranda, M., D. Chambers and C. Coumans (2005) 'Framework for Responsible Mining: A Guide to Evolving Standards'; www.frameworkforresponsiblemining.org, accessed 20 August 2008.

O'Faircheallaigh, C. (1996) 'Negotiating with Resource Companies: Issues and Constraints for Aboriginal Communities in Australia', in R. Howitt, J. Connell and P. Hirsch (eds.), *Resources, Nations and Aboriginal Peoples* (Melbourne: Oxford University Press): 184-201.

—— (2000) *Negotiating Major Project Agreements: The 'Cape York Model'* (Canberra: Australian Institute for Aboriginal and Torres Strait Islander Studies).

—— (2002) 'Implementation: The Forgotten Dimension of Agreement-Making in Australia and Canada', *Aboriginal Law Bulletin* 5 (October 2002): 14-17.

—— (2003) *Implementing Agreements between Aboriginal People and Resource Developers in Australia and Canada* (Research Paper No. 12; Brisbane: Aboriginal Politics and Public Sector Management, Griffith University).

—— (2004a) 'Denying Citizens their Rights? Aboriginal People, Mining Payments and Service Provision', *Australian Journal of Public Administration* 63.2 (June 2004): 42-50.

—— (2004b) 'Evaluating Agreements between Aboriginal Peoples and Resource Developers', in M. Langton, M. Tehan, L. Palmer and K. Shain (eds.), *Honour among Nations? Treaties and Agreements with Aboriginal People* (Melbourne: Melbourne University Press): 303-28.

—— (2006a) 'Aborigines, Mining Companies and the State in Contemporary Australia: A New Political Economy or "Business as Usual"?', *Australian Journal of Political Science* 41.1 (March 2006): 1-22.

—— (2006b) 'Mining Agreements and Aboriginal Economic Development in Australia and Canada', *Journal of Aboriginal Economic Development* 5.1 (October 2006): 74-91.

—— (2007) 'Environmental Agreements, EIA Follow-up and Aboriginal Participation in Environmental Management: The Canadian Experience', *Environmental Impact Assessment Review* 27.4 (May 2007): 319-42.

—— (2008) 'Negotiating Cultural Heritage? Aboriginal–Mining Company Agreements in Australia', *Development and Change* 39.1 (January 2008): 25-52.

—— and T. Corbett (2005) 'Indigenous Participation in Environmental Management of Mining Projects: The Role of Negotiated Agreements', *Environmental Politics* 14.5 (November 2005): 629-47.

O'Reilly, K., and E. Eacott (1999) *Aboriginal Peoples and Impact and Benefit Agreements: Report of a National Workshop* (Yellowknife, NT: Canadian Arctic Resources Committee).

Senior, C. (1998) *The Yandicoogina Process: A Model for Negotiating Land Use Agreements* (Regional Agreements Paper No. 6; Canberra: Australian Institute of Aboriginal and Torres Strait Islander Studies).

Smith, M. (2007) 'Environmental Militarism: Burma's Extractive Industries', *Greener Management International* 52: 47-61.

Sosa, I., and K. Keenan (2001) *Impact Benefit Agreements between Aboriginal Communities and Mining Companies: Their Use in Canada* (Ottawa: Canadian Environmental Law Association).

Strelein, L. (2006) *Compromised Jurisprudence: Native Title Cases Since Mabo* (Canberra: Aboriginal Studies Press).

Weitzner, V. (2006) *'Dealing Full Force': Lutsel K'e Dene First Nation's Experience Negotiating with Mining Companies* (Ottawa: North-South Institute).

5

Indigenous peoples, corporate social responsibility and the fragility of the interpersonal domain

Richie Howitt

Department of Human Geography, Macquarie University, Australia

Rebecca Lawrence

Department of Sociology, University of Stockholm, Sweden

This chapter draws on experience at the interface of engagement between indigenous cultures and corporate cultures in Sápmi,[1] and indigenous domains in Australia and the Asia-Pacific. We highlight the fragility of corporate actors' and indigenous peoples' dependence on **personal relations** as the basis for the performance of the formal processes of coexistence and engagement on the ground. We argue that personal relations can provide both a basis for engagement and an arena in which to improve and refine good process in intercultural and place-based relations. In particular, we seek to explore the significance of the interpersonal domain in the translation of corporate cultures and corporate policies into the social sphere, and the *retranslation* of the interpersonal back into the corporate sphere. In doing so, we pay attention to forms of interaction and engagement that *leak* through the walls of 'formal process' and exist beyond officially recognised frameworks of engagement. Without seeking to dismiss the importance of good structure and process in formal engagement, we argue that mean-

1 Sápmi, or Saamiland, is the traditional homeland of the Saami and covers the northern parts of Norway, Sweden, Finland and the Kola Peninsula of Russia.

ingful engagement between corporations and indigenous communities remains rooted in the interpersonal domain and the strengths and weaknesses of this need to be acknowledged by all involved. However, given the contingency of the 'interpersonal', we see great cause for caution. Rather than being naïvely optimistic about the integration of higher standards of corporate social responsibility into corporate culture in the resources sector through the interpersonal domain, we suggest that this dependence on the interpersonal makes good process vulnerable to staff turnover, changes in corporate leadership and share-market takeovers, and ultimately fails to address structural incapacities to recognise, acknowledge and address social realms within many resources corporations. It therefore needs to be matched with stronger structural and strategic embedding of good process in corporate structures, and stronger external oversight by civil society and the state.

In the Australian context we suggest that resource corporations' relations with indigenous groups are often challenged in seeking to reconcile the structural orientation of increasing formal performance indicators (both for staff and functional areas and for activities including agreement-making with indigenous groups) with the widespread orientation of indigenous relations in the sector to interpersonal relations and community-scale engagement and relationships. At the same time, in Scandinavia debates about corporate/indigenous engagement are only just beginning to take hold in the mainstream. As Saami communities and organisations call for a more structured approach to agreement-making and negotiation processes, new questions about the dynamics between *process*, *structure* and *people* are arising.

The chapter highlights how these two different contexts can shed light on the significance of the fragility of a personal/process nexus of engagement, albeit from rather different angles. Increased formalisation of corporations' cross-cultural engagement in Australia, involving processes of 'stakeholder engagement', 'indigenous communities relations programmes' and formal agreement-making, can often serve to sterilise and limit the freedom of people to negotiate in good faith. They can constrain the issues that can be negotiated, restrict the time-frame for issue resolution to a negotiation timetable, or exclude issues (or groups) that do not fit within the terms of negotiation set by the corporation or by relevant legislation. In addition, resource corporations tend to fit stakeholder engagement and community relations into a much larger strategic picture than the site-based contingencies that focus the attention of place-based stakeholders such as indigenous groups (Burton 2001).

On the other hand, in Scandinavia, where stakeholder engagement and negotiated agreement processes are not yet standard praxis, indigenous people in the Nordic states are finding themselves expected to conform to historical precedents that assumed (indeed, insisted on) the absence of collective Saami interests and opened scope for a limited range of 'national interests'. 'National interests' (*riksintressen*) are identified by Swedish County Administrations, and framed in the Swedish Environmental Act 1998, whereby Saami reindeer-herding areas may be classified as areas of 'national interest'. Those same areas, however, may also be legislatively recognised as of 'national interest' for the mining, wind power or tourist industries. In this system, Saami are expected to settle for 'consultation' alongside other national interests, in the absence of a legal framework and an ethic of corporate best practice that promotes 'negotiation' with indigenous people based on free, prior and informed consent. In this light, Saami claims are reduced to some sort of equivalence with the claims of any other recognised interest group, and, while the various 'national interests' are theoretically equal before the law, Saami communities are without the financial resources necessary to participate 'equally'.

There is in fact not a single example where the Swedish Environmental Court has found that reindeer-herding rights, as a legislatively recognised 'national interest', should prevent the construction of a mine, road or large infrastructure project. In an attempt to work outside this system, and its inherent structural discrimination, Saami communities are increasingly seeking to establish direct dialogue with companies over their planned and ongoing natural resource development activities on Saami lands. Yet, when Saami communities *are* successful in getting corporations to the negotiating table, the fragility of negotiations becomes acute, as personalities drive processes and no legal framework regulates corporate compliance with appropriate standards or responses to key issues of collective interest, customary entitlements or implications of resource projects for the cultural landscapes of Saami geography. In this context, demands by Saami communities to have their ancestral rights recognised may be met with feigned personal offence from company representatives as they express their personal disappointment over failed meetings. For many Saami communities this has contributed to their experience that project proponents are often prepared to be a 'friend' to the Saami community, *until* the Saami community starts placing demands. Without appropriate structural or legal drivers requiring accountable oversight of the activities of corporations in relation to Saami rights, Saami communities become bound to the fragility of their personal relationships with local corporate representatives.

Scaled geographies of engagement and responsibility

In any resource sector activity, questions of responsibility extend well beyond the actual project sites, to include the actions, influence, capacities and values of investors, financiers, customers and others. As a result, the boundaries between a site-constrained project and the wider domains of the corporation, corporate culture and the resource sector as a whole become blurred. This often renders key sites of contestation of indigenous claims on place-based projects and their impacts and consequences remote and inaccessible to conventional community relations processes. Similarly, the exclusion of indigenous institutions (both representative and customary) from wider scales of decision-making (the board, the industry, the processing and consumption sectors) brings into play a **scale politics** that reimposes conditions of absence and erasure on indigenous interests whose political actions have re-inscribed their presence in the landscapes of regional and national politics with varying levels of effectiveness (Howitt 2006).

To a great degree, it is the invisibility of longer chains of responsibility in the impact and opportunity processes around a place-based resource project and its social, cultural and environmental footprint on indigenous peoples that produces this erasure and invisibility. The focus of much impact assessment work, for example, is on what are considered first- or second-level impacts, direct construction and operational impacts and opportunities and directly related secondary effects (Goldman 2000; Barrow 2000). No impact assessment studies that we know of have grappled explicitly with questions arising, for example, from financial, corporate or industry linkages in order to draw longer lines of responsibility for negative impacts or for delivery of positive opportunities. Yet the rhetoric of the corporate social responsibility (CSR) literature has been to make precisely these longer lines of responsibility more visible, accountable and effective (Elkington 1997, 2006; Warhurst 1998; Wilson 2001). Similarly, issues of cumulative impacts on wider communities of interest, or the political implications of policies and programmes to attract and enhance resource sector investment, or resource industry campaigns on political issues, are often significant in their effect on the standing of indigenous issues in the politics of the day. They also affect consideration of indigenous concerns in wider policy debates about, for example, economic, environmental or social legislation (Langton 2001).

The place of the corporation in these wider debates is also a matter of concern. Indigenous contesting of state claims to sovereignty over lands, waters and resources challenges the property rights from which corporate claims to real and potential wealth are derived. In the domains of CSR, it is the corporation, not the state, that is seen as taking responsibility, while the property and resource rights that are granted by the state as the basis for the corporate stake at any negotiating table are rendered as naturalised, unproblematic and not open to challenge. Yet the experience of many indigenous groups is that a particular company's acquisition of such rights, and the privileges and power that they support, occurs as an incidental consequence of share market transactions (including takeovers), currency fluctuations and shifts in global supply and demand. They do not result from any credible process of state evaluation of development options for a major national resource, or from corporate engagement with the source and implications of the rights. In this way, the vested interest of the corporation is quickly transformed into an interest that is representative of the national interest against parochial, self-interested local players whose efforts to secure inclusion are reconstructed as beyond and hostile to the national interest (Howitt 1991).

The mining and forestry sectors have very different histories of engagement with indigenous people (Lawrence 2005, 2007; Lertzman and Vredenburg 2005; Mabee and Hoberg 2006; O'Faircheallaigh and Corbett 2005; Trigger 1997). Does engagement have to be located 'in place', in real time, between people on the ground in order to be meaningful? Or can it be mediated through distant chains of relations? Miners and mining executives are often engaged directly with indigenous peoples impacted by resource-based projects such as a local mine. A pulp and paper manufacturing company may relate to indigenous people as a result of logging by its suppliers of timber or it may, under the pressure of NGO and customer criticism, be forced to engage directly with indigenous communities. Bankers and banking executives investing in companies mining or logging on indigenous peoples' territories have a different kind of engagement with indigenous communities (Lawrence 2008). In what ways do the local, national and global geographies of place, scale and connection matter in the performance and realisation of these relations? Is one set of relations (for example, those focused on site-specific production, downstream processing, or boardroom decision-making) less meaningful than another? Is one less accountable than another? What does it mean for an indigenous community to engage with a multinational enterprise whose head office might be located in a different national jurisdiction? Is that more significant than the geographical distance to head office in the metropolitan capital of a people's home state? What does it mean for bankers, managing risks and investments from their offices at Canary Wharf in London, to think about questions of indigenous rights? Is the distance between their realities so great that the idea of meaningful engagement becomes farcical, unachievable? Clearly, a single day of cross-cultural training cannot equip executives to deal with complex issues of indigenous rights, histories of colonialism, the specificities of particular indigenous cultures and people–people and people–place relations, and the impacts of resource development projects on subsistence livelihoods.

Equally, what does it mean for an NGO to transport indigenous people from the Philippines to a bank's headquarters in Sydney? What happens when indigenous concerns are brought to a company's annual general meeting, or a board meeting is held on-site with indigenous people given access to board members? These constellations of engagement at various scales involve not only different kinds of ethical question, but also different kinds of corporate culture. The answer to each depends very significantly on context, on timing and on the tensions between the structure-based and agency-based elements of the engagement.

Changing corporate cultures of responsible and sustainable engagement with indigenous peoples

Corporate cultures are multifaceted, plural and diverse, as are the people who work for 'corporations'. In the discourses of CSR the responsibility and role of individuals *within* companies is often too easily dismissed. CSR practices are embedded in a larger web of social and political relations. Our point here is not to reinvent the cult of the CSR guru. On the contrary, pseudo-religious narratives that celebrate CSR 'converts' should be read with caution. The case for developing corporate cultures that foreground issues of sustainability, triple-bottom-line reporting and CSR generally emphasise a 'business' case rather than a case founded in morality or religious or pseudo-religious values (Crawley and Sinclair 2003; Lertzman and Vredenburg 2005; Whitehouse 2006).

Some CEOs and managers, in espousing the CSR case, may refer to a moment of enlightenment, or a series of events, that first led them to realise the importance of sustainability issues and indigenous rights issues. Within this narrative, personal transformation becomes a catalyst for the individual to seek, support or initiate corporate transformation. For some, this 'transformation' comes through personal engagement with indigenous people on the ground. For others, a particular event, such as the experience of many mine employees in the Bougainville crisis in Papua New Guinea or the Marcopper tailings system collapse in the Philippines, generates a mismatch between their understanding of their own circumstances and how they are seen outside the company. For others, the shift is primarily a pragmatic one, fuelled by professional standards and a perceived reputational risk of not engaging with issues of indigenous rights. But, while these shifts and transformations are often, in hindsight, reconstructed as life-changing—both for the company and the individual—these shifts are often in reality much less dramatic and are more simply based on everyday interactions and relationships between *people*. Specific events might in part contribute to shifts in corporate responses to indigenous claims. Yet it is the realisation that the claims of a community are based on extreme need rather than political opposition, the shared moment fishing with an elder, sharing a family victory or tragedy, the acknowledgement of equality at a human level that provides the 'shock of understanding' (York and Pindera 1991), that reshapes executive understanding of the relationship between corporate activity and community concern, fear or aspiration. It is the complex and everyday interaction (or lack of interaction) between corporations and indigenous people that constitutes the 'nitty-gritty', the hard work of cross-cultural community engagement.

Moreover, while policies, procedures and 'corporate values' may contribute to an overall framework for engagement and while there is an increasing attempt to embed social and moral distinctions into a 'technical' framework' (O'Malley 2004: 8), there is a need to acknowledge that some things continue to exist in a space outside policies and procedures. For example, the formal corporate discourse of community engagement, community relations, stakeholder engagement, corporate social responsibility or social licence to operate can serve to erase the complicated and often *unequal* power relations between corporations and indigenous communities. We argue this not to give a necessarily negative value to the space of engagement outside formal frameworks. In some cases, non-formal engagement is a strategic way for indigenous communities to draw corporate actors into *their* worlds and *their* spheres of action. What does it mean for a corporate representative to spend time with indigenous communities on their traditional lands and waters during cultural activities or during a reindeer round-up? To understand this complexity it is necessary to recognise the messiness and stickiness of indigenous–corporate relationships. It is this messiness and stickiness that constitutes engagement between indigenous peoples and corporations and, in turn, makes them inescapably dependent on and reflective of both the strength and weakness of personal relationships. It is to this 'fragile dependency' that we now turn.

As the anthropologist Marilyn Strathern has argued, relations bind parties to one another in various ways (Strathern 1988). Relations may include obligations, responsibilities, dependencies, expectations and forms of engagement that challenge both cultural boundaries and are in some cases 'counterintuitive'. What does it mean for an indigenous community to engage with corporations already operating on, or proposing to operate on, their traditional territories? Does engagement necessarily mean an acceptance of development, and, if so, is it development on terms defined by others? What does it mean for indigenous communities to foster a personal relationship with developers proposing development on indigenous territories? What does it mean to a community when a previously hostile, antagonistic or indifferent corporation wants to engage, or, as Burton puts it, when corporations want to cuddle (Burton 2001)? Many communities experience this as a dilemma. Engaging with development proponents may be seen as a 'sell-out' if communities are opposed to the development, but without engagement any opportunity to benefit from or avoid pauperisation by a project (or worse) may be lost. What does it mean to get 'too close' or 'too friendly' with development proponents?

At the same time, without a deeper engagement through dialogue and the building of relations, corporate representatives are often left without the tools to understand what traditional lands *means* for local indigenous people. Is it, however, the responsibility of communities to *teach* corporate representatives about indigenous ways of belonging? How is corporate learning possible in a context in which indigenous claims, ways of seeing the world and cultural connections to land are often not respected or given space in a broader political context? In many ways, this raises questions of the capacity of corporate cultures to respond to cultural information and to retain learning about local specificity, history and meaning-making. In many situations, companies make relationships, commitments and even agreements as a foundation for ongoing engagement with communities affected by their activities, only to forget the details, abandon the learning or lose the records that become woven into the fabric of groups more reliant on oral transmission and continuity of relationships. Indeed, Doohan (2006) reports finding the entire archival documentation of one mining company's relationships with affected indigenous communities in a humid, dusty warehouse in boxes labelled as ready for disposal!

The challenge of changing corporate cultures in the resources sector to better ground CSR in sustainable, just and equitable relationships with diverse (and dynamic) interests of affected communities is not simply a matter of changing policies, nor of changing staff attitudes, although both are necessary. The challenge, to paraphrase David Natcher and his colleagues, is to focus on the relationships rather than either the resources or the values (Natcher *et al.* 2005). This can be illustrated in a series of brief case studies that offer windows on the CSR processes as they are experienced in affected communities.

Placer Dome's sustainability reporting and CSR discourses in the 1990s[2]

In the mid-1990s, Canadian-based international gold mining company Placer Dome initiated internal and strategic changes that placed it at the forefront of resources sector companies grappling with questions of CSR, sustainability and corporate strategies for growth, survival and competitiveness. The drivers for Placer Dome's shift can be identified in a variety of sources from corporate mergers, long history in mining in intercultural locations in Papua New Guinea, and the contingencies of personalities. The most significant markers of this period of change on Placer Dome were the development of a sustainability report, implementation of community-scale project monitoring and integration of industry leadership on these issues into the corporation's global strategies:

> Placer Dome is proud to lead the way in the mining industry in adding a further consideration to our bottom line—the social impact of mining. This 'triple bottom line' approach led us to become the first mining company in Australia to produce a Sustainability report (Loney and Williams 2000).

The company's foray into industry leadership on these issues did not survive changes in corporate leadership and a takeover in 2006 by Barrick Resources. Key elements of the strategies, however, continue to be reflected on some Placer sites.

2 This section is based on research supported by the Australian Research Council Large Grant (A00000132) (Changing Corporate Culture), which was funded 2000–2004. The cooperation of staff and host communities of Placer Dome's mining operations in Australia and Papua New Guinea is gratefully acknowledged, as is the exceptional generosity of the company's Australian and international leadership group at that time in making themselves available for in-depth interviews.

The contingencies of the share market and personality both come into play in assessing the significance of these shifts nearly ten years later. A series of CEOs took Placer further along the CSR track than other gold miners were willing to go in the mid to late 1990s. Placer Dome Asia Pacific (PDAP), the company's regional subsidiary, took a leadership role, undertaking stakeholder engagement that extended to including a former Greenpeace leader as a business consultant and developing innovative programmes for managing mine closure, community health, sustainability reporting and other matters. PDAP included staff who had been on-site on Marinduque Island in the Philippines where a tailings system failure in 1996 produced an environmental and social disaster that added to the ongoing calamity of poor environmental systems at the Marcopper site since 1975 (Macdonald and Southall 2005). It is not our purpose here to review events at Marcopper, but to consider how the Marcopper catastrophe was understood within Placer Dome, and how it shaped the corporate culture of Placer Dome in general and PDAP in particular. As one insider put it, PDAP's managing director was deeply affected by events at Marcopper:

> His feeling was that if we were to get out of there believing that we had done the right thing we had to address the magnitude of the issues . . . He had been there in the early days. When he went back, he was horrified to see what had happened because, in the early days there was a great deal of pride and respect and the Marcopper operation was seen as the jewel in Placer's crown and then obviously it lost that status and people forgot about it. It was just an asset on the books but it wasn't a cost so no one worried about it and it ended up being a place that they sent people when they weren't sure what else to do with them and you know so it completely changed from sending the best.[3]

PDAP's regional leadership team drew on personal experience at Porgera and Misima in Papua New Guinea as well as Marcopper, and insisted PDAP take new directions that would make repetition of the catastrophe at Marcopper virtually impossible. For many of the team, experiences of mining in cross-cultural settings shaped their thinking about key issues of social and environmental impact, the local responsibilities of a mining operation and the responsibility of corporate agents in the processes of social and economic change:

> This experience in Papua New Guinea sort of changed the way [people were] thinking. I can remember [one team member] talking about going and engaging with the women's' committees and you know the sort of responses that he got from the women about what was happening to the men when they came home from work and all this sort of stuff, what was happening to the money and they were the same symptoms of what you saw in other . . . mining communities in the developed world . . . he immediately . . . looked for . . . solutions to those problems to stop . . . I think the point is that these sort of projects attracted people who perhaps had . . . a different way of thinking to the traditional projects in Canada or the US or Australia . . . I mean a lot of the Canadian remote mines are still very much mining communities: they build the towns, they attract the people; it's the remote isolated parts of the communities but it wasn't quite the same.[4]

3 Field interview with former Placer Dome senior manager, Washington, DC, USA, September 2001.
4 *Ibid.*

It is in these cross-cultural mining experiences that we can identify key shifts in personal values. People form relationships and understandings of the mine site that reflect their experience of indigenous and local values, local priorities and local experience—mediated by their own sense of what is right and reasonable. As one former mine manager reflected:

> I was the General Manger running the business, I had a budget and the aim was to produce as much gold as possible for the least possible cost and obviously looking forward we had to ensure that we minimised corporate liability when mine closure occurred. [We were very conscious] that what was happening with Ok Tedi at the time really emphasised the problems. And the problems that we had at Marcopper really emphasised the need to be proactive in minimising long-term corporate liability. So I put those things such as minimising corporate liability alongside adding social value as part of the triple-bottom-line thing; all those sorts of things tied in if you like. [In PDAP this] coincided with an expanded response to stakeholder consultations, which had us talking to villagers, talking to WWF and a whole range of others. This was happening at the same time as the company was going through its internal debate, developing a sustainability policy and debating what it meant. Then the Placer Dome President instituted his annual president's award for sustainability, and Misima won the President's Award for Sustainability two years in a row on the basis of the closure planning process, the community consultation process and all those other programmes that interrelated into it. So, Misima was much held up as the example within [Placer Dome] of what to do, and so maybe what we did at Misima will help contribute to that corporate change process.[5]

This overlay of personal experience, personal values and the shift in corporate culture took Placer Dome to a leadership role in the mining industry that drew strongly on middle and senior management seeking to align personal values with corporate value. For example, the former Misima mine manager felt that the company's contribution to treatment and control of Filariasis (see, for example, Sapak *et al.* 2004; Wynd *et al.* 2007a, 2007b) was a marker of how integration of social responsibility and sustainability into a robust business model could be achieved in practice. For him, the Marcopper catastrophe

> occurred in 1996 at the time of our awakening—if you like, environmental and social awakening. First, many of the senior people within the company that had worked at Marcopper were actually personally affronted by what occurred there. Second, [we were challenged] by the way it got thrown back at us by civil society [and] we really didn't have a defence . . . they found a large part of the community on Marinduque were actually disenfranchised by the mine because they hadn't benefited from the mine but they were impacted by the mine. It's . . . the same thing that Ok Tedi went through in dealing with the people around the mine, but didn't deal with the people down river who felt themselves to be impacted and felt themselves to be stakeholders by definition but weren't adequately consulted . . . at Porgera and Misima . . . we were trying to do the right thing . . . but we'd just taken our eye off the ball at Marcopper and then of course we found ourselves in a, in a political situation where we were just on a hiding to nothing.[6]

5 Field interview with Placer Dome Asia Pacific senior manager, Brisbane, Australia, November 2001.
6 *Ibid.*

Placer Dome's second Vice President, Sustainability reflected on the demands of industry leadership:

> my role is providing strategic direction for the overall corporation in the areas of sustainability. It's somewhat similar to [the previous Vice President, Sustainability leadership role], although he had the word 'Leadership' right in his title. I do not. I guess . . . my role . . . is . . . to find ways to implement sustainability policy within our business . . . I guess it's our belief and our hope that sustainability will be a business advantage in the share price of Placer Dome. In the short term we see that being achieved through what we can actually do at the individual operations, getting support for our communities, having good environmental protection, getting good risk control and management and getting the support of our stakeholders . . . We hope that the mining industry will get there and [Placer Dome Inc.] will move . . . towards that with the rest of the world, but perhaps we won't be quite as far in advance as we were before . . . I think we still are viewed as leaders in sustainability and I think we certainly want to be among the leaders of sustainability and be recognised by ethical investment funds. We see advantages to doing that. But, again, do we want to be . . . so far out in front? I use the analogy of a 'V' of geese. In Canada geese mean a lot to us, and you see those birds flying around. I guess what we want to do is to be among that leading group. We may be the head goose sometimes, [and at other times] we may be a little bit further back in the 'V'. But the way I envision that is that we'll always be in that leader group, but we may not be that lone goose that's way out there pioneering in all sorts of directions.[7]

While the commitment to leading the industry from the front might have weakened, Barrick Mining has retained much of the rhetoric of CSR as the framework for its long-term global strategy,[8] but it is notable that it provides no site-specific sustainability report for Porgera, where the PDAP strategies to integrate sustainability and CSR thinking into the structure of the operations were perhaps most influential. The Marcopper catastrophe still provides a case study of environmental injustice (e.g. Coumans 2002; Gregory 2000; Macdonald and Southall 2005). The corporate leadership whose personal experience and relationships within a leadership team reshaped Placer Dome's profile moved on, either within or beyond the company that has developed from the 2006 merger with Barrick. The opportunities provided by those contingencies have significant echoes: for example, in the execution of closure planning at Misima and the ongoing Filariasis work. But the execution of these practices and opening of opportunities to engage across the whole group seems to have been rather more fragile than the optimism of the leadership group in the mid-1990s might have hoped. As each subsequent CEO and leadership group has marked its managerial territory, the balance between value and values seems to have changed to restrict the sort of openness that saw PDAP invite critics of riverine tailings disposal at Porgera to join monitoring and evaluation activities in order to improve practice and impacts if there was improvement to be found.

7 Field interview with Vice President, Sustainability, Placer Dome Inc., Vancouver, Canada, September 2001.

8 See, e.g., www.barrick.com/CorporateResponsibility/OurCommitment/default.aspx, accessed 15 August 2008.

Social impact assessment in Scandinavia[9]

While social impact assessment (SIA) has been accepted practice for developments affecting indigenous peoples in Australia, Canada and the USA for the last two to three decades (Downing *et al.* 2002; Joyce and MacFarlane 2001), and is established practice in developing parts of the world, it has received relatively little attention in Scandinavia. Saami communities have long experienced frustration over the limited way in which natural resource developments are assessed, yet industry and governments of Scandinavia have been slow to pick up on international developments in SIA practice. This is not limited to Scandinavian companies alone. Canadian and Australian corporations active in Sápmi have failed to undertake SIAs for their recent mining developments, and have been slow to show respect for Saami rights. This is despite the fact that both SIA and corporate engagement with indigenous peoples' rights is the established 'norm' in Australia and Canada. Dragon Mining AB, for example, a subsidiary to the Australian mining company Dragon Mining, is currently being sued by the Vapsten Saami community for damages caused by the Svartliden mine to traditional reindeer-herding pastures. Throughout the environmental impact assessment (EIA) process in 2005–2006 the Vapsten Saami community made requests that the socioeconomic and cultural impacts of the mine be taken into account. However, no SIA was ever undertaken.

Part of the reason for this 'gap' between international best practice and corporate practice in Scandinavia may be attributable to personal relations on the ground. While Canadian and Australian mining companies pride themselves on healthy community relations with indigenous communities in their respective countries, and Scandinavian companies equally so with the indigenous inhabitants of Africa and other developing parts of the world, the rights of Saami people somehow 'slip through the cracks'. Both multinational mining companies and Scandinavian companies often fail to make the link between the rights of indigenous peoples and the fact that Saami people inhabit areas of natural resource exploitation in Scandinavia. While Saami communities feel that they are at times successful in communicating their concerns to the management of Canadian and Australian companies, they have also experienced frustration when local Scandinavian management teams of those companies, or local government planning officials or politicians, deal with issues of Saami rights on the ground. Conflicting personal relations, historical antagonisms and more general challenges of systematic and structural racism in post-colonial Scandinavia have contributed to poor communication and a general 'freeze' in dialogues. That is not to argue that Australian or Canadian companies are unequivocally good, nor that breaches of Saami rights are simply attributable to a Scandinavian management style that fails to recognise Saami rights as special rights. On the contrary, Australian and Canadian companies may benefit from turning a blind eye to the fact that standards in their home countries are not properly implemented in Saami territories.

9 This section draws on Rebecca Lawrence's experience as an adviser to Saami communities throughout Sápmi, and the support of those communities and Saami organisations (in particular the National Swedish Reindeer Herding Association and Saami Council) is gratefully acknowledged. The sources of direct quotes and company names have not been identified in those cases where negotiations are ongoing or of a confidential nature.

While Saami communities have long expressed frustration over the lack of consideration give to socioeconomic and cultural impacts of developments, it was not until the beginning of 2007 that Saami organisations and communities in Swedish Sápmi began articulating concrete demands on companies to undertake SIAs as such. Several communities have been successful in getting companies to commit to undertake SIAs that take into account Saami perspectives. However, considerable challenges remain. These mainly concern more general problems that have pervaded the EIA process, and which are now being reproduced in the SIA process.

One major challenge concerns the reduction of Saami claims to 'stakeholder interests'. Within the Scandinavian stakeholder discourse Saami people are given the right to express concerns over development in the same way as *any other* affected citizen group. Any claims to special rights are dismissed. So while Saami people have a right of appeal over development as citizens or as an interest group through the County Administration appeal process and the Environmental Court, their claims to traditional Saami lands, and the protection and rights that connection entails, are largely ignored. This structural discrimination manifests itself also at the personal level. The majority of development proponents and SIA/EIA consultants, while often demonstrating sympathy for the plight of Saami individuals trying to protect their 'livelihoods', show little real respect for the individual or collective rights of Saami people as indigenous people. Similarly, a reindeer herder's livelihood is considered a 'day job' and little more. While lip-service is given to the reindeer-herding 'industry', there is little corporate or personal engagement with how a Saami right to self-determination would actually fundamentally change the way in which both EIAs and SIAs are undertaken. The cultural landscapes that bring Saami language, history and economy together into a coherent whole are easily elided when the social and cultural domain is simply defined as absent from the realm of 'environment'. But the solution to this is not to simply add the social and cultural as if it was an appendix to an environmental or economic material 'reality'. The danger of this approach—both theoretically and practically—presumes that we can get around the problem by simply adding a 'cultural filigree to what in the end is still held to be an economic core' (Amin and Thrift 2004: xiv).

Some Swedish civil servants and company representatives have expressed confusion over why Saami communities would argue for a socioeconomic assessment of the impacts of a mine when the only logical conclusion (in their eyes) to such an assessment would be that the benefits of mining for a local (non-reindeer-herding) community, in terms of employment, are greater than the negative impacts for a Saami community, in terms of loss of income from reindeer herding. But so long as SIA is treated as an appendix to what is held to be the 'nitty-gritty' of the impact assessment process—that is, traditional environmental reporting and quantitative economic feasibility studies—companies will continue to fail to engage with the Saami experience of an inextricably connected landscape and social structure that continues to bear the costs of natural resource development projects. Environmental and social impact reporting and decision-making within companies cannot proceed in any meaningful way if the affected landscapes are not simultaneously recognised as homelands and cultural landscapes for Saami people.

For many Scandinavian companies the relationship with Saami communities is conceptualised as a simple, one-way relationship that is not essential to framing their decision-making. Both EIA and SIA are considered as processes whereby the company *informs* the affected Saami community of its proposal and the Saami community in turn *provides facts* on how the reindeer use the land. Several Saami communities in Swedish Sápmi are in the process of negotiating the first SIAs in Saami areas, and these Saami communities have had to deal with the company's notion that the Saami community should 'do the work for the company' and unconditionally provide 'information' on their land-use practices for the company's disposal. This concern has been expressed by Saami communities throughout both the EIA and the SIA processes.

Where companies are often keen to present their relationship with indigenous communities in a positive light, indigenous communities themselves may have a very different understanding. But, where legislative recognition of rights does not exist, and communities are dependent on favourable personal relations with company representatives, communities may be more or less forced to present their relationship with a company more positively than is actually the case. As one Saami Council politician noted:

> When I was 22 and started actively in Saami politics, I followed an older reindeer herder who used to say in every public meeting he attended how good the cooperation was with his cooperative and Metsähallitus [Finnish state logging company]. And he always said that we'll get along just fine as long as the person that is in charge in Metsähallitus is still in there. In a way, using these tactics (flattering, keeping Metsähallitus close) he managed to buy almost 20 years of different benefits for his cooperative, and, at the same time, the rest of the cooperatives had to pay the price.[10]

In the absence of structures and processes that ensure a recognition of rights, Saami reindeer-herding cooperatives in Finnish Sápmi are particularly vulnerable to the varied and often unpredictable decisions by the Finnish logging company, Metsähallitus, over logging of forests on traditional Saami lands. Moreover, a lack of recognition of traditional knowledge and the experiences of Saami over the impacts of these activities are not included in any assessment of Metsähallitus's activities, nor is there any consistent monitoring of cumulative impacts. SIA, while legislatively required in Finland for development projects, does not feature in the toolbox of Metsähallitus. Ironically, forestry is quite possibly the greatest contributor to loss of grazing lands for Saami throughout both Sweden and Finland, yet there are no SIA requirements or even a basic assessment of cumulative impacts of logging on Saami communities and landscapes.

Moreover, there is a fundamental difference in the way the resource sector corporations and Saami communities view development. For the corporate sector environmental governance decisions such as the decision to reopen a mine or develop a wind-power farm concerns fine-tuning a proposal. The location of a tailings dam can be adjusted to suit reindeer herding, or fences can be erected around wind turbines to govern the reindeer's movements.[11] At the same time that industry commonly argues for mitigation measures in order to achieve 'sustainable development', development per se is taken as the status quo. As such, reindeer herding and other traditional Saami land-use practices are often framed by the natural resource sector as 'doomed', or at best in need of serious modernisation in order to survive. The notion that these may be ecologically sustainable cultural and economic practices that warrant support and reinforcement rather than modernisation is not entertained at all. The impacts that arise from resource devel-

10 Personal communication.

11 Paine offers a coherent account of the delicate balance between human and reindeer behaviours that provides the ecological foundations of reindeer pastoralism (Paine 1982).

opments are seen by industry merely as speeding up an inevitable progression towards a different way of life for Saami people. In this construction, Saami loss is naturalised and the loss of the sustainable ecological relations of Saami pastoral economies rendered not only as inevitable, but also as already under way and therefore not a current issue. On the other hand, for the affected Saami communities, mitigation measures are considered a 'drop in the ocean' when the overall impact of a proposed project is considered to have catastrophic effects for reindeer herding and Saami ways of life.

Such epistemological differences lie at the heart of many personal misunderstandings between company management and Saami communities. One popular saying used in Saami communities expresses this dilemma: 'It's no use explainin' if they just don't get it.'[12] This frustration at being heard but not listened to, of being 'at the table' but unable to enter meaningful negotiation is a reflection of a wider marginalisation in Scandinavian discourses of CSR (Lawrence 2007).

Newmont and Aboriginal employment in Central Australia[13]

The mining industry in Central Australia, by virtue of its exploration on freehold Aboriginal land, requires the identification of traditional owners in order to allow negotiations for grant of exploration and mining leases. Indeed, there is an 'assertion of a special status for "traditional owners" of resource-rich areas' (Howitt *et al.* 1996: 2). The legislative recognition of Aboriginal peoples' rights under the Aboriginal Land Rights (Northern Territory) Act 1976 (ALRNTA) and the legal structures and processes in place require that companies directly engage with traditional owners. In the Central Australian mining industry, Newmont Mining Corporation has made a commitment to employ indigenous people at Newmont's Granites Gold Mine in the Tanami Desert, and a comprehensive agreement between Newmont and the local Warlpiri Traditional Owners governs this commitment, among others (Lawrence 2005). These commitments, a result of negotiations between the company and local community under the ALRNTA, put in place structures for the sharing of benefits to local communities. But formal commitments can prove empty if they are not supported by a place-based engagement by Newmont staff. This is recognised by the company itself in the following quote from Newmont's 2001 information package:

> Tanami Operations has worked hard to build respectful, long-term relationships with its Indigenous landlords. It provides benefits to local Indigenous people, in recognition of the Company's appreciation of the use of their land (Newmont Mining Corporation 2001: 6).

12 'Det går int att förklara för den som inte begrip'.'

13 This case study is based on work undertaken by Lawrence (with Howitt as a co-supervisor) towards an honours degree (Lawrence 2002). The support of Warlpiri people and organisations is acknowledged, as are the valuable cooperation of Newmont and Central Land Council and financial support from Macquarie University.

Yet, despite the existing agreement, and the commitment by the company to long-term relationships with its 'indigenous landlords', the employment levels of Warlpiri people at the Granites Gold Mine have historically remained below a level acceptable to Newmont and the Central Land Council. The majority of indigenous mine employees are sourced from regional centres and cities rather than local indigenous communities. This challenge, among others, indicates that, despite formal comprehensive agreements, and a relatively strong protection of rights in law, CSR commitments to provide employment and training benefits to local indigenous people can prove difficult to realise.

Reasons offered for this failure vary within and across Newmont, the CLC and among Warlpiri Traditional Owners themselves. A whole variety of discourses describe Warlpiri people as either too culturally different, indifferent to 'white-fella' employment, or just plain 'lazy' (Lawrence 2005). These views are articulated both by the company itself and by some Warlpiri people. At the same time, Newmont's policies and the internal rhetoric of Newmont suggest that the Warlpiri people are powerful traditional owners. For the company, this sets the stage for the relationship whereby 'we're paying for the use of the land, and we're working for them [the Warlpiri]'.[14] Various everyday discourses concerning the status of Warlpiri people posit them not only as Newmont's 'landlords',[15] but also as 'the Sheiks of the Tanami'.[16] For Newmont, the company is only operating 'at the liberty of the traditional owners'.[17] The relationship between Newmont and local Warlpiri communities is thus decidedly ambivalent. As one corporate executive observed, 'it's all very personality-driven'.[18]

The Central Land Council, representing the Warlpiri people, argues that the depiction of Warlpiri Traditional Owners as powerful needs to be placed within an historical context. For example, the Granites leases at Tanami Operations were historical leases that pre-dated enactment of the Aboriginal Land Rights Act in 1976. Negotiations over those leases did not include a right to veto; therefore the Warlpiri people never had a legally recognised right to refuse mining development. Negotiation of the mining agreements then was mainly concerned with exploration licences over land surrounding the mine, with the aim to find further deposits. Furthermore, while consent to a larger area for the new mineral lease was needed, it is likely that the mine development was inevitable and refusing to consent was not a realistic possibility.[19]

It is interesting that, within these discourses of traditional ownership, power is conceived as an entity held by Warlpiri people *over* Newmont, failing to acknowledge the productive and heterogeneous relations of power between actors and agencies. Furthermore, there is a clear vested interest on Newmont's behalf to present its relationship with

14 Field interview with Mine Manager, Granites Goldmine, Newmont Mining Corporation, Australia, January 2002.

15 Field interview with Indigenous Relations Manager, Granites Goldmine, Newmont Mining Corporation, Australia, 2002.

16 Field interview with Group Executive External Relations, Newmont Mining Corporation, Sydney, Australia, December 2001.

17 Field interview with Director Community Relations, Newmont Mining Corporation, Adelaide, Australia, January 2002.

18 Field interview with Group Executive External Relations, Newmont Mining Corporation, Sydney, Australia, December 2001.

19 Field interview with Manager, Mining Section, Central Land Council, Australia, January 2002.

the Warlpiri in positive terms. It is important to Newmont and its success as a company, that it be seen to enjoy a stable relationship with its 'indigenous landlords' (Newmont Mining Corporation 2001: 6). This is emphasised at both a corporate and personal level:

> One of Newmont's major focus has been providing employment and training opportunities for local [indigenous] communities and more broadly. There's a whole history to that and our previous CEO, Robert Champion de Crespigny . . . part of it was his personal commitment but also part was being an astute business man . . . if you're going to get access to land then you're going to need to engage with Aboriginal people, that's a competitive edge.[20]

Community relations is thus seen to be driven by both a personal commitment of corporate leaders, and to ensure strategic access to indigenous lands. But does engagement bind indigenous people to permitting 'access' to their lands? As the manager of Central Land Council's Mining Section notes:

> Normandy came as a company with a corporate philosophy of involving indigenous communities . . . not a social welfare attitude, but as responsible corporate citizens . . . but with these rights such as land rights, there's more pressure to be getting along with communities . . . they were talking the talk, and then there was pressure to walk the walk.[21]

In the context of corporate cultures oriented towards performance, this idea of 'walking the walk' has considerable leverage, particularly where there is also a shift toward inclusion of specific performance indicators that address the domains of community relations, sustainability, human rights and intercultural communication (Harvey 2002). This also shows that an opposition between 'structure' and the interpersonal is misleading, and that indigenous groups need to develop strategies that incorporate a focus on both. The value of structure is realised only so far as companies and communities are actually able to engage in meaningful ways to produce real benefits for local communities. For example, comprehensive agreements with commitments to employ indigenous people and share benefits of mining are vulnerable to unintended interpretations by individual mining professionals. Could the Central Land Council have known, when negotiating the agreement on behalf of Warlpiri people, that indigenous employment would later be interpreted to mean employment of *any* indigenous person, and not necessarily Warlpiri, as originally intended?

Discussion: the social footprint of community relations

Community relations is often a marginal functional area of resource company operations. In many situations, it is left up to individuals who 'get on well' with the locals and becomes a haven of minor corruption and major disruption. In many ways, it has been one of the well-trodden roads to hell that has been paved with good intentions and

20 Field interview with Director Community Relations, Newmont Mining Corporation, Adelaide, Australia, January 2002.

21 Field interview with Manager, Mining Section, Central Land Council, Australia, January 2002.

appalling social and cultural misunderstanding. In some cases, the intent to open opportunities has been constructed only in the limited cultural terms of the dominant cultures of resource exploitation, with companies mistaking the goals, ambitions and priorities of their own industry cultures for universal values that make 'common sense'. In the cases discussed in this chapter, we see the emergence of an internal critique that seeks to align shareholder value with human values and to secure sustainable benefits from resource exploitation for the host communities, often within legal jurisdictions that offer limited standing to indigenous rights, or challenge to basic values, well-being and livelihood of affected indigenous groups. Whatever internal recognition of difference and diversity might be secured in such settings, corporate cultures of tolerance and engagement are fragile and fleeting. They are subject to the personal intervention of senior leaders, the innovative management of middle-level leaders and the site-based interaction with open-minded locals. The community partners of resource corporations Placer Dome and Newmont discussed in the case studies above faced significant shifts in corporate priorities and cultures, as well as personnel, after share market transactions that saw takeovers of the parent companies of locally innovative subsidiaries. In the Saami case, the blurred jurisdictional context between corporation, state and dominant society leads to ambiguity in the invitations to Saami to engage.

In the short-sightedness of the corporate strategy and the blurred edges of colonial artefacts, the hybrid spaces of corporate engagement are potentially dangerous places of deep colonising (Rose 1999). In spite of good intentions, the prevailing structural, political, legal and economic constraints limit the capacity for the corporations to commit to people and place in the ways that indigenous cultures generally understand such notions. There is no 'dwelling' for the resource company (e.g. Ingold 1995). Its purpose, as put so succinctly by one of our interviewees: '[Our] aim was to produce as much gold as possible for the least possible cost and obviously looking forward we had to ensure that we minimised corporate liability when mine closure occurred.'[22]

The risk for indigenous groups is that there is also, ultimately, no co-existence, no shared spaces, only possession and its opposite. Rose's powerful image of the deep colonising pole of the Western (corporate) self engaging in monologue and masturbation that parades as conversation and productive intercourse (Rose 1999) haunts the fragile terrains of community relations for many community groups. It is a terrain that is deeply littered with incomplete projects, partly realised partnerships, intercultural miscues and misunderstandings, and expectations that others will simply, ultimately, be just like us, whoever and whatever 'we' are.

Howitt (1991) wrote of the confusion between vested interests of corporations seeking resource rights on indigenous lands, which self-interested governments transformed into representatives of the 'national interest'. This conflicts with the representative interests of Aboriginal publics, which experienced deeply scarring social, economic and cultural exclusion from the 'national interest'. They found themselves transformed into 'vested interests' whose arguments for recognition and rights threatened the 'national interest' that was conveniently represented by the state-sanctioned resource company. Such inversions continue to challenge relations between states, resource companies and indigenous groups. Saami forest territories, for example, continue to be subsumed into

22 Field interview with Placer Dome Asia Pacific senior manager, Brisbane, Australia, November 2001.

the national interest, which is constructed as a simple and singular entity by Scandinavian states whose recognition of diversity within the nation is limited by ideologies on national unity and public good.

Inevitably, such constructions are often underpinned by deeply etched values of racism and the legacies of colonial occupation that has left issues of identity, diversity and legality of land, water and resource tenures somewhat ambiguous—or at least morally questionable—in many jurisdictions. While discourses of CSR and community engagement have opened some spaces from which to challenge the notion that such questions are long settled in favour of the corporate and colonial entity, those discursive spaces are fragile, contingent and almost always short-lived. They rely on the contingent relations of the interpersonal domain in settings such as the state and corporation, which are structured to devalue the uncertainties and ambiguities of personal difference and interpersonal relations in favour of structural, legislative and codified arrangements.

Conclusion

Changing corporate culture is an important element of CSR practice and its implications for indigenous groups and their relations with resource sector corporations. The rate of change, the directions in which particular corporations move, and the expectations such changes place on community partnerships are all quite unpredictable, because so much of the performance of CSR is constructed in top-down strategies within companies, reflecting the current leadership's priorities and demands. Looking from the bottom up, from the vantage point of on-the-ground relationships, changing corporate cultures needs to be less a matter of describing the nature and processes of change, and more about influencing and achieving changes in corporate culture as it is performed and experienced. In changing corporate cultures indigenous groups often seek to use the interpersonal domain as the only accessible spaces of intercultural dialogue. Yet the capacity of individual corporate employees to carry the responsibilities of such partnership, indeed their capacity to even apprehend how their presence in indigenous domains implicates their employer corporation in culturally constructed roles, obligations and activities, is often effectively non-existent. While such cross-cultural experiences have significant implications for personal values, integration of such experiences into corporate memory, and transferring the experiential learning of individuals into the collective understanding of the corporation, is both problematic and fragile.

Longstanding expectations of 'direct dialogue' between local indigenous people and resource corporations without mediation ('interference') from NGOs or indigenous representative bodies have been justified by arguments that suggest that negotiations should take place between *people* directly involved. Even if this involves the CEO of a mining company negotiating directly with the head of a local indigenous community, they will always be embedded in historically structured power relations that implicate states that delegate property rights over indigenous peoples' territories and domains. These relations are far from equal and need to be recognised as such. They reflect the structural racism and deep colonisation of national culture that assumes the primitiveness and fragility of indigenous culture as a reflection of the human past, rather than a vigorous and robust indigenous present and future continuous being. In most cases, the delegation of specific responsibilities in the mediation of corporation–community relations to specific staff at different stages of feasibility, development, operation and closure, each with specific performance indicators, time-frames and career linkages, means that the corporation's singular legal identity is untranslatable into the language of intercultural discourse. Whether it is transactions on the share market or decisions to dispose of corporate archives, the integration of culturally literate strategies at the local scale into corporate memory and culture at the institutional scale is not a task that has been prioritised as the basis for ongoing relations with indigenous peoples affected by resource industry activities. Natcher *et al.*'s analysis that sustainable, just and equitable land and resource management regimes in indigenous domains rely on a more sophisticated cultural (and, crucially, intercultural) literacy than is commonly available in resource industry professionals (Natcher *et al.* 2005) remains reflected in corporate practice.

Within CSR discourses, the notion of stakeholder engagement poses a particular challenge to indigenous claims because here a corporation often posits itself as a *mediator* of interests in the everyday engagement between those representing corporations and those representing indigenous people. The corporation is often constructed in this situation as mediating (often in collaboration with NGOs or representative organisations) the recognition of indigenous interests when more hostile state or dominant society values seek their continuing denial or further erasure. In other cases, representative organisations are constructed as meddling in relations between well-intentioned resource companies and indigenous communities affected by their operations, and corporations represent themselves as more reliable sources of support and goodwill in the politically unstable settings of NGO rights advocacy. In both such settings, paternalism and deep colonising represents the indigenous domain as unable and unsuitable to exercise decisive influence over environmental and social governance.

Yet, at the scales of indigenous self-governance, it is precisely these domains that should be constructed as both decisive and accountable in mediating the interface between indigenous and corporate cultures as it is performed in specific resource projects. Acceptance of tribal regulation, for example, in the form of the Navajo Nation Environmental Protection Agency, has allowed development of coal, power generation and other resource sector activities in the Navajo Nation within a context of tribal-to-federal governance arrangements that major resource corporations have integrated into their strategic planning, investment and development processes. In places that deny even this level of indigenous self-governance and regulation, the emphasis on the interpersonal allows valuable intercultural understanding to be lost and ignored, and for institutional accountability to be avoided. It is this lack of external oversight and accountability, and the fragility of the influence of any set of personal commitments in the wider corporate setting, that reinforces the colonial patterns of relations in resource sector relations with indigenous groups.

Focusing on good process in mediating corporate cultures, CSR and indigenous cultures, rights and values must, then, move towards a critical consideration of institutional arrangements. These must provide both recognition and capacity to affected indigenous groups through self-determination and self-governance, and require the performance of CSR to shift away from the fragile and deep-colonising domains of intention towards the more accountable domains of governance.

References

Amin, A., and N. Thrift (2004) 'Introduction', in A. Amin and N. Thrift (eds.), *The Blackwell Cultural Economy Reader* (Oxford, UK: Blackwell): x-xxx.

Barrow, C.J. (2000) *Social Impact Assessment: An Introduction* (London: Arnold).

Burton, B. (2001) 'When Corporations Want to Cuddle', in G. Evans, J. Goodman and N. Lansbury (eds.), *Moving Mountains: Communities Confront Mining and Globalisation* (London/New York: Zed Books): 109-24.

Coumans, C. (2002) *Placer Dome Case Study: Marcopper Mines* (Sydney: Minerals Policy Institute; www.miningwatch.ca/updir/PD_Case_Study_Marcopper.pdf, accessed 2 October 2008).

Crawley, A., and A. Sinclair (2003) 'Indigenous Human Resource Practices in Australian Mining Companies: Towards an Ethical Model', *Journal of Business Ethics* 45.4: 361-73.

Doohan, K. (2006) ' "Making Things Come Good": Aborigines and Miners at Argyle' (PhD dissertation, Macquarie University, Sydney).

Downing, T.E., J. Moles, I. McIntosh and C. Garcia-Downing (2002) *Indigenous Peoples and Mining Encounters: Strategies and Tactics* (London: International Institute for Environment and Development).

Elkington, J. (1997) *Cannibals with Forks: The Triple Bottom Line of 21st Century Business* (Oxford, UK: Capstone Publishing).

—— (2006) 'Governance for Sustainability', *Corporate Governance: An International Review* 14.6: 522-29.

Goldman, L.R. (2000) *Social Impact Analysis: An Applied Anthropology Manual* (Oxford, UK: Berg).

Gregory, C. (2000) 'Environmental Justice Case Study: Marcopper in the Philippines', *Environmental Justice Case Studies by University of Michigan Students*; www.umich.edu/~snre492/cases.html, accessed 2 October 2008.

Harvey, B.E. (2002) *New Competencies in Mining: Rio Tinto's Experience* (Melbourne: Rio Tinto).

Howitt, R. (1991) 'Aborigines and Restructuring in the Mining Sector: Vested and Representative Interests', *Australian Geographer* 22.2: 117-19.

—— (2006), 'Scales of Coexistence: Tackling the Tension between Legal and Cultural Landscapes in Post-Mabo Australia', *Macquarie Law Journal* 6: 49-64.

——, J. Connell and P. Hirsch (eds.) (1996) *Resources, Nations and Indigenous Peoples: Case Studies from Australia, Melanesia and Southeast Asia* (Melbourne: Oxford University Press).

Ingold, T. (1995) 'Building, Dwelling, Living', in M. Strathern (ed.), *Shifting Contexts: Transformations in Anthropological Knowledge* (London/New York: Routledge): 59-79.

Joyce, S.A., and M. MacFarlane (2001) *Social Impact Assessment in the Mining Industry: Current Situation and Future Directions* (London: International Institute for Environment and Development Mining, Minerals and Sustainable Development Project).

Langton, M. (2001) 'Dominion and Dishonour: A Treaty between our Nations?', *Postcolonial Studies* 4.1: 13-26.

Lawrence, R. (2002) 'Governing Warlpiri Subjects: Indigenous Employment and Training Programs in the Central Australian Mining Industry' (honours dissertation, Departments of Human Geography and Sociology, Macquarie University, Sydney).

—— (2005) 'Governing Warlpiri Subjects: Indigenous Employment and Training Programs in the Central Australian Mining Industry', *Geographical Research* 43.1: 40-48.

—— (2007) 'Corporate Social Responsibility, Supply-chains and Saami Claims: Tracing the Political in the Finnish Forestry Industry', *Geographical Research* 45.2: 167-76.

—— (2008) 'NGO Campaigns and Banks: Constituting Risk and Uncertainty', *Research in Economic Anthropology* 28: 241-69.

Lertzman, D.A., and H. Vredenburg (2005) 'Indigenous Peoples, Resource Extraction and Sustainable Development: An Ethical Approach', *Journal of Business Ethics* 56.3: 239-54.

Loney, J., and I. Williams (2000) 'Towards True Triple Bottom Line Reporting', presentation by Placer Dome Asia Pacific Ltd to CPA (Certified Practising Accountants), *Centering on Excellence Conference*, Hilton Hotel, Sydney, March 2000.

Mabee, H.S., and G. Hoberg (2006) 'Equal Partners? Assessing Comanagement of Forest Resources in Clayoquot Sound', *Society and Natural Resources* 19.10: 875-88.

Macdonald, I., and K. Southall (2005) *Mining Ombudsman Case Report: Marinduque Island* (Sydney: Oxfam Australia).

Natcher, D.C., S. Davis and C.G. Hickey (2005) 'Co-management: Managing Relationships, Not Resources', *Human Organization* 64.3: 240-50.

Newmont Mining Corporation (2001) 'Our People', in Newmont Australia, *Now and Beyond 2001* (Adelaide, SA: Newmont Mining Corporation).

O'Faircheallaigh, C., and T. Corbett (2005) 'Indigenous Participation in Environmental Management of Mining Projects: The Role of Negotiated Agreements', *Environmental Politics* 14.5 (November 2005): 629-47.

O'Malley, P. (2004) *Risk, Uncertainty and Government* (London: Glasshouse Press).

Paine, R. (1982) *Dam a River, Damn a People? Saami (Lapp) Livelihood and the Alta/Kautokeino Hydro-Electric Project and the Norwegian Parliament* (Copenhagen: International Work Group for Indigenous Affairs).

Rose, D. (1991) 'Indigenous Ecologies and an Ethic of Connection', in N. Low (ed.), *Ethics and Environment* (London: Routledge): 175-87.

Sapak, P., W. Melrose, D. Durrheim, F. Pawa, S. Wynd, P. Leggat, T. Taufa and M. Bockarie (2004) *Evaluation of the Lymphatic Filariasis Control Program, Samarai Murua District, Papua New Guinea*, August 2004; www.jcu.edu.au/school/sphtm/documents/lfmisima/LFMisima04.pdf, accessed 2 October 2008.

Strathern, M. (1988) *The Gender of the Gift* (Los Angeles: University of California Press).

Trigger, D. (1997) 'Mining, Landscape and the Culture of Development Ideology in Australia', *Ecumene* 4.2: 161-80.

Warhurst, A. (1998) *Corporate Social Responsibility: A Discussion Paper Prepared for the Working Group on Corporate Social Responsibility of the WBCSD* (Warwick, UK: Mining and Environment Research Network).

Whitehouse, L. (2006) 'Corporate Social Responsibility: Views from the Frontline', *Journal of Business Ethics* 63.3: 279-96.

Wilson, R. (2001) 'Corporate Social Responsibility: Putting the Words into Action', speech to the *RIIA-MMSD Conference on Corporate Citizenship*, Chatham House, London, 16 October 2001.

Wynd, S., J. Carron, W. Selve, P.A. Leggat, W. Melrose and D.N. Durrheim (2007a) 'Qualitative Analysis of the Impact of a Lymphatic Filariasis Elimination Programme Using Mass Drug Administration on Misima Island, Papua New Guinea', *Filaria Journal* 6.1; www.filariajournal.com/content/6/1/1, accessed 2 October 2008.

——, D.N. Durrheim, J. Carron, W. Selve, J.P. Chaine, P.A. Leggat and W. Melrose (2007b) 'Socio-cultural Insights and Lymphatic Filariasis Control: Lessons from the Pacific', *Filaria Journal* 6.3; www.filariajournal.com/content/6/1/3, accessed 2 October 2008.

York, G., and L. Pindera (1991) *People of the Pines: The Warriors and the Legacy of Oka* (Boston/ London/ Toronto: Little, Brown & Co.).

6

Corporate engagement with indigenous women in the minerals industry

Making space for theory

Ginger Gibson

Norman B. Keevil Institute of Mining Engineering, University of British Columbia, Canada

Deanna Kemp

Centre for Social Responsibility in Mining, University of Queensland, Australia

Indigenous women have always been involved with mineral development through activities such as artisanal mining, direct or indirect employment in larger-scale mining operations and (rarely) through direct ownership of mining operations.[1] In addition, women are family members and care providers to miners. Until recently there has been limited recognition of women's contributions and interaction with the mining industry, with even less attention paid to indigenous women and processes of exclusion and inclusion in mineral development.

1 There is no simple definition of the concept of indigenous people. The discussion offered by Connell and Howitt (1991: 3) is helpful as it conceives indigenous identities in a dynamic sense and indigenous people as active participants in development processes, offering resistance to forms of domination.

This chapter focuses on corporate engagement[2] of indigenous women in and around large-scale mining.[3] The chapter explores empirical data on this subject and relates this to several different theories. We invoke theory in two main ways: as a frame to help make sense of the world, and as a predictive tool to understand what might happen some time in the future. As with many other domains of knowledge, contributors to the literature on corporate engagement with indigenous women are often not explicit in their use of theory. Consequently, in exploring the use of theory in relation to our topic, we have 'read between the lines' of much of the empirical work we reviewed when the theoretical perspective was not made explicit. This chapter aims to open up space for debates about theory, in order that thinking is extended beyond particularised contexts and experiences, and so that individual cases can contribute to theory-building in our quest to understand how and why indigenous women are marginalised, or not, in the large-scale mining context. The theory that we review includes Marxist or class-based analysis, with Marxist–feminist theory as an outgrowth of this general approach, and cultural theory, with postmodern cultural theory as an outgrowth.

Before moving on to explore theory, the chapter outlines how the minerals industry considers indigenous women at a policy level and then provides some empirical findings about corporate engagement with indigenous women.

Indigenous women in the corporate policy domain

With the global expansion of mineral development into remote areas, the impact of mining on indigenous peoples has grown. The mining industry, largely through industry bodies and leading multinational mining companies, has undertaken research to better understand its impacts on indigenous peoples (Cruz and Marcos 2002; Downing *et al.* 2002). There has been some adjustment to industry and corporate policy regarding mineral development and indigenous people (ICMM 2003; MCA 2005; Mining Association of Canada 2006). However, these policy positions tend not to have an explicit gender dimension,[4] despite acknowledgement in *Breaking New Ground*, the final report of the Mining Minerals and Sustainable Development initiative (IIED and WBCSD 2002), that the impact of mining on women has been exacerbated by the failure to identify women as a distinct group of stakeholders.

Indigenous women are in a 'double blind'—a term we use to describe corporate engagement that tends to overlook the nexus between gender and indigeneity—a fact

2 Our interpretation of engagement is broad, encompassing any form of interaction, effective or not, between mining companies and indigenous women.

3 This chapter does not explore concepts of CSR in great depth, other than to agree with other authors in this volume that CSR is under-conceptualised. Perceptions and judgements of whether CSR initiatives are adequate or not, or indeed whether CSR is an adequate concept in and of itself, will depend on definitions of CSR, and will thus influence interpretations both in the empirical and theoretical realms.

4 We consider gender to be a socially constructed relationship between men and women, which is often unbalanced and unequal. Like Lahiri-Dutt (2006: 163) we also see this relationship as fluid rather than fixed in nature.

reflected in the literature. Indigenous women are rarely referred to in corporate documentation; instead, they are usually obliquely captured under the category 'women' (read 'white women') or 'indigenous' (read 'indigenous men'). However, there are several drivers that may prompt the mining industry to strengthen its focus on this group. The corporate social responsibility (CSR) agenda is increasingly focusing on gender: for example, the Global Reporting Initiative's recently proposed mandatory gender indicators for corporate sustainability reports (Global Reporting Initiative 2006). Companies adhering to the Global Reporting Initiative are now required to report gender-disaggregated data for workforce composition, employee turnover and salary. Some mining companies now relate their social performance to the Millennium Development Goals,[5] one of which is gender equality and the empowerment of women. In guidance notes from the International Finance Corporation gender is referred to, requiring that companies 'be inclusive of both women and men and of various age groups in a culturally appropriate manner' (IFC 2006: 29). And, in recognition of the gendered distribution of risks and benefits of extractive industry projects, the World Bank has proposed a guideline that aims to improve the sustainability and development impact of such projects (World Bank 2006).

Change in the mining industry's position on gender is also being driven by social movements. The International Women's Mining Network, for example, brings together networks of women internationally and nationally to focus on gender issues in mining. A 2004 conference in India led to a statement heavily focused on the rights of indigenous women, invoking action on economic liberalisation, mining impacts, consultation structures and employment (International Women and Mining Network 2004). Such statements are also being made along geographic lines, with women from the Pacific Region of the network calling for governments and all mining companies operating in the region to protect human rights and ensure that mining occurs only with the consent of local communities and with no negative impacts (Pacific Region International Women and Mining Network 2007).

While the drivers for a change in approach are strong, within the large-scale minerals industry there is still a gap at the level of policy, as indigenous women are not explicitly identified as a key stakeholder group. This disconnect becomes obvious when examining the research record of site-level corporate engagement with indigenous women.

Empirical data on indigenous women and engagement

The exclusion of indigenous women at a policy level within the minerals industry also plays out in practice, as evidenced within the empirical literature. Three broad aspects of this exclusion emerged from a literature review of indigenous women in mineral development:

5 The Millennium Development Goals, adopted by the world's heads of government/states at a special UN summit in 2000, include quantified goals for development and poverty eradication by 2015 (UN 2008).

- Engagement during negotiations
- Environmental, social, economic and cultural impacts
- Distribution of labour and employment

Engagement during negotiations

Historically, engagement with indigenous peoples about economic benefits of mining has often been absent (Connell and Howitt 1991). Lack of engagement with indigenous women in the early stages of mineral development often reflects a legal framework that either is not enforced or does not establish indigenous or gender rights in the first place. In some countries, indigenous people are provided with a legal framework that provides the right to negotiate, such as through the Australian Native Title Act 1993 and the Papua New Guinean Mining Act 1992. However, legal frameworks do not guarantee better outcomes. For example, in a review of the cases arbitrated by Australia's National Native Title Tribunal, it was found that every mining lease in dispute between a mining company and indigenous people was granted, and no substantive new conditions were imposed on any of the mining companies (Corbett and O'Faircheallaigh 2006).

Neither do legal frameworks guarantee equitable representation on the basis of gender. In Papua New Guinea (PNG), women are rarely represented in formal negotiations, though they are consulted by men (Macintyre 2002), and their inclusion on negotiation teams does not mean they are free to contribute equally (Byford 2002). The adequacy of indigenous women's involvement in negotiations during the development of gold mining in Central Australia has been called into question, with 'Land Council business' in this region perceived to be men's domain (Howitt 1991). There are several examples where indigenous women have been excluded from negotiations over mineral development, and consequently failed to gain benefits such as compensation payments and jobs, as occurred at Misima in PNG (Byford 2002), in Freeport in Indonesia (Simatauw 2002) and in India (Ahmad and Lahiri-Dutt 2006). The involvement or non-involvement of indigenous women in negotiations about mineral development is a fertile area for further research.

Environmental, social, economic and cultural impacts on indigenous women

The empirical literature suggests that, while there are a number of dynamics at play such as patriarchal customary practices and inadequate legal and market frameworks, exclusion from engagement in negotiations with large-scale miners contributes to inequitable outcomes for indigenous women, including negative impacts and lack of benefit (IIED and WBCSD 2002; Macdonald 2002; World Bank 2006). While social, economic and cultural impacts for indigenous women are felt in development- and location-specific ways, there are common themes. Studies show that the advent of mining transforms the economic and cultural life of indigenous communities, particularly when there is a change from subsistence cultivation to cash economies, or where resettlement is involved (Ahmad and Lahiri-Dutt 2006; Bonnell 1999). Whether in PNG (Gerritsen and MacIntyre 1991), Indonesia (Robinson 1986, 1996) or Fiji (I. Macdonald 2004), the sudden influx of cash through compensation, royalties and wages through direct or indirect employment (usually of men) in mining sees radical change and the renegotiation of gender roles, usually with a more negative impact on indigenous women.

It is indigenous men rather than women who tend to gain employment in large-scale mining operations, even when jobs are part of a compensation agreement. With indigenous men displaced into wage labour, the division of labour in subsistence agriculture and social routines are quickly altered. In some countries, indigenous women face increased workloads as they manage households and family responsibilities alone (Bhanumathi 2002; Carino 2002). Mining often sees an influx of migrant labour, usually men, hoping to secure mining employment, either directly with the mine or within other related industries (Nyame and Grant 2007). Such migration can change the dynamics of communities and impact on community life. Women and families can become dependent on male wages from mining as their primary source of livelihood rather than collecting subsistence resources to sustain the family's overall livelihood, stripping them of their traditional means of acquiring status and wealth (Macdonald 2006). Some indigenous women face an increase in alcohol-related abuse, domestic violence and general social disruption once men find employment in mining. Some men are also known to spend wages on short-term consumables and status items (Trigger 2003) and on prostitutes, which in turn exposes women to diseases such as HIV/AIDS (Silitonga *et al.* 2002). With men controlling financial and other resources, a new and unequal redistribution of power often occurs (Connell and Howitt 1991; Gerritsen and MacIntyre 1991).

Environmental impacts can affect indigenous women uniquely, as women often rely on the environment to provide for their family's livelihood. McGuire (2003) discusses the gendered impacts of mines, one in Australia and the other in Indonesia, to highlight that exclusion of women in the consultation phase of mine development can result in significant gendered environmental impacts. McGuire shows that mining can result in the destruction of traditional lands and can see the removal of forests, which can prevent women from accessing traditional medicines, foods and cultural materials. The pollution of water by poorly managed mining discharge can also contaminate drinking and bathing water, and result in toxic levels of metals in local fish. Resettlement and relocation has also been shown to disproportionately affect women. Displacement from original homes often means not only physical displacement, but also social, cultural and economic displacement, including loss of livelihoods derived from subsistence resources from the local area (Ahmad and Lahiri-Dutt 2006).

When the impacts of mining operations are felt, women often actively seek to engage with the company to address them. Emerson-Bain (1994) notes that women have been active in community-based opposition to mining and records worker struggles in Fiji and landowner struggles in Bougainville, where women were on the front line of the earliest clashes with the mining company. The Porgera Women's Association (PWA) in PNG, established to give voice to women's concerns, dealt with issues of road traffic, dust from the roads, healthcare services and standards, inclusive of maternal and child health. Bonnell (1999) argues that such associations have the potential to provide support for women to take risks and enter the public debate, although she does not clarify the degree of influence that the PWA actually had over changing operations to address women's concerns, or the processes of negotiation. Women have also called in third parties to assist them. For example, since 2000, Oxfam Australia's Mining Ombudsman has provided indigenous (and non-indigenous) women and men with a third-party mechanism through which to raise concerns about the impact of Australian mining companies on their community (Oxfam 2007). The Ombudsman does not adopt a pro- or anti-mining position. Rather, its human rights-based approach attempts to level the often unequal power dynamics between community and company, particularly along gendered lines. The Ombudsman aims to resolve disputes through participatory processes, capacity-building, involvement of local organisations and the use of local knowledge, including women's knowledge. There is limited research about the degree to which such third-party dispute-resolution mechanisms adequately represent the views of indigenous women, achieve long-term outcomes or function to empower women, even if disputes remain unresolved.

Although the adverse impacts of mining on indigenous women are more extensively covered in the literature, positive outcomes from mining are occasionally highlighted. Researching the impact of the activities of a Canadian company, PT INCO in Indonesia, Robinson (2002: 43) mentions increased access to education and the chance to travel outside the community. Byford (2002) notes that expenditure on infrastructure at the Misima Mine in PNG has indirectly aided women in negotiating a new role within the community, with roads and transport providing accessibility to new markets. Similarly, MacIntyre (2003) notes dramatic improvements in transport, health services, water supplies and education facilities at the Lihir mine, also in PNG. However, that facilities exist does not guarantee access to them: they may be inside the mine compound to guarantee security but this may prohibit access to them, or exorbitant user fees may be charged (Gibson 2003). Corporate community development programmes may target women, but there is little publicly available qualitative or quantitative data available on programme outcomes from the perspective of indigenous women. Given recent corporate commitments within the mining industry to the achievement of the Millennium Development Goals, this situation appears neglectful and disappointing.

Empirical research confirms the widely accepted view that indigenous women have not been adequately engaged about the impacts and opportunities of mining, and experience adverse impacts differently and often more deeply than men. There is increasing evidence to suggest that social impact and gender assessments are being undertaken by mining companies, but the results of these rarely sit in the public domain, and thus it is not currently possible, from the research record, to appreciate how companies understand and respond to impacts on indigenous women. Empirical research relating to corporate responses to concerns articulated by indigenous women is virtually non-existent.

Labour and mining employment

Employment is an important avenue for indigenous people to gain benefits from mineral development. Some mining operations employ small numbers of indigenous women. However, there are few empirical studies about indigenous women and employment in large mines, and it is difficult to tell from publicly available data just how many indigenous women are involved. In Australia it is estimated that indigenous women make up 0.4% of the total mining workforce (compared with a 15% representation of non-indigenous women), which represents 16% of the indigenous mining workforce (Parmenter and Kemp 2007). Such figures suggest that indigenous women, many of whom live in remote areas and are Traditional Owners with custodial rights and cultural knowledge of lands, are one of the groups least likely to be working at Australian mines (Bryant and Tedmanson 2005).

Bryant and Tedmanson cite Moreton-Robinson (2000), who sees the neglect of indigenous women as code for privileging white women's experience or situations. Responding to this gap, Parmenter and Kemp (2007) re-analysed data across three different studies that involved, but did not originally focus on, indigenous women at mines in north-west Queensland in Australia, to profile indigenous women's employment experience in large-scale mines, including factors that attracted them to mining, barriers to entering mainstream employment and their experience of mining work. The data indicated that, in comparison to non-indigenous women, the following factors were more important in attracting indigenous women to mining: the opportunity to work with family or friends; the availability of study assistance; the chance to be a role model for other indigenous people; and to pursue personal development and financial benefit. However, in relation to the last point, very little is known about the career aspirations of indigenous women in mining (Parmenter and Kemp 2007). Indigenous women from one of the mines with a relatively high percentage of indigenous female employees indicated that their employment experience was generally positive, notwithstanding dissatisfaction expressed about specific aspects of the workplace.

The South African context supports the assertion that while the mining workplace is characterised by structural gender divisions, racial divisions also persist in terms of the types of jobs held. Ranchod (2001) explains that, in South African mines, professional women are likely to be white, while unskilled women in mining are likely to be black (Ranchod 2001: 6). This follows the general historical trend of indigenous labour being concentrated in unskilled, temporary and underprivileged positions (Tiplady and Barclay 2007) such as in the case of the exclusion of the lower castes (*Adivasis*) in the Indian collieries (Sinha 2002).[6]

Literature focusing on the role of women in mining in developing countries is growing, although the focus tends to be on artisanal and small-scale mining, where women are better represented as a percentage of the overall workforce. Macintyre and Lahiri-Dutt (2006) bring together a range of case studies from the Asia-Pacific, African and Latin American regions documenting the histories and experiences of women, often indigenous, working in mining. The case studies aim to make visible the roles and contributions of women as miners and stimulate debate about how gender and inequality are constructed and sustained within the industry. A key finding is that numbers of women employed fall as mining moves to the large scale, particularly in developing countries.

There is a small but notable concentration of articles that relate to the experiences of female employees in large-scale mining in PNG. For example, Bonnell (1999) documents the experiences of local women who gained employment at the Porgera mine; Macintyre (2003) does the same at the Lihir mine. These authors suggest that, while women generally enjoyed their employment, they experienced problems with sexual harassment, male backlash and limited childcare options. In their study of Australian mines, Kemp and Pattenden (2007) concluded that indigenous women face additional challenges such as systemic social disadvantage, complex family responsibilities (including pressure to stay at home and look after children and family members) and issues associated

6 The Indian context is particularly complex as it sees both class- and caste-based divisions.

with holding positions of authority over indigenous men. Combining family and child-care responsibilities with shift work has been cited as a key reason for women leaving work in large-scale mines (Lahiri-Dutt 2006; Kemp and Pattenden 2007).

In summary, the empirical evidence indicates that indigenous women have largely been excluded from negotiations about benefits from mineral development, including employment, and often experience adverse impacts of mining more substantial than those felt by men.

Theoretical approaches

Theoretical treatments of gender and the extractive mode of production explain and frame how gender roles and identities operate in society, and how structures and processes work to reinforce them. A range of theoretical perspectives are examined here. Feminist perspectives on Marxist political economy describe the influence capitalism has on the identities, roles, households and bodies of women. Postmodern cultural theory offers distinct methodologies and insights on impact areas and power relations. There is substantial overlap and common ground in these theoretical perspectives which hint toward the potential for hybrid or concurrent use of theory.

Marxist–feminist theory

Feminists who employ a Marxist political economy perspective suggest that segregated gender roles and a division of public and private life are required to maintain the capitalist mode of production. Change is hard to trigger, because the ruling class protects its ability to accumulate capital by presenting gendered roles, identities and places as normal, natural and universal (Miewald and McCann 2004; Robinson 1986). However, markets can require new labourers so that market transformation can lead to gender role shifts (Kideckel 2004). For example, as in World War II, with men gone in a wartime economy, economic necessity brings women into the workforce, leading to a softening of perceptions of who might occupy a job (Williamson 2003).

Gender, according to this perspective, has operated as a key driver of exclusion from the economy and in the mines (Emerson-Bain 1994; Robinson 1996). Mining communities and sites are constructed as male landscapes (Collis 1999; Robinson 1996), preserves of masculinity maintained through 'misogynistic forms of differentiation' (Eveline and Booth 2002: 11), such as nude pin-ups and segregated social activities. The sexualisation of work and space reinforces the gendered division of labour (Tallichet 1995), and, even as women enter the job hierarchy, they are expected to occupy subordinate positions in deference to men. Roles and identities are conditioned by unequal power relations (Miewald and McCann 2004) and the power struggles that emerge in mines and at home (over, for example, what constitutes a woman's role) are predicted to have outcomes on the bodies, formation of households and kinship networks of women (Miewald and McCann 2004).

A key relationship of struggle in the extractive economy is marriage, and feminist theorists suggest the social environment of public patriarchy operates to reinforce private patriarchy, creating unequal power relations. Power, conceptualised as the ability to gain a desired end for oneself or thwart that of another (Collis 1999), is most frequently in the hands of men due to their economic and physical power. Many of these feminists predict little or subtle change in responsibilities at home and at work as women become engaged in the economy (Hall 2004; Williamson 2003). Even with public authority, private patterns of authority and power are not changed, so that women's wages remain ancillary to men's (Miewald and McCann 2004; Williamson 2003), and outside work does not reshape the sexual division of labour at home (Measham and Allen 1994; Moretti 2006). New roles for women are predicted to invite aggression, so that 'violence and depression that accompany threats to traditional notions of masculinity in particular constitute a major concern and play a significant part in women's attempts to assert new ideals of gender relations' (Miewald and McCann 2004: 1,056). Some of these feminists theorise a shift in power in this relationship through the emergence of the 'partnership marriage' (Collis 1999; Miewald and McCann 2004), role switching or divorce (Kideckel 2004). Collis suggests that strategies such as bargaining, manipulation, supplication, autocracy, bullying and disengagement are used by women to attain desired outcomes in marriages involving a worker in the extractive industries (Collis 1999). As women work in mines, relationships outside the nuclear family are predicted to become harder to maintain, leading to the reshaping of social relations (Miewald and McCann 2004).

Marxist–feminists also theorise that strategies of resistance to male power also emerge in trade unions, as women creatively engage as mothers and sisters through dual auxiliaries (Merithew 2006) and through alternative neighbourhood forums (Measham and Allen 1994). Where 'unionism encompassed both male workers in the coal pits and female labour in the homes . . . [women] gained power [in the union] through motherhood' (Merithew 2006: 81). A parallel union structure, the auxiliary, has been framed as a site of limited power (Merithew 2006). Union organisers, according to Marxist–feminists, have marginalised women. For example, Lahiri-Dutt (2006) demonstrates how Indian female labourers of a particular caste were marginalised from the workforce as a new class of male immigrants entered the region. While a Marxist–feminists perspective prioritises the union as a site of resistance, this perspective has also considered how women mobilise in neighbourhoods and on family and home-based issues (Measham and Allen 1994). This leads to a call for a 'broader definition of women-centred politics and activism [which] locates women's political activism within familial and familiar support networks' (Measham and Allen 1994: 43).

Marxist–feminist approaches have addressed the symbolic dominance of the extractive mode of production, as well as the challenge the industry poses to normative values of indigenous peoples (Robinson 1996). Wages embed a worker in the capitalist mode of production, but also cause tension in pre-capital structures and relations. An ethnography of Bolivian tin mines illustrates how capital and labour become embedded in existing systems of reciprocity and exchange, so that economic goods are not things in themselves but determinants of reciprocating human relations (Taussig 1980). The wage-based economy is transformed and apprehended through the local network of reciprocal relations. Further, Bolivian tin miners protect themselves through invocations to the devil they feel lives in the mine in order to ensure success in mining, which is ultimately believed to be destructive of life (Taussig 1980: 14). Macintyre (2003) also surfaces this concept of reciprocity in the study of Petztorme ('Working Together'), a women's organisation funded by a mine in the Lihir region in PNG. In that case, a well-meaning philosophy of self-help for women failed in part owing to the lack of funds or goods for distribution by women leaders who are expected to 'extend largesse' to others in the immediate lineage and clan (Macintyre 2003: 128). The Marxist–feminist approach instructs us that failure to read the symbolic function of wages, as illustrated by Macintyre (2003), and the historical shift of economies (Taussig 1980) may well neglect the core impact on indigenous belief systems.

Postmodern cultural theory

Cultural theory, and more specifically postmodern cultural theory, has focused attention on how culture, stories and language operate at least partly to reproduce unequal power relations and thereby exclude women from the mining economy. For example, study of pollution stories and myths in Bolivia reveals how men believe women, except for tourists and widows, destroy men's ore-finding luck (Absi 2006) and thus prohibit women from entering the economic domain of mining. In PNG, mining's gendered employment can be linked to mythology associated with women in mining that constructs women as polluting and dangerous (Clark 1993). Western mining company beliefs of appropriate roles for women (Moretti 2006) have also helped to maintain certain jobs as primarily male. Further, values and traditions of certain cultures (for example, Lihirian) prohibit men from consulting women publicly, with women required to rely on the ability of men to represent them (Macintyre 2003; Macintyre and Lahiri-Dutt 2006).

Postmodern cultural theory focuses on how language operates to preserve power and reinforce gendered constructions of work and knowledge, all of which form identity that is partial, interpretive and contingent (Lahiri-Dutt 2006). When work is described as dirty and difficult, language operates to protect the work for men and sanction their behaviour; Somerville (2005: 19) describes how 'violence and aggression were accepted as a "natural" accompaniment to the hegemonic masculinity required by the job'. The attention to language reveals how mining has become associated with the male gender. Lahiri-Dutt (2006: 164) argues that, by mining work in India being marked as 'dangerous' and 'dirty', it has become associated with men, thereby excluding women (who are labelled as clean and requiring protection). Women were then 'protected' from engagement in this profession by legislation that functions to strengthen male solidarity, bonding and militancy in mining.

Postmodern cultural theory, which emphasises the social construction of culture and nature, has challenged images of 'miner's wife' or 'country woman' constructed through early Marxist–feminist research, yielding new understandings of power and identity. Hall (2004) suggests identity is more complex than these archetypes of women might suggest, which describe women as baby-makers, activists or drudges. Instead, through postmodern interpretations, women are written of as having agency and multiple subjectivities (Gibson-Graham 1995). Power is inscribed everywhere (by women and by men) and constitutes the subjectivities of people through everyday practices. Power is shown to exist in many relations, such as older–younger, urban–rural, wealthy–poor and researcher–researched. The task of postmodernity in research on mining and gender has been to 'constitute alternative sites of power and places of political intervention'

(Gibson-Graham 1995: 15), so that identity is seen to be variable and women embrace different and new ideologies to express their identities (Hall 2004) and influence power.

The focus and language of postmodern cultural work yields new conceptual tools for analysis of the role of women in mining. One of these is the concept of deconstructing 'performances', a term used to explain critical workplace occurrences or events and associated language, to understand their root causes and how they influence subjectivities. Eveline and Booth (2002) analyse 'gendering episodes': that is, male-dominated performances, which function to harass women and exclude them from mines, such as the tangible example of a female miner's picture spliced onto a picture of a man's body and labelled 'dyke' for a mine-wide pin-up. Highly public support of a female worker is introduced as a possible counter-performance, an act that might change the dynamics of power if 'critical acts' are accompanied by the support of those in authority (Eveline and Booth 2002).

Even as the range of strategies analysed seems to expand with this turn toward postmodern cultural theory, women's opportunities within the mines continue often to be in 'zones of domesticity' (Robinson 1996: 139). When women move into non-traditional roles, they apparently 'normalise' the workforce so that pin-ups and other sexually explicit material behaviour become inappropriate (Miewald and McCann 2004; Ranchod 2001), but even as they do this they are still marginalised and often become the minders of the site, the men and the machines (Eveline and Booth 2002). Indeed, Lahiri-Dutt (2006: 366) points out that women are hired in Indonesia precisely because they are more 'docile and easier to control' than men. However, they are expected to behave like men and they cannot afford to be too different.

Indigenous theorists, in the postmodern era, have brought new issues to light that escape the cultural preconceptions of the 'other'. For example, concepts of family tend to be defined by the models feminist theorists live in themselves. Notions of mothering and nurturing have come to be narrowly defined within the dichotomies of public and private. These binary models have imposed often inappropriate redefinitions on indigenous family and gender roles (Edmunds 1996). Indigenous perspectives reveal impact pathways that dislocate transmission of gender-based knowledge. In Australia, Kopusar (2002) points to three aspects of this issue: legislation protective of free entry systems; corporate practice allowing men to enter sacred areas; and the cultural construct of gender. Kopusar explains that legislation allowing free entry into indigenous territory allows men to transgress women's sacred land, breaking local laws of exclusive female occupation of sacred spaces. As special places become unavailable to women through mining and exploration, women are unable to then transmit sacred knowledge to children. Knowledge transmission, the meaning of spaces for knowledge access and the significance of spaces in a cultural landscape are important issues to bring to life in impact assessments. As Kopusar (2002: 13) suggests: 'The land isn't empty. The bushes and trees are teeming with food for the children . . . every part of the land has a message for the people.'

Linking theory and empirical work

While these theories are helpful as a lens through which to view the world, a key question is whether these theories are also useful for predicting what is found in the empirical realm. The discussion so far suggests that Marxist–feminist and postmodern cultural theory foreshadow the complex reality of lived experience; our next section tackles this question more directly.

Engagement during negotiations

Marxist–feminist approaches predict the capitalist mode may displace other modes of production, but empirical research often fails to recognise forms of non-capitalist relations, such as when goods become entangled in symbolic exchanges in networks and challenge the normative values of indigenous peoples (Robinson 1996). Macintyre (2003), relying on postmodern cultural theory, suggests that a corporate design for a women's organisation failed because of the lack of understanding of non-capitalist relations. Explanations of why indigenous women fail to be engaged in negotiations tend to ring hollow unless the culturally defined importance of symbolic exchanges and power relations in and between men and women are acknowledged. And these failures to get a seat 'at the table' cannot be overcome without this prior acknowledgement.

Very few empirical studies show women to be at the negotiating table, other than through consultations outside the formal engagement process. Marxist–feminists might presuppose this exclusion because they see male- and class-based privilege overwhelming negotiations before they even begin. Postmodern cultural theory might predict that women will be excluded from negotiations through the use of male-dominated and female-denigrating language and performative acts. While Macintyre (2002: 28) suggests the 'best opportunity for women to get around the bargaining table is during the initial negotiation phase, when mining companies are most susceptible to local demands', she offers no explicit discussion of why this is so. However, implicit in the argument is that stakes, perceptions and outcomes can be negotiated at early stages as identities are just being formed and before gendered structures of workers' unions emerge to exclude women, or occupational structures (with sexualised language and spaces) control entry.

Postmodern cultural theory reveals multiple levels of gendered exclusion. Cultural relativism has been employed to provide defences of male privilege, because mining companies may not risk offence to indigenous men by suggesting women need to be included, or these men may use corporate funds and wages to reinforce their own predominant status and roles (Macintyre 2003). Indigenous women may have to combat mining companies' preference for negotiating with indigenous men rather than women, a corporate preference which combines to give male issues and discourses of masculinity (Eveline and Booth 2002) a priority as the natural way of things. Thus, indigenous women may be excluded by non-indigenous and indigenous men alike. Human rights literature suggests that women's engagement to make decisions about what affects them is a basic and inalienable right (Macdonald 2006), but moral arguments are yet to be reflected in a growing number of indigenous women negotiators. This gap between principle and practice needs to be better assessed.

Marxist–feminist approaches tend to incorrectly predict the underlying pathways, patterns and intensity in how indigenous women mobilise or assert agency; as a result, the structures for public participation introduced by unions are often entirely inappropriate. These theorists predict that women organise in solidarity through class-based organisations, but empirical literature reveals other motivations and mechanisms (Measham and Allen 1994). Some empirical literature suggests that the initial organising structures of local community representation tend to be introduced by corporations, state agencies or consultants and are often modelled on similar structures at other projects (Ballard and Banks 2003; Macintyre 2003). When the village is presumed to be the model of organisation, the negotiation and consultation model of the corporation is inappropriate if the social unit in which people organise and identify is the family, clan or church. The mismatch in modes of organisation has led to a call for organising based on familiar and family-based networks at the local level (Macintyre 2003).

Impacts and processes of resistance

While some corporate statements of CSR call for greater inclusion of women in stakeholder engagement and impact assessment, it is challenging for theoretical Western constructs of environment and nature to discern the subtleties encoded in indigenous epistemologies. Even gender itself may be an insufficient construct to understand indigenous women's knowledge, as illustrated by Edmunds (1996). Rather, a combination of public and private knowledge, secret and sacred knowledge, knowledge based on gender and age as well as other perspectives are required to bring forth new understandings of gendered impacts of mining (Edmunds 1996).

Postmodern cultural theory suggests struggles over language, theory and terrain will aim to contain women as 'stakeholders', keeping them from active negotiations with mining companies, away from the mine site, and under-utilised in the effort to identify impact. The company will emphasise, along with international lenders, the necessity of the extractive industries for regional development, and CSR work may serve to entrench further the 'neo-colonial relations that manage and control aboriginal life' (Banerjee 2000: 14). Postmodern cultural theory might suggest that women will be contained from active negotiation over resources.

Empirical and theoretical explorations of resistance of indigenous women to mining and its impacts are under-developed. Moments of resistance can be vexing for mining corporations as they represent the possibility that communities will not support the extractive endeavour. Yet it is these very moments that some indigenous theorists characterise as anti-colonial and symbolic struggles (Banerjee 2000). These are likely to be the spaces that many theorists easily overlook: for instance, in conferences, in dialogues and in story-telling (Macdonald and Rowland 2002). These are also the moments that are under-theorised, and tend to be viewed as the minority discourse in postmodern cultural theory (Tully 2000), if they are viewed at all. Subaltern perspectives, which would include indigenous theorists or people rendered without agency due to their social status, can reveal that a white woman can influence perspectives in the corporations more than an indigenous man (Banerjee 2000; Ranchod 2001).

Work and employment

The motivations, roles, networks and identities of indigenous women are notably absent in theoretical literature by Marxist–feminists and postmodernist cultural theorists. Empirical work with white women reveals different motivations and organisational forms than those predicted through Marxist–feminist research. In 1984–85 coal miners' strikes, British women mobilised through kin networks, not grassroots unions (as predicted by Marxist–feminists), over the fear of job loss (rather than the impulse of solidarity), and supported the movement through the continuance of critical roles of feeding, cleaning and clothing (Measham and Allen 1994). Thus, one might expect indigenous women to also reject a simplistic Marxist–feminist explanation of motivations and identities. Indeed, international women's mining movements appeal to both miners and those that oppose the mine and do not sustain the class-conflict analysis or gendered division of labour (Gier and Mercier 2006). A recently organised international women's organisation on mining allows for the participation of both those who support capitalism and those who oppose it, countering the class divisions and economism inherent in Marxism (International Women and Mining Network 2004). However, ethnographic approaches guided by Marxist–feminism can capture distinct motivations of women mine workers, who are able to 'develop outside of the kinship-based forms of power that had previously structured their lives and had limited their personal movement outside the home' (Robinson 2002: 42).

While employment data on indigenous women is sparse, indications are that indigenous women are employed at extremely low levels and numbers within mining corporations. Marxist–feminist theory predicts that capitalism depends on the gendered division of labour, holding that even as women become involved in the industry, they are trapped in a limited range of jobs and activities by virtue of the hydraulic sexuality of mines (Robinson 1996) and gendered identities and roles associated with work. The seemingly 'normal' male process of extraction has naturalised gendered codes in everyday practice (Robinson 1996). Large-scale mining is likely to remain deeply masculine, with gendered concepts of nature, forms and practices held in place, until these premises are challenged. In the mines, women appear as site minders and in positions without authority, and they are challenged, directly and indirectly, if they behave unlike a woman apparently ought to, if they begin to advance, or when they challenge male authority. Outside the mine, women appear as wives, domestic workers and in the sex trade. Just as with negotiations, women can be excluded not only through the productive mode, but also by the lens and frame of the endeavour itself.

Exclusions born of distance and geography have had little play in the literature. With a significant proportion of mine sites now developed as fly-in, fly-out operations, indigenous women have many home and family roles to leave and perceptions to overcome in order to engage in the occupational site of a mine. Given that white women's early mobilisation was based in the auxiliary, and later in the union, remote mines with no 'home space' present challenges for conventional modes of female mobilisation, making any grassroots, home-based women's movement much more difficult. Also, unions may appeal to indigenous men but not specifically to indigenous women (Merithew 2006) in that unions may organise labourers in the pit, processing plants or machine shops, very likely missing the feminised 'zones of domesticity' (Robinson 1996) where indigenous women might be employed. Thus, postmodern cultural theory predicts that a company's 'double-blindedness' to the nexus between gender and indigeneity will continue to hold, as unions may only appeal to, or more importantly, be available to, the relatively small number of women working in the pit or process-plant, not to women in the zones of domesticity in the mines, where they are much more likely to be.

CSR may aim to engage indigenous women with new policy and programmes, yet the empirical research to demonstrate effectiveness of strategies has not been done. While affirmative action has been shown to make good business sense (Eveline and Booth 2002), it is possible that CSR initiatives will function to replicate existing class, raced and gendered structures, privileging white women or placing enormous pressure on a few indigenous woman to represent all. Organisational tools such as gender mainstreaming, critical acts in support of women in the mines and gaining a new critical mass of indigenous women in organisations may hold out the possibility of small wins for indigenous women. Parmenter and Kemp (2007) explained how the Century Mine in north-west Queensland in Australia attracted and retained a notable and relatively stable number of indigenous female employees using the critical mass hypothesis, suggesting that once a threshold of women is achieved the corporation is able to retain women. However, CSR theorists argue that mainstreaming requires activities and guidelines within corporations buttressed by political support and policy change, rather than simply identity- and group-based organising (Grosser and Moon 2005).

Marxist–feminists and postmodern cultural theorists do not explain how indigenous women working at a mine site may continue to face multiple pressures, harassments and difficulties of balancing home and work life. If the stories of male identity protection revealed through postmodern accounts can teach anything, it may be that, if indigenous women begin to spin yarns about themselves, with new narrative threads and characters, then subtle shifts may occur in sites. However, gendering and 'racing' episodes are likely to continue to oppress and exclude indigenous women, a fact white woman may continue to overlook or even ignore.

Each approach, Marxist–feminism and postmodern cultural theory, brings alternative perspectives on power, identity, domestic and work roles, gender and agency. The nuances of postmodern cultural theory are helpful in deconstructing the archetypes that were built up with Marxist–feminist analyses.

Space for further research

Our closing comments relate to how theory can be used to better inform empirical work about corporate engagement with indigenous women. First, postmodern cultural theory holds potential to chart new territory, following the lead of some of the best research from ethnographic studies based on a strong theoretical model (Lahiri-Dutt 2006; Macintyre 2003; Robinson 1996). Postmodern cultural theory, and the turn to linguistics and studies of power, may well serve to deconstruct the large-scale mining method and the language of impact assessment, surfacing assumptions of how indigenous women's bodies, identities, environments and knowledge are impacted through mining. This theory can be brought to bear to highlight similarities across case studies about the impacts of mining development on indigenous women and how they engage with the mining industry. From these foundations, empirical work and the development of principles for industry have a stronger platform from which to drive greater female engagement in negotiation and implementation of agreements. Both postmodern cultural and Marxist–feminist theory are helpful for framing the argument that adequate engagement of women has the potential to lead to more equitable outcomes.

Second, there is space to understand how and when indigenous women mobilise, resist and engage with the industry. Under-theorised spaces, such as the mobilisation apparent in the International Women and Mining Network, are ahead of both theory and empirical research, appealing across class, ethnicity and age ranges, and to miners and non-miners alike (International Women and Mining Network 2004). However, while postmodern cultural theorists suggest that women may rely on men to represent them (Macintyre 2003), further work on the structure and construction of gender roles in indigenous societies will be helpful. Macintyre's important work (2003) on how volunteerism in mining-based organisations frustrates women's engagement, as well as the symbolic and material impact of introduced funds, is a notable start.

Third, given the low numbers of indigenous women in the corporation, more robust understandings of how race and gender act together to exclude women in the mines are needed. Class and ethnicity have been combined to understand the exclusion of women in small-scale mining (Lahiri-Dutt 2006), but have yet to be applied to the large-scale mining corporation. Marxist–feminist approaches might also observe age- and role-based exclusions associated with child-rearing and care. They must also be brought to bear in the corporation, testing whether performance acts, critical mass and gendering episodes operate to exclude indigenous women and whether CSR can overcome the multiple barriers indigenous women face, following Render's (2004) observations that white women progress faster than black women in industry. Lastly, we have only just begun to understand the gendered and perhaps even aged sources of knowledge of women. All theorists and applied researchers need to 'come out of the closet' (Nicoll 2000) to more fully explore the power relations between women, indigenous and non-indigenous, and the large-scale mining economy.

References

Absi, P. (2006) 'Lifting the Layers of the Mountain's Petticoats: Mining and Gender in Potosí's Pachamama', in J. Gier and L. Mercier (eds.), *Mining Women: Gender in the Development of a Global Industry, 1670–2005* (New York: Palgrave Macmillan): 58-70.

Ahmad, N., and K. Lahiri-Dutt (2006) 'Engendering Mining Communities: Examining the Missing Gender Concerns in Coal Mining Displacement and Rehabilitation in India', *Gender, Technology and Development* 10: 313-38.

Ballard, C., and G. Banks (2003) 'Resource Wars: The Anthropology of Mining', *Annual Review of Anthropology* 32: 287-313.

Banerjee, S.B. (2000) 'Whose Land is it Anyway? National Interest, Indigenous Stakeholders, and Colonial Discourses: The Case of the Jabiluka Uranium Mine', *Organization and Environment* 13.1: 3-38.

Bhanumathi, K. (2002) 'The Status of Women Affected by Mining in India', in I. Macdonald and C. Rowland (eds.), *Tunnel Vision: Women, Mining and Communities* (Melbourne: Oxfam Community Aid Abroad): 20-24.

Bonnell, S. (1999) 'Social Change in the Porgera Valley', in C. Filer (ed.), *Dilemmas of Development: The Social and Economic Impact of the Porgera Gold Mine, 1989–1994* (Pacific Policy Paper 34; Canberra: Asia Pacific Press): 19-87.

Bryant, L., and D. Tedmanson (2005) 'Drilling Down—Diversity in the Mining Industry: Exploring the Barriers to Gender and Indigenous Diversity in the Australian Mining Industry', *International Journal of Knowledge, Culture and Change Management* 5.3: 157-68.

Byford, J. (2002) 'One Day Rich: Community Perceptions of the Impact of the Placer Dome Gold Mine, Misima Island, Papua New Guinea', in I. Macdonald and C. Rowland (eds.), *Tunnel Vision: Women, Mining and Communities* (Melbourne: Oxfam Community Aid Abroad): 30-34.

Carino, J.K. (2002) 'Women and Mining in the Cordillera and International Women and Mining Network', in I. Macdonald and C. Rowland (eds.), *Tunnel Vision: Women, Mining and Communities* (Melbourne: Oxfam Community Aid Abroad): 16-19.

Clark, J. (1993) 'Gold, Sex, and Pollution: Male Illness and Myth at Mt Kare, Papua New Guinea', *American Ethnologist* 20: 742-57.

Collis, M. (1999) 'Marital Conflict and Men's Leisure: How Women Negotiate Male Power in a Small Mining Community', *Journal of Sociology* 35.1: 60-76.

Connell, J., and R. Howitt (eds.) (1991) *Mining and Indigenous Peoples in Australasia* (Sydney: Sydney University Press).

Corbett, T., and C. O'Faircheallaigh (2006) 'Unmasking the Politics of Native Title: The Native Title Tribunal's Application of the NTA's Arbitration Provisions', *University of Western Australia Law Review* 33.1: 153-76.

Cruz, O., and A. Marcos (2002) *Indigenous Peoples, Mining and International Law* (London: International Institute of Environment and Development).

Downing, T.E., J. Moles, I. McIntosh and C. Garcia-Downing (2002) *Indigenous Peoples and Mining Encounters: Strategies and Tactics* (London: International Institute for Environment and Development).

Edmunds, M. (1996) 'Redefining Place: Aboriginal Women and Change', in R. Howitt, J. Connell and P. Hirsch (eds.), *Resources, Nations and Indigenous Peoples: Case Studies from Australasia, Melanesia and Southeast Asia* (Melbourne: Oxford University Press): 121-36.

Emerson-Bain, A. (1994) 'Mining Development in the Pacific: Are We Sustaining the Unsustainable?', in W. Harcourt (ed.), *Feminist Perspectives on Sustainable Development* (London: Zed Books): 46-59.

Eveline, J., and M. Booth (2002) 'Gender and Sexuality in Discourses of Managerial Control: The Case of Women Miners', *Gender, Work and Organization* 9.5: 556-78.

Gerritsen, R., and M. MacIntyre (1991) 'Dilemmas of Distribution: The Misima Gold Mine, Papua New Guinea', in J. Connell and R. Howitt (eds.), *Mining and Indigenous Peoples in Australasia* (Sydney: Sydney University Press): 34-53.

Gibson, G. (2003) *Community Perspectives on the Extractive Industries* (Vancouver, BC: CoDevelopment Canada).

Gibson-Graham, J.K. (1995) 'Identity and Economic Plurality: Rethinking Capitalism and Capitalist Hegemony', *Environment and Planning D: Society and Space* 13.3: 275-82.

Gier, J., and L. Mercier (eds.) (2006) *Mining Women: Gender in the Development of a Global Industry, 1670–2005* (New York: Palgrave Macmillan).

Global Reporting Initiative (2006) 'G3 Guidelines'; www.globalreporting.org/ReportingFramework/G3Online, accessed 20 August 2008.

Grosser, K., and J. Moon (2005) 'Gender Mainstreaming and Corporate Social Responsibility: Reporting Workplace Issues', *Journal of Business Ethics* 62: 327-34.

Hall, V.G. (2004) 'Differing Gender Roles: Women in Mining and Fishing Communities in Northumberland, England, 1880–1914', *Women's Studies International Forum* 27.5–6: 521-30.

Howitt, R. (1991) 'Aborigines and Restructuring in the Mining Sector: Vested and Representative Interests', *Australian Geographer* 22.2: 117-19.

ICMM (International Council on Mining and Metals) (2003) 'Sustainable Development'; www.icmm.com/our-work/sustainable-development-framework, accessed 9 August 2008.

IFC (International Finance Corporation) (2006) 'International Finance Corporation Performance Standard 7: Indigenous Peoples'; www.ifc.org/ifcext/enviro.nsf/Content/PerformanceStandards, accessed 10 August 2008.

IIED (International Institute for Environment and Development) and WBCSD (World Business Council for Sustainable Development) (2002) *Breaking New Ground: Mining, Minerals and Sustainable Development. Final Report of the MMSD Project* (London/Sterling, VA: Earthscan; www.iied.org/mmsd/finalreport/index.html, accessed 20 August 2008).

International Women and Mining Network (2004) 'Statement of the International Women and Mining Network: Third International Women and Mining Conference, Visakhapatnam, India, 1–9 October 2004'; www.mmpindia.org/womenminingstatement.htm, accessed 20 August 2008.

Kemp, D., and C. Pattenden (2007) 'Retention of Women in the Australian Minerals Industry', in Minerals Council of Australia and Australian Government Office for Women, *Unearthing New Resources: Attracting and Retaining Women in the Australian Minerals Industry* (Canberra: Australian Government Office for Women; www.minerals.org.au/__data/assets/pdf_file/0016/20275/MCA_Women_In_Mining_WebVersion.pdf, accessed 20 August 2008).

Kideckel, D.A. (2004) 'Miners and Wives in Romania's Jiu Valley: Perspectives on Postsocialist Class, Gender, and Social Change', *Identities: Global Studies in Culture and Power* 11.1: 39-63.

Kopusar, P. (2002) 'An Australian Indigenous Women's Perspective: Indigenous Life and Mining', in I. Macdonald and C. Rowland (eds.), *Tunnel Vision: Women, Mining and Communities* (Melbourne: Oxfam Community Aid Abroad): 12-15.

Lahiri-Dutt, K. (2006) 'Mining Gender at Work in the Indian Collieries: Identity Construction by Kamins', in M. Macintyre and K. Lahiri-Dutt (eds.), *Women Miners in Developing Countries: Pit Workers and Others* (Aldershot, UK: Ashgate): 163-79.

Macdonald, I. (2002) 'Introduction: Women's Rights Undermined', in I. Macdonald and C. Rowland (eds.), *Tunnel Vision: Women, Mining and Communities* (Melbourne: Oxfam Community Aid Abroad): 4-7.

—— (2004) 'Case Summary: Vatukoula, Fiji'; www.oxfam.org.au/campaigns/mining/ombudsman/cases/vatukoula/docs/fullreport.pdf, accessed 21 May 2007.

—— (2006) 'Women Miners, Human Rights and Poverty', in M. Macintyre and K. Lahiri-Dutt (eds.), *Women Miners in Developing Countries* (Aldershot, UK: Ashgate): 307-29.

—— and C. Rowland (eds.) (2002) *Tunnel Vision: Women, Mining and Communities* (Melbourne: Oxfam Community Aid Abroad).

Macintyre, M. (2002) 'Women and Mining Projects in Papua New Guinea: Problems of Consultation, Representation and Women's Rights', in I. Macdonald and C. Rowland (eds.), *Tunnel Vision: Women, Mining and Communities* (Melbourne: Oxfam Community Aid Abroad): 26-30.

—— (2003) 'Petztorme Women: Responding to Change in Lihir, Papua New Guinea', *Oceania* 74.1–2: 120.

—— and K. Lahiri-Dutt (eds.) (2006) *Women Miners in Developing Countries: Pit Women and Others* (Aldershot, UK: Ashgate).

MCA (Minerals Council of Australia) (2005) 'Enduring Value: The Sustainable Development Framework for the Australian Minerals Industry'; www.minerals.org.au/enduringvalue, accessed 20 August 2008.

McGuire, G. (2003) 'Environmental Impacts of Mining on Women in Indonesia and Northern Australia'; siteresources.worldbank.org/INTOGMC/Resources/336099-1163605893612/mcguireenvironmentalimpacts.pdf, accessed 20 August 2008.

Measham, F., and S. Allen (1994) 'In Defense of Home and Hearth: Families, Friendships and Feminism in Mining Communities', *Journal of Gender Studies* 3.1: 31-45.

Merithew, C. (2006) ' "We Were Not Ladies": Gender, Class, and a Women's Auxiliary's Battle for Mining Unionism', *Journal of Women's History* 18.2: 63.

Miewald, C.E., and E.J. McCann (2004) 'Gender Struggle, Scale, and the Production of Place in the Appalachian Coalfields', *Environment and Planning A* 36.6: 1,045-64.

Mining Association of Canada. (2006) *Framework on Mining and Aboriginal Peoples*; www.mining.ca/www/Towards_Sustaining_Mining/index.php, accessed 20 August 2008.

Moreton-Robinson, A. (2000) *Talkin' up to the White Woman: Indigenous Women and White Feminism* (St Lucia, QLD: University of Queensland Press).

Moretti, D. (2006) 'The Gender of the Gold: An Ethnographic and Historical Account of Women's Involvement in Artisanal and Small-Scale Mining in Mount Kaindi, Papua New Guinea', *Oceania* 76.2: 133-49.

Nicoll, F. (2000) 'Indigenous Sovereignty and the Violence of Perspective: A White Woman's Coming Out Story', *Australian Feminist Studies* 15.33: 369-86.

Nyame, K., and J. Grant (2007) 'Implications of Migration Patterns Associated with the Mining and Minerals Industry in Ghana'; www.imi.ox.ac.uk/pdfs/Nyame%20Grant%20Ghana%2007.pdf, accessed 22 September 2008.

Oxfam (2007) 'The Mining Ombudsman Project'; www.oxfam.org.au/campaigns/mining/ombudsman/index.html, accessed 22 September 2008.

Pacific Region International Women in Mining Network (2007) 'Statement of the Pacific Region', Regional International Women in Mining Conference Meeting, Madang, Papua New Guinea, 24–26 October 2007; www.oxfam.org.au/campaigns/mining/docs/Statement-of-the-Pacific-Region-RIMM-meeting-October-2007.pdf, accessed 20 August 2008.

Parmenter, J., and Kemp, D. (2007) 'Indigenous Women in Mining Employment in Australia', paper presented at the *Minerals Council of Australia Sustainable Development Conference*, 29 October–2 November 2007; www.csrm.uq.edu.au/docs/Indigenous%20women%20and%20mining%20employment%20in%20Australia.pdf, accessed 20 August 2008.

Ranchod, S. (2001) 'Gender and Mining. Workplace: Input to the Mining Minerals and Sustainable Development Southern Africa Regional Research', August 2001; www.natural-resources.org/minerals/cd/mmsd_saf.htm#SocialIssues, accessed 20 August 2008.

Render, J. (2004) *Mining and Indigenous Peoples Issues Review* (London: International Council on Mining and Metals).

Robinson, K. (1986) *Stepchildren of Progress: The Political Economy of Development in an Indonesian Mining Town* (New York: State University of New York).

—— (1996) 'Women, Mining and Development', in R. Howitt, J. Connell and P. Hirsch (eds.), *Resources, Nations and Indigenous Peoples: Case Studies from Australasia, Melanesia and Southeast Asia* (Melbourne: Oxford University Press): 137-51.

—— (2002) 'Labour, Love and Loss: Mining and the Displacement of Women's Labour', in I. Macdonald and C. Rowland (eds.), *Tunnel Vision: Women, Mining and Communities* (Melbourne: Oxfam Community Aid Abroad): 40-43.

Silitonga, N., A. Ruddick and F.S. Wignall (2002) 'Mining, HIV/AIDS and Women: Timika, Papua Province, Indonesia', in I. Macdonald and C. Rowland (eds.), *Tunnel Vision: Women, Mining and Communities* (Melbourne: Oxfam Community Aid Abroad): 44-48.

Simatauw, M. (2002) 'The Polarisation of the People and the State on the Interests of the Political Economy and Women's Struggle to Defend their Existence: A Critique of Mining Policy in Indonesia', in I. Macdonald and C. Rowland (eds.) *Tunnel Vision: Women, Mining and Communities* (Melbourne: Oxfam Community Aid Abroad): 35-39.

Sinha, S. (2002) 'Colonialism and Capitalism: Unearthing the History of Adivasi Women Miners of Chotangapur', in M. Macintyre and K. Lahiri-Dutt (eds.), *Women Miners in Developing Countries: Pit Women and Others* (Aldershot, UK: Ashgate): 89-106.

Somerville, M. (2005) 'Working Culture: Exploring Notions of Workplace Culture and Learning at Work', *Pedagogy, Culture and Society* 13.1: 5-26.

Tallichet, S.E. (1995) 'Gendered Relations in the Mines and the Division-of-Labor Underground', *Gender and Society* 9.6: 697-711.

Taussig, M. (1980) *The Devil and Commodity Fetishism in South America* (Chapel Hill, NC: University of North Carolina Press).

Tiplady, A., and M. Barclay (2007) *Indigenous Employment in the Australian Minerals Industry* (Brisbane: Centre for Social Responsibility in Mining, University of Queensland).

Trigger, D. (2003) 'Mining Projects in Remote Aboriginal Australia: Sites for the Articulation and Contesting of Economic and Cultural Futures', paper presented to the *Mining Frontiers Workshop*, Max-Planck Institute for Social Anthropology, 16–18 April 2003.

Tully, J. (2000) 'The Struggles of Indigenous Peoples for and of Freedom', in P. Ivison, P. Patton and W. Sanders (eds.), *Political Theory and the Rights of Indigenous Peoples* (Cambridge, UK: Cambridge University Press): 36-59.

UN (United Nations) (2008) 'The Millennium Development Goals Report 2008'; www.un.org/millenniumgoals/pdf/The%20Millennium%20Development%20Goals%20Report%202008.pdf, accessed 22 September 2008.

Williamson, M. (2003) '"I am Going to get a Job at the Factory": Attitudes to Women's Employment in a Mining Community, 1945–65', *Women's History Review* 12.3: 407-21.

World Bank (2006) 'Mainstreaming Gender into Extractive Industries Projects: Proposed Guidelines'; siteresources.worldbank.org/EXTEXTINDWOM/Resources/ttl_ei_gender_guidance.pdf?andresourceurlname=ttl_ei_gender_guidance.pdf, accessed 20 August 2008.

7

Archaeological heritage and traditional forests within the logging economy of British Columbia

An opportunity for corporate social responsibility

Bill Angelbeck
Department of Anthropology, University of British Columbia, Canada

Logging is and has been a primary industry for the forested province of British Columbia (BC), Canada, expanding with colonial settlement. Clear-cutting remains a widely used practice, involving 74% of all cut-blocks, and, as the industry spreads into new watersheds, the majority of the cutting (78%) shears old-growth forests (David Suzuki Foundation 2005). Environmentalists point to the effects forestry has on salmon-spawning streams and on species at risk, but this industry's practices have had substantial consequences for First Nations people, as their traditional territories consist predominantly of forest areas. Industrial logging has impacted their heritage, damaging and eroding archaeological sites, including the logging of culturally modified trees (CMTs), which are living examples of that heritage. In addition, logging has reduced the available areas for the continuation of traditional practices. This includes areas for gathering cedar bark and for felling monumental-sized cedar logs for carving poles and canoes, pools for spiritual bathing and areas for placing specialised dance regalia, among other traditional practices.

Until recently, companies conducted logging without much communication with First Nations groups about their claims to the land or where traditional and sacred areas needing protection might be located. This is an important issue since almost no treaties had been established with BC First Nations. They still claim an unbroken ownership and use of their traditional territory and few claims regarding land use have been settled. In fact, virtually all forests under government tenure are subject to ongoing native claims (Nathan 1993). Despite the widespread belief that land-claim issues are relatively recent, First Nations began such claims when colonisation began and have continued to press their case for well over a century. As Paul Tennant (1990) has argued, First Nations have shown a resiliency that has persisted since contact despite repeated and ongoing attempts to repress their culture and limit their sovereignty.

In this chapter I discuss the impact of industrial logging on First Nations' heritage and traditional territories and highlight some cases where conflicts have erupted in litigation and direct action. I indicate how corporate social responsibility (CSR) can work for forest companies to avoid such conflicts in the future as well as enhance the protection of indigenous heritage sites. CSR needs to involve four primary principles, including human rights, environmental sustainability, economic efficiency and the social licence to operate (Ali and O'Faircheallaigh 2007). Each of these is part of the First Nations' concerns regarding how logging is conducted. It is a human right, as defined by the United Nations (OHCHR 1997), to engage in one's cultural traditions and religious practices. For BC First Nations, this means being able to access old-growth forest areas, specifically those areas traditionally used in the past, as indicated by archaeological heritage. These First Nations also advocate long-term approaches to the forests, such as the 'thousand-year strategy' of the Haida, not only to maintain environmental sustainability but also to ensure that large or ancient cedars will continue to be available for the carving of canoes, totem poles or house posts for their ceremonial longhouses. First Nations also advocate for efficient industry practices that would leave less of an imprint on the landscape, favouring selective logging over clear-cutting scars and heli-logging or horse logging over the lasting marks of logging roads. Lastly, if corporations are to maintain their social licence to operate, they need to be cognisant of the historical and social context in which they work, that they are logging within the traditional lands of First Nations, within territories that have not been ceded and that, for most groups, are still a part of ongoing treaty negotiations.

In this chapter I focus on culturally modified trees (CMTs) and how those lasting marks of indigenous heritage should be respectfully handled to engender support from First Nations communities, as part of the social licence to operate. An opportunity exists for CSR as First Nations generally are not opposed to logging. In fact, they are the original loggers in the region, and historically they have been employed in the industry. Common ground can be found in pursuing practices that are consistent with their long tradition of sustainable stewardship of these forests.

Traditional forests

When Captain Cook first arrived on the west coast of Vancouver Island, his men began cutting trees down for lumber and fuel. The Nuu-chah-nulth, who lived on that coast, confronted them for taking trees from their traditional territory without having obtained permission, which would usually involve gifts to their chiefs, essentially buying or trading for such a right. Cook wrote in his journal on 22 April 1778 that he had not encountered indigenous peoples 'who have such high notions of everything the Country produced being their exclusive property as these' (cited in Gough 1984: 14). Cook found that the peoples of the Northwest Coast have customs of stewardship, and even ownership, regarding their traditional territories, both coastal and mainland.

In the ethnographies of the Northwest Coast, anthropologists have often focused on the marine orientation of First Nations and their reliance on canoe travel, whale hunting or salmon fishing along narrows of rivers. However, forests have always been a substantial focus, and this can be most readily seen in the sacredness of western red cedar (*Thuja plicata*), which provides multiple uses with its bark, withes, roots and wood. Cedar's acidic properties make it quite rot-resistant, an important feature in the temperate rainforests of the North Pacific, where its wood has been employed for house posts and beams, which were roofed and walled with long cedar planks. The bark is peeled from living trees to make clothing, cordage, matting and water-tight baskets. With its withes and roots, First Nations weave ropes, lashes and ties. Many of the famous carvings in Northwest Coast art style—totem poles, plank poles, welcoming figures, masks, canoes and whaling shrines—use cedar. Its use is so broad that many First Nations, such as the Haida and Coast Salish, have stories of a sacred origin for the tree. Long ago, during the original time of metamorphosis, when most species were essentially human, a man named Cedar happened to be a very giving individual and the Transformer changed him into a cedar tree for his generosity (Reid 1971, 1984).

Such was cedar's importance, Garibaldi and Turner (2004) referred to it as a 'cultural keystone species', meaning that, without the existence of western red cedar, the cultures of the Northwest Coast would be altogether quite different. In fact, the spatial distribution of Northwest Coast cultures approximates the range of western red cedar (Pojar and Mackinnon 1994: 42). Temporally, archaeologists have determined that the development of the distinctive 'Northwest Coast pattern' (see, for example, Matson and Coupland 1995) occurred only after western red cedar and yellow cedar became commonplace in regional forests, 4,000–6,000 years ago (Hebda and Mathewes 1984; Mathewes 1991).

For all the importance of cedar to Northwest Coast cultures, it has only recently become of interest to archaeologists. For the most part, this is due to the highly acidic nature of regional soils, within which wooden and other organic objects disintegrate rapidly. Even bone tools and faunal remains are largely absent unless disposed within shell middens, wherein the alkalinity of the shells acts to preserve hardier organic material. With the advent of wet-site excavations in the area, predominantly occurring in the last 30 years, the recovery of wooden artefacts is possible, due to the anaerobic environments (Bernick 1995). Wet sites are those contained within water tables or swamp areas, or encased within a mudslide such as occurred at the village of Ozette, described as the region's 'Pompeii' (Samuels and Daugherty 1991: 3). These wet-site excavations have shown that over 90% of the artefacts are made of wood and fibre, while only 5–10% are stone, bone or shell—the typical materials recovered from most excavations in the region (Croes 2003).

A recent development has been the archaeological documentation of CMTs, which has largely occurred since the mid-1980s. Before then, archaeological excavations focused on shell-middens, burial sites, or sites with lithic debris. This shows that archaeologists found what they searched for, and they were not looking at the trees until relatively recently. CMT surveys have become a routine practice for archaeologists since the late 1980s, and especially so since the Forest Practices Code came into effect in 1995, establishing protocols for the documentation of CMTs in proposed forestry cut-blocks (Klimko *et al.* 1998).

The most common type of CMT is one caused by bark-stripping, generally in either tapered or rectangular forms, which is done to acquire the inner bark, used for clothing, diapers, baskets and many other items. The cedar, after its bark is stripped, withstands the removal, and healing lobes form around the scar (see Fig. 7.1). Archaeologists perform dendrochronological studies, which involve the microscopic counting of annual tree rings to determine a specific year of use, and many are several centuries old. Some have shown that the lobes of ancient trees have often grown completely over older CMT scars, leaving the cultural mark only within the tree-ring patterns, which can be seen only when the tree is cut down. Therefore, the documentation of CMTs is biased towards the last several centuries—the oldest date recorded is AD 1467 (Mobley and Eldridge 1992), but ancient trees potentially contain older scars. It took political activism on the part of First Nations and archaeologists to pressure the province to include CMTs as 'legitimate' archaeological sites (Bernick 1984; Hicks 1984; Stewart 1984).

CMTs represent an expansion in archaeological practice, documentation and analysis in the Northwest Coast. They also represent a nexus of interaction between archaeology and ethnography, requiring archaeologists to seek anthropological records as well as community elders to understand the ancient woodworking evidence they have encountered (Arcas Associates 1986; Stryd and Eldridge 1993). Just as distinctive wooden artefacts at wet-site excavations have promoted such cooperation (Foster and Croes 2004), the presence of CMTs has also encouraged collaboration to understand how these scars were made or what the materials were for. In fact, some of the earlier investigations of CMTs were commissioned at the behest of First Nations communities (see, for example, Ham and Howe 1983).

In bringing about such collaborations, CMT studies and surveys have brought the past into a tangible, contemporary presence. These are living trees that still bear the marks of past use. The documentation of thousands of living CMTs are a testament to the philosophy espoused by the Northwest Coast peoples that the cedar is sacred and should not be needlessly killed to attain what one needs. For instance, if one needed lots of bark, one could peel a full block of bark, girdling the tree and effectively killing it. Instead, a typical CMT exhibits one narrow strip, or a pair of strips, on trees that are still standing and living, with lobes of growth slowly covering the scar face. Furthermore, the trees are not just documenting use from centuries ago, since there are stands of CMTs with scars only a decade or two old, or younger. Northwest Coast peoples continue their cedar practices today, and have concerns for untouched areas for future uses.

The documentation of CMTs has also invigorated First Nations in their resistance to development and logging in their traditional territories as these CMT sites document the history of their own forestry traditions, and add to their claims of Aboriginal title to

FIGURE 7.1 A culturally modified tree (CMT) from the Kwoiek Valley, just south of the Stein Valley

Photo by author

Note cut marks at top and bottom of scar and huge lobes indicating centuries of growth since scarification. The CMT flagging is used to clearly indicate their presence to loggers.

the land. Environmental groups that oppose clear-cut logging have come to their support arguing for the protection of old-growth areas which contain CMTs, and for ending clear-cut logging practices. In some cases discussed below, it is clear that CMTs have become a tool in the political power plays regarding provincial forests.

The continuity of tradition

Past: effects on heritage

According to most First Nations in BC, they were the original loggers in the territory, and when the commercial loggers came to the region, many First Nations men became loggers, as it was one of the primary jobs available to them. Researchers have challenged a common view that native people were bystanders in the economic development and industrialisation of BC, documenting that First Nations people helped in the establishment of settlements, worked to build railways, participated in agriculture, fishing, trapping and other resource industries and especially in logging (Lutz 1992; Menzies and Butler 2001). While logging provided employment, First Nations were never the employers, nor did their local communities receive many benefits. In fact, existing 'land policies ensured that Native people would participate in industrial forestry as labourers rather than as owners or managers' (Harris 2002: 307). The areas that they had used for years to gather forest resources were being tenured to logging companies, mining companies and others. For most of the time since logging started, there was not even consultation with the First Nations about logging within their traditional lands. As the logging industry expands throughout the province, it unleashes a broad array of effects on the heritage and traditions of the indigenous groups, impacting evidence of their past, limiting their present abilities to conduct many of their traditional practices, and restricting the potential for the continuation of these activities.

The destructive nature of most logging activities undoubtedly has an impact on archaeological sites. Huge trees are felled, often impacting other trees as they drop, causing trees to uproot. Fallen trees are dragged across the surface. When clear-cutting occurs and destroys surface vegetation, any remains are subject to high degrees of erosion, especially considering the amount of rain in the Pacific Northwest and the steepness of mountain slopes. Impacts to surface soils subsequently disturb any subsurface evidence of camps, structures, hearths and so on.

Regarding CMTs, the evidence is not below ground at all, but rather primarily contained in living trees. Prior to the 1980s, all CMTs within cut-blocks were logged, and entire archaeological sites were often destroyed. The passage of the Forest Practices Code in the 1990s did help protect CMTs. However, under the terms of the provincial Heritage Conservation Act, logging companies can still apply for a 'site alteration permit' from the provincial archaeology branch to log CMTs once archaeologists document the trees' cultural traits, and perhaps arrange to perform dendrochronological studies on thin slices of the harvested tree trunks or 'cookies'.

Present: the alteration of contemporary practices

Many First Nations are currently engaged in the treaty process in regards to their land claims and Aboriginal rights and title. As their sites have been eroded or literally cut down, logging practices contribute to the destruction of their heritage and their traditional ties to the land. This deletes their marks of past use, evidence that can be used in treaty negotiations to help document their longstanding use of the landscape. Therefore, the logging industry has contributed to weakening their case for stewardship of their traditional territories.

Past and ongoing logging practices not only affect First Nations' evidence of past use but also constrain their ability to continue their traditional practices. Clear-cutting affects wildlife and First Nations' ability to hunt in traditional spots, and the run-off from erosion affects salmon streams and fishing spots. Whereas First Nations used to rely on hunting and fishing, they are no longer able to do so as their ability to hunt and fish even in areas unaffected by logging is licensed and controlled by federal and provincial regulations. Essentially, the First Nations are prevented from continuing their traditional practices in their traditional territory, much of which has been tenured to multinational corporations. Under the terms of the Indian Act, which gives the federal government control over reserve land in trusteeship (Tennant 1990: 45), the Ministry of Forests has control over logging on reserve lands and many have been clear-cut (Nathan 1993: 139).

This situation is in contrast to the general anthropological discourse regarding the Northwest Coast peoples, in which these were heralded as 'wealthy' hunter-gatherers in the resource-rich temperate rainforest environments: streams were plentiful with salmon, forests bore cedar and deer, oceans contained whales and seals, river deltas annually teemed with flocks of geese, and intertidal zones contained numerous forms of shellfish, the valves of which accrued in coastal village middens. So plentiful was their environment and their use of it that their society and economies were oriented around surplus resources and goods. The potlatch ceremony involved elaborate feasting and the giving away of items in a networked gift economy. Stores of smoked or dried salmon, packed berries or barrels of fish oil also furnished well-developed trading networks. The effects of colonialism have reversed those conditions and, in many cases, legally prevented the continuation of past practices. As one Stó:lo Chief recounted,

> You look at the way housing is right now [on the reserves] . . . [you need] money to build houses. In the past, we didn't have to borrow money. If you needed to build a house, you just went out into the forest, cut the trees down and built a house . . . We need to be able to do that again . . . There's always some policy that infringes on our rights and title . . . Every reserve has to look for money to build houses, but in the past we just went out and got them . . . [When the] commercial logging started up in the valley, that's the first [thing] they did was clear out the valley (McHalsie 2003: 67).

Others point out the irony of housing shortages on reserves, most of which are surrounded by forests in which 'cutting trees is illegal without a provincial permit, which is usually only provided to private commercial loggers and logging companies' (Rude and Deiter 2004: 23).

Logging also has effects on the continuation of religious traditions. In the Coast Salish area, many still engage in the tradition of winter dance or spirit dancing ceremonies. These entail solo quests deep into the forests to find solitude, a connection with nature and territory—all to encourage the attraction of spirit powers. Spirit dancing has seen a revival since the 1960s (Amoss 1978; Bierwert 1999; Jilek 1982; Kew 1970; Robinson 1963) as groups of initiates travel a circuit from longhouse to longhouse across southern BC and in north-western Washington State. In conducting several interviews with spirit dancers and elders, I have encountered repeated complaints about how development and logging hinder their efforts at continuing the spirit dance tradition. First, old-growth areas are preferred for spirit questing areas. Old-growth forests are sought precisely for ancientness—the trees are often hundreds and some can be over a thousand years old. As families revisit questing spots over generations, those trees can be the same trees their ancestors quested near. More specifically, however, elders state that spirit powers are drawn to such areas. Due to the degree of logging, many dancer initiates have increasingly had to settle for second-growth forests.

Second, of particular interest for questers are pools along mountain streams. These pools are used for spiritual 'bathing', or 'swimming'. Logging may affect sacred bathing pools even if the immediate area is not logged. Clear-cutting upstream produces high rates of soil erosion that clouds and pollutes areas downstream, including bathing pools. When this happens, the traits that initially made the site suitable for 'swimming' are ruined; moving stream water is preferred for its coldness, clarity and movement, as opposed to the murkier conditions of ocean and lake waters, stagnant ponds, or even broad rivers (Angelbeck 2003).

Third, the extent of logging has had effects on the ability for many First Nations groups to continue to use cedar bark, roots, withes or logs. Elders and community members have related instances where they have not been able to take bark from cedar because the stand was located on private land or tenured to a logging company (Angelbeck 2003). Because of the diminishing ranges of old-growth areas, it is difficult for carvers to find a tree suitable for carving a house beam, plank pole or canoe, each of which is carved wholly from a single log. In order to carve such immense pieces, they require an old-growth tree simply to have a log of sufficient size. The trees also need to be straight and branchless for over 10 metres in length and more than a metre wide. Even when such trees are available, they are located in remnant old-growth stands that typically are in harder-to-access areas—the primary reason they are still left intact. The cost of getting such logs from remote areas can be prohibitive. It is no longer the case where one can fell a tree near the shore or a river's edge and float it back to the village, or simply carve a large log in the forests surrounding one's house.

Future: maintaining potential for traditions

If cedar availability is currently a problem in BC, what does that foreshadow for future uses of cedar? With contemporary rates of logging, particularly of old-growth stands, the areas suitable for traditional use will continue to shrink by the year. While archaeologists conduct surveys to record 'past use' areas and ethnographers conduct 'traditional use' studies to document historic and contemporary areas for hunting, gathering or spiritual use, there are no areas designated as 'potential use areas' that contain accessible stands of suitable cedar for *future* cultural use. One example of such long-term thinking has been advanced by the Haida, who aim to implement a thousand-year strategy for cedar throughout Haida Gwaii, or the Queen Charlotte Islands. Such long-term thinking considers sustainability of contemporary practices, and is at odds with prac-

FIGURE 7.2 Map of south-western British Columbia indicating traditional territories contested due to logging threats

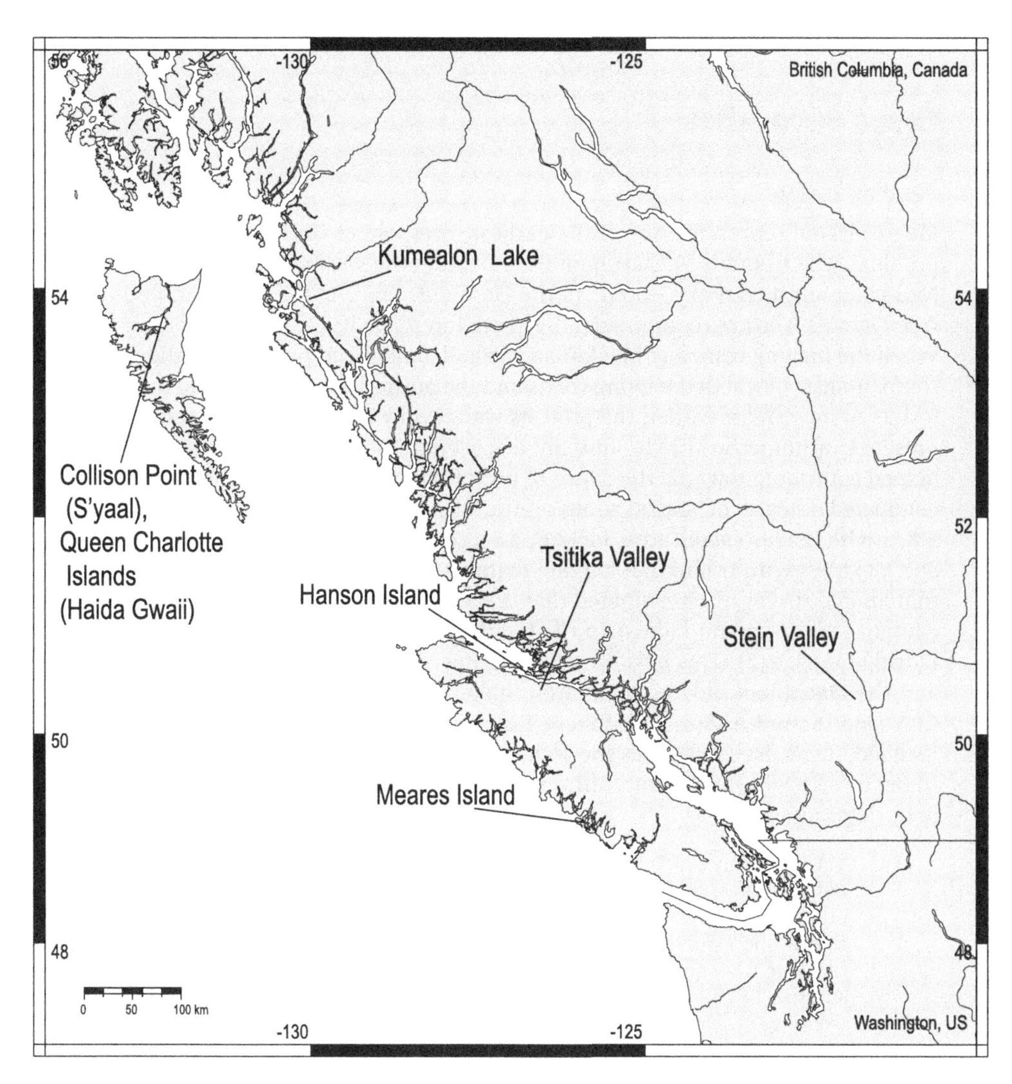

tices oriented at profit in the short-term. This situation has understandably led to conflict.

Cases of conflict

Meares Island

Meares Island represents one of the earliest cases where CMTs surfaced as a political problem. The island is in Clayoquot Sound along the west coast of Vancouver Island, Nuu-chah-nulth territory—home to the Hesquiat, Tla-O-Qui-aht and Ahousaht bands (see Fig. 7.2). The area contained old-growth forests on its coast and numerous islands. When the logging company MacMillan Bloedel announced plans to log the area, environmentalists blockaded logging roads and the authorities jailed several hundred protesters in 1993. Years earlier, the protests were led by First Nations concerned with the presence of numerous CMTs. Initially discovered during a survey in 1982, the CMTs helped contribute towards the cause of preservation. In 1984 and 1985, archaeologists conducted a survey of coastal zones on the island and encountered 1,779 CMTs within a 8,500 ha (22,000 acre) area, including 1,334 trees with one or more bark-strip scars and 445 trees with traditional logging features (planked, felled or tested trees). From that limited survey, they estimated that the area along the coastal margin alone would contain 13,800–25,600 bark-strip CMTs and 2,700 to 5,350 logged CMTs (Stryd and Eldridge 1993: 212). This bolstered the arguments of the protesters and First Nations, and the MacMillan Bloedel plans were halted in 1993; eventually UNESCO declared Clayoquot Sound a World Biosphere Reserve. As Braun and Wainwright (2001: 59) detailed, the map generated by the archaeologists was 'key' to their success in saving the island from logging at the time, although the island's ultimate status of protection is still to be determined:

> The significance of this [CMT] map was that it challenged the erasures, or 'cognitive failures,' that had underwritten forestry since the 1940s. Where forestry officials had always seen a 'natural' landscape without political-cultural claims, the Tla-o-qui-aht interjected a landscape worked-over by many centuries of Tla-o-qui-aht land use. With this map the exclusions that were constitutive of colonial forestry on the coast returned to disrupt the smooth workings of colonial power (Braun and Wainwright 2001: 59).

Hanson Island

Hanson Island to the north was the focus of similar protests in the early 1980s. Archaeologists initially surveyed for CMTs on Hanson Island in 1982 (Ham and Howe 1983), but the battle by the Kwakwa̱ka̱'wakw bands to protect Hanson Island became much more protracted, extending through the 1990s. The original survey attempted to document and test the location of traditional Kwakwa̱ka̱'wakw Nimpkish settlements and other sites. It documented 29 sites, including traditional trails, pictographs, shell midden sites and CMT sites. The island was under threat for clear-cutting, as planned by a multinational timber company, Crown Zellerbach. Nearly half of the sites that Ham and Howe (1983) encountered were forest utilisation sites, or CMT sites. At that time, there was not much awareness of CMTs, and the evidence for intense past forest use on the island had not been archaeologically documented. Ham and Howe's report did temporarily stall logging operations, and they called for more traditional use studies, but no funds were provided for such research.

Crown Forest, which acquired Crown Zellerbach's licence, later hired their own archaeologist to conduct studies of portions of the proposed cut-blocks. In that study, 75 CMTs were found along their survey transects, in addition to 30 others (Lackey 1987). Based on those results, Crown Forest claimed that 'It can be seen that the actual concentration of CMTs on Hanson Island is relatively low' (cited in Garrick and Waldon 1999). Subsequently, David Garrick, an anthropologist, led a team of volunteers using their own time to begin a broader inventory of CMTs on the island (Garrick 1998). They documented 1,878 CMTs on the island, arguing for evidence of use for over 700 years. With those results, the Nimpkish band blocked logging operations on the island once again, with protests delaying logging five more years. Meanwhile, Garrick continued to lead surveys for CMTs on the island, ultimately documenting 2,600 trees, apparently despite threats from the Port McNeill Forest District to burn his research camp (Garrick and Waldon 1999). Yet the third owner of the island's forestry licence, TimberWest, submitted proposals for logging on the island, which again met with protests.

By 2003, the actions of First Nations were successful in protecting the area, with a proclamation by the province that no timber, mining, or development activities would occur on the island. The Kwakwa̱ka̱'wakw bands that bear claim to Hanson Island would share in management of the head lease to share revenue for any tourism or recreation (*Courier-Islander* 2003).

Stein Valley

The case of the Stein River Valley is an example where archaeologists were not seen as contributing to the preservation of sites, but viewed as being pro-development. The Stein Valley is located along the Fraser River canyon and it contains old-growth forests. It had been protected from logging due to difficulties of access, in part due to narrows at the lower canyon, or the 'Devil's Staircase'. But, in the late 1970s, after the surrounding watersheds had been logged, proposals were made to log the Stein River Valley as well. The logging company, BC Forest Products Ltd, hired an archaeological consulting company, I.R. Wilson Consultants, to conduct the survey of the proposed logging haul road (Wilson 1985). The valley contained numerous examples of pictographic rock art, which the First Nations maintained were the result of vision quests by their ancestors, specifically shamans. The area also contained CMTs, mostly tapered and rectangular bark-strip scars; it also included one example of a pictograph inscribed on the scar of a CMT (Glavin 1989). The local Nlaka'pamux First Nation, the Lytton Indian band, were unsatisfied with I.R. Wilson's survey and its recommendation that the logging proceed. The band hired its own archaeologists to conduct another survey of the area and those encountered new pictograph sites, but also found serious problems with the initial survey, which Arnett (1993) described as having a 'pro-development bias'. He argued that it 'demonstrates the inability of "impact assessment archaeology" to understand the significance of the rock writing sites in their full setting. This omission seems obvious to aboriginal elders, as well as to other people who understand that a forest is composed of much more than marketable old growth trees' (Arnett 1993: 20).

To keep the pictographs or CMTs intact but allow destruction of the environment they are situated within is to destroy part of what makes them culturally valuable. As Lepofsky (1988) stressed in her report:

> According to Wilson's recommendations, avoiding impact of falling rocks at specific pictograph site locations is enough to maintain the sites' heritage value. The natural setting is integral to the heritage of a pictograph site—it cannot be separated from it. To destroy the wild nature of a pictograph location would be to destroy the heritage value of that spot. It is absurd to talk about the heritage significance of a pictograph resulting from a spirit quest, when the pictograph would overlook a logging road. This would, in fact, be the fate of several pictographs if a road were constructed in the Stein (Lepofsky 1988: 54-55).

Ultimately, the band's efforts were successful. The province imposed a moratorium on logging in the area in 1989, and in 1995 declared that the valley would be protected as a provincial park, one that would be co-managed by BC Parks and the Nlaka'pamux bands.

Sy'aal, or Collison Point

In 1990, Guujaaw, the prominent leader of the Haida of the Queen Charlotte Islands, or Haida Gwaii, made a visit to the *Sy'aal* site, after a quarter of its CMTs were logged. Guujaaw considered those CMTs to be 'living archaeology', wherein '[these represent a] unique opportunity to get close to the activities of our ancestors in a live and dynamic setting . . . To our people the spiritual effects of being in the footsteps of our ancestors is the primary significance of the site' (Guujaaw and Wanagun 1991: 2). Guujaaw himself is a renowned carver, and he spent time learning the methods of his ancestors by studying CMTs. At the *Sy'aal* site, there are remains of incomplete canoes carved from trunks of ancient cedar trees, as well as bark-stripped trees. Guujaaw remarked that the 'The method being used is regular clear cutting, which, if allowed to continue, will totally and completely destroy the cultural values of the site' (Nelson 2003: 17).

The Haida served an injunction against Stejack Logging for damaging the site that year, which halted all logging activities in the immediate area, while the issue went to court (*Massett Band et al. v. Stejack Logging*). The Haida criticised the BC government's Archaeology Branch (located in the Department of Tourism, Sports and the Arts), which handles preservation and implementation of the Heritage Conservation Act. One of the Haida's lawyers stated that 'It would seem that their job would be to protect archaeological sites, but in this case they [the Archaeology Branch] refused because of the money and politics' (Dunphy 1990). Guujaaw also remarked on the province's emphasis on the 'scientific value', as if it were the only type to consider beyond an economic or monetary value:

> The Heritage [or Archaeology] Branch . . . thought that really what is important at these [CMT] sites is the information; it's the archeological stuff that's important. They were quite willing to come up and count rings and give us an idea of when this site was used and basically give it back to us in a little folder

> that we could stick on the shelf. We just told them that we're not interested in that. And the Heritage Branch has been like that consistently, not just with the forest but with the land. The first priority has been how can we get the information that we need in order that development can proceed (Guujaaw 1996: 47).

This case was important as it established that not only is the scientific value of trees significant, but that there is a cultural context and relationship to the surrounding forest, a connection between the CMTs and the stand of unmodified trees surrounding them.

Tsitika Valley

In the case of Tsitika Valley, logging companies fought back against protests over CMTs and clear-cutting. In 1990, the Ministry of Forests approved clear-cut logging within the Tsitika Valley on Vancouver Island, within Kwakwa̲ka̲'wakw territory of the Tlowitsis-Mumtagila bands. Another alliance of First Nations and environmentalists led to protests and a blockade. This time, the logging company, MacMillan Bloedel, would not tolerate the road block and acquired an injunction against the protesters to stop all delaying tactics, which the protesters ignored. About 150 of the loggers formed their own blockade at the Tsitika River Bridge, which prevented the protesters from leaving their camp. Fifteen of the environmentalists were arrested. The First Nations charged that the Royal Canadian Mounted Police (RCMP) were technically supporting the logging company as they denied their own protests yet allowed the illegal blockade by the loggers. The RCMP allowed only two archaeologists through the roadblocks to check for the presence of CMTs.

Chief John Smith of the Tlowitsis-Mumtagila argued that one tree with a tapered scar appeared to be cultural in origin, which the archaeologists attempted to verify. If proven cultural, the band hoped to save the valley from logging, arguing that—as noted in the *Vancouver Sun*—'A single cedar or hemlock tree found stripped of its bark or with a deep scar from a chiselled-out plank might clinch the Tlowitsis-Mumtagila Indian band's bid to obtain a logging injunction pending a full land-claims case' (Pynn 1990). Tired of such protests, one logger responded that 'If they tie up this, then what are we all going to do? . . . Are we all going to become hippies and move to Denman Island?' A timber industry spokesman stated that they were 'concerned that [the] discovery of CMTs could ultimately result in a smaller land base for BC's forestry sector' (Fournier and Scott 1990).

Unfortunately for the Tlowitsis-Mumtagila, the archaeologists determined the marking on the tree to be natural (Eldridge 1990) and the valley was logged. MacMillan Bloedel even harvested the tree in question for analysis, and the archaeologists analysed a stem-round slice sample from the tree's trunk and verified that it was natural scarring (Eldridge 1991).

Kumealon Lake

Kumealon Lake is in Tsimshian territory, just south of Prince Rupert in north-western BC. In 1994, Interfor had plans to log the forests on karst landscape that contained larger-than-usual cedars. Several CMTs were known to be in the area, but Interfor had acquired a site alteration permit, which allowed them to log the CMTs once those were documented archaeologically. The Kitkatla band went to court challenging the Heritage Conservation Act itself and its claim to be able to 'permit' the destruction of their heritage. The main issue revolved about the constitutionality of the Act. The band argued that the province did not have a right to rule against matters that regarded their Aboriginal rights as guaranteed by the Constitution.

In 1998, the Kitkatla were successful in getting the court to establish a moratorium on logging. The Justice had found the company's notifications and consultations with the Kitkatla had not been sufficient (*The Province* 1998; Bell 2001: 247). Significantly, this was aided by the *Delgamuukw* decision the previous year, which determined that Aboriginal groups that had not signed treaties still have a constitutional right to their traditional territories; therefore, governments must consult with Aboriginal groups regarding resource extraction or other developments in their traditional territories (Thom 2003a, 2003b). This Kitkatla injunction raised alarms that this would be the first of many such injunctions that would slow development in the province. The Toronto *Globe and Mail* ran the headline 'Courts block logging in BC: Miners, loggers fear a wave of lawsuits as Indians invoke Delgamuukw decision' (Matas 1998).

The Kitkatla's victory, ultimately, was short-lived. In 2002, the Supreme Court of Canada ruled 7–0 against the Kitkatla, finding that, while the trees had 'ethnic' and 'scientific' significance, they were commonplace in BC. A newspaper account claimed that the Kitkatla failed to show a genuine association with 'trees with patches of missing bark' (*Daily Commercial News and Construction Record* 2002). Chief Matthew Hill countered such a notion, however, by describing the CMTs as 'living museums of Aboriginal culture' (Bell 2001: 248). Since they had experienced epidemic levels of disease and decades of cultural repression through residential schools, and a general separation from their own territory, much of their traditional knowledge about forestry management had been lost. Therefore, Chief Hill explained:

> CMTs become even more important as a way to know our ancestors and their way of life. By studying the location and the details of how the forest was used in the past, we can learn about our own culture even when there are no people alive to tell us about those places or those uses. It is like the spirit of our dead ancestors still have a way to speak to us and our children (Bell 2001: 248).

Despite failing to stop the site alteration permit, the Kitkatla did create a lot of public awareness about the issue of CMTs, which can now be seen as a success of another sort (Nelson 2003: 16).

A role for CSR

Conflicts between timber companies and First Nations in BC will continue to occur if logging continues to be scheduled for forests with indigenous heritage sites. To address their concerns, First Nations have predominantly sought settlement and negotiation with the province. They envision, perhaps rightly, that, to have any footing with the corporations extracting resources from their territory, they need to change the province's laws to gain the leverage to constrain the extent of logging and the methods used in the industry. An assumption here, driven by past experience with extractive resource corporations, is that without such policies and laws in place there is no need for a company to engage with them. That is, the First Nations have not experienced and do not expect to experience many actions that accord with a sense of CSR. They have interacted with resource-extracting companies, for example with the early fur-trading companies since the late 1700s, and gold-mining companies in the mid-1800s, which have all operated without much sense of obligation to local groups. Resources and monies were extracted outward to industrial centres, from the 'periphery' to the 'core'. The modern structure of logging companies, most of which are multinational, perpetuates such externally directed operations, with timber heading to other states and monies to generally non-local corporate centres. The multinational scope of the industry does not lend itself to an emphasis on local community involvement. Drushka (1993) has recommended government action to increase diversification and decentralisation of the forest industry in BC. Perhaps such a move would contribute to a greater sense of local involvement and social responsibility. Instead, with the Liberal Party in power in the province, a wave of deregulation has been instituted in the forest industry (West Coast Environmental Law 2004).

Logging corporations could do much on their own to improve relationships with First Nations. They should engage in genuine efforts at consultation, not prolong the 'inadequate consultation and effort' of logging industry efforts in the past (Flahr 2002: iii). Consultation should involve shared decision-making regarding the forests of traditional territories, one that allows First Nations to participate in the design of decision-making protocols (NAFA 1995: 6). Joint ventures and other collaborative projects should also be encouraged with local First Nations communities to direct economic benefits to them from the resources extracted from their territories. Just as stumpage fees are paid to the province for cutting trees on public land, forms of revenue-sharing should be commonplace for trees extracted from First Nations territories.

Companies could also contribute to better relations by changing the methods used in the industry. Reducing or eliminating clear-cutting would be more compatible with First Nations patterns of traditional forest use, as well as with their values. Forms of selection logging and other sustainable forms of logging can work in combination with First Nations' traditional use of the forests. A good sign is that ecosystem-based management (see, for example, Slocombe 1993) is being implemented by some smaller companies along the coast. Ecosystem-based management explicitly aims for the protection of the long-term health of the forests and encourages diversification of forest practices and respect for the traditions of First Nations. For instance, such an approach allows First Nations to generate income from non-timber forest products, as the Siska have successful done along the Mid-Fraser Valley, generating a line of 'Siska Traditions' jellies, jams, soaps and teas from their traditional foods and medicinal plants (Lavoie 2007; Turner and Cocksedge 2001). Such practices would also allow for traditional uses of forest areas in ways that clear-cut areas completely eliminate. By adopting ecosystem-based management, timber companies would take a positive step toward the principles of CSR.

Yet corporate action does not appear to be enough on its own. Action must be taken by the provincial government as well. As part of treating First Nations heritage as valued, it should be governed by an Archaeology Branch that is not administrated within the Ministry of Tourism, Sports and the Arts. As the Archaeology Branch maintains the provincial inventory of what are predominantly indigenous archaeological sites (over

23,000 sites), it should be located within the Ministry of Aboriginal Relations and Reconciliation, as urged in a letter to the premier from the Union of British Columbia Indian Chiefs (Phillip *et al.* 2005). Such a move would shift away from an emphasis on protection of only 'extraordinary' sites worthy of profitable tourism, to a setting that is institutionally inclined to consider all heritage sites that might be of value to First Nations.

Furthermore, to better facilitate protection of their cultural heritage, First Nations should be given a role in decision-making over the management of their own heritage sites, perhaps after the model suggested by Budhwa (2005), in which the First Nations select the archaeologists to handle reviews, rather than have firms contracting directly to forest companies, which can lead to accusations of 'pro-development' archaeology. In addition, for those instances when it is agreed that CMTs are to be protected, there is no systematic protocol for verifying that CMTs have been protected once logging is completed. A form of sampling should be employed to survey logged areas. Otherwise, the bulk of the effort is put into the front end of heritage compliance, without a system of checks to ensure whether heritage is actually protected.

First Nations have both suffered and proven resilient since the onset of colonialism. Just as in the cases discussed above, it is likely that more defensive measures to protect their heritage will take place through the direct actions of protests, blockades and other political statements. Through court actions, First Nations will also continue to serve injunctions to stop logging in areas threatening heritage without adequate protection or compensation. In the past, the law has mostly served as 'a tool of colonisation' since it requires participation in a manner that accepts the legitimacy of provincial and federal rule (Toovey 2005: 87). Yet, as Guujaaw noted after they achieved an injunction in the Stejack case: 'it certainly makes it a little bit more interesting that the law could work with us, considering we've had nothing but an adversarial relationship with the law' (Dunphy 1990). Compared with direct actions and litigation, negotiations with the province and logging companies have seemed to work poorly for First Nations. However the Nisga'a Final Agreement negotiated in 1998 may reveal a change of course as it allows for Nisga'a ownership of a portion of their traditional territory as well as management responsibilities within their whole traditional territory. Furthermore, negotiation may yet prove of value as direct actions and court actions show the resolve of First Nations in their defence of traditional territory. In so doing, the multi-pronged approach of BC First Nations may serve as a model for other cultures that may face related heritage threats as the forest industry continues to expand in old-growth forests.

Conclusion

The forests of BC have been primarily treated as if their only value is their economic value as timber. The trees are regarded as a form of agriculture, and the industry's language reflects this ('crops', 'yields' and 'harvests'). There is in reality significantly more value to these forests than just the crop of trees. First Nations value the forests of their traditional territories as integral to the continuation of their cultural practices. In the Northwest Coast, much has been written about the potlatch and, as Mauss (1990) points out, the gifting in a potlatch is more than an economic transaction of a gift or item.

Rather, the gift is emblematic of a relationship between the giver and receiver. In many ways, this potlatch pattern of relationships in the Northwest Coast extends to other relationships First Nations maintain. For instance, the First Salmon Ceremony celebrates their relationship with salmon that return semi-annually and they praise the salmon people and offer gifts to them to continue their relationship in good standing. Similarly, a ceremony is done for cedar prior to bark-stripping or logging. Given this dialogical nature of relationships, First Nations retain a need to maintain this obligation of stewardship, not just philosophically, but within a long tradition of ecologically sustainable practices.

Since colonialism, their own cultural traditions and practices have been repressed, with the outlawing of the potlatch and other ceremonies, and claims of Crown ownership of their lands. The first colonialism was political and militaristic, Winona LaDuke (1984) has argued, accomplished through the conquest of territory and domination by a state. The second colonialism essentially was corporate, occurring through claims of mining and timber tenures and the machinations for the extraction of those resources by multinational corporations. She writes: 'Is it possible to heal the wounds of the people, of whatever sort, caused by the process of separating them from the land, while keeping them separated by virtue of a process which literally consumes the land itself?' (LaDuke 1984: iii). There is a need to 'decoloniz[e] British Columbia's forests', as Boyd and Williams-Davidson (2000: 130) have stated. First Nations should be allowed and encouraged to participate in the traditional management and role with the forests in accordance with their traditional role and relationship with their territories.

From a heritage standpoint, perhaps the most significant aspect of the logging and removal of CMTs is simply the erasure of their imprint on the landscape of their traditional territories. As discussed in the examples above, First Nations have fought to keep the heritage that these forests contain in order to maintain their cultural and spiritual relationship with their ancestral areas. The erasure removes a clear reminder of indigenous presence in the landscape, markers that these are traditional territories. The destruction of these trees is essentially the destruction of records. The logging of these heritage trees can readily be seen as a rewriting of history—the removal of the evidence of Aboriginal forestry in the wake of a corporate one. Given the context of unsettled land claims, this erasure takes on greater importance.

Logging companies can either contribute to this cultural loss, and face continued resistance from First Nations, or begin to engage in socially responsible practices that can help perpetuate First Nations traditions and preserve their heritage.

References

Ali, S.H., and C. O'Faircheallaigh (2007) 'Introduction', *Greener Management International* 52 (theme issue on 'CSR, Extractive Industries and Environment: Values and Principles'): 5-16.

Amoss, P. (1978) *Coast Salish Spirit Dancing: The Survival of an Ancient Religion* (Seattle, WA: University of Washington Press).

Angelbeck, B. (2003) *Stó:lo Forests: Traditional and Contemporary Values, Resources, and Uses of Forests* (Report on file; Sardis, BC: Stó:lo Nation).

Arcas Associates (1986) *Native Tree Use of Meares Island, BC* (Report on file; Victoria, BC: Archaeology Branch of British Columbia).

Arnett, C. (1993) 'The Archaeology of Dreams: Rock Art and Rock Art Research in the Stein River Valley', in A. York, R. Daly and C. Arnett (eds.), *They Write their Dreams on the Rock Forever: Rock Writings in the Stein River Valley of British Columbia* (Vancouver, BC: Talonbooks): 1-26.

Bell, C. (2001) 'Protecting Indigenous Heritage Resources in Canada: A Comment on Kitkatla v. British Columbia', *International Journal of Cultural Property* 10: 246-63.

Bernick, K. (1984) *Haida Trees: Remains of Canoe Manufacture in the Forests of Southern Masset Inlet* (Report on file; Victoria, BC: Archaeology Branch of British Columbia).

—— (ed.) (1995) *Hidden Dimensions: The Cultural Significance of Wetland Archaeology* (Vancouver: UBC Press).

Bierwert, C. (1999) *Brushed by Cedar, Living by the River: Coast Salish Figures of Power* (Seattle, WA: University of Washington Press).

Boyd, D.R., and T.-L. Williams-Davidson (2000) 'Forest People: First Nations Lead the Way toward a Sustainable Future', in D. Salazar and K. Alper (eds.), *Sustaining the Forests of the Pacific Coast: Forging Truces in the War in the Woods* (Vancouver: UBC Press): 123-47.

Braun, B., and J. Wainwright (2001) 'Nature, Poststructuralism, and Politics', in N. Castree and B. Braun (eds.), *Social Nature: Theory, Practice, and Politics* (Malden, MA: Blackwell Publishers): 41-63.

Budhwa, R. (2005) 'An Alternate Model for First Nations Involvement in Resource Management Archaeology', *Canadian Journal of Archaeology* 29: 20-45.

Courier-Islander (2003) 'Management agreement means First Nations will be involved in planning future opportunities for Hanson Island', *Courier-Islander*, 6 August 2003: C9.

Croes, D. (2003) 'Northwest Coast Wet-Site Artifacts: A Key to Understanding Resource Procurement, Storage, Management, and Exchange', in R. Matson, G. Coupland and Q. Mackie (eds.), *Emerging from the Mists: Studies in Northwest Coast Culture History* (Vancouver: UBC Press): 51-75.

Daily Commercial News and Construction Record (2002) 'Band loses high court bid to protect old trees', *Daily Commercial News and Construction Record* 75.4 (4 April 2002): A16.

David Suzuki Foundation (2005) *Canada's Rainforest Status Report* (Vancouver: David Suzuki Foundation).

Drushka, K. (1993) 'Forest Ownership and the Case for Diversification', in K. Drushka, B. Nixon and R. Travers (eds.), *Touch Wood: BC Forests at the Crossroads* (Vancouver: Harbour Publishing): 1-22.

Dunphy, M. (1990) 'Haida move to block logging on cultural site: Injunction granted against company', *Vancouver Sun*, 31 May 1990: D8.

Eldridge, M. (1990) *A Culturally Modified Tree Survey of Cutblock 101, Tsitika River* (Report on file; Victoria, BC: Archaeology Branch of British Columbia).

Foster, R., and D. Croes (2004) 'Joint Tribal/College Wet Site Investigations: A Critical Need for Native American Expertise', *Journal of Wetland Archaeology* 4: 125-37.

Fournier, S., and O. Scott (1990) 'Tsitika questions make MB wonder: "If we get the OK to log, I wonder what is next?" ', *Vancouver Sun*, 1 November 1990: 5.

Garibaldi, A., and N. Turner (2004) 'Cultural Keystone Species: Implications for Ecological Conservation and Restoration', *Ecology and Society* 9: 1-18.

Garrick, D. (1998) *Shaped Cedars and Cedar Shaping: A Guidebook to Identifying, Documenting, Appreciating and Learning from Culturally Modified Trees* (Vancouver: Western Canada Wilderness Committee).

—— and R. Waldon (1999) 'CMT Politics', in *Yukusam-Hanson Island Urgent Conservation Plea, Wilderness Committee Educational Report* (Vancouver: Western Canada Wilderness Committee; www.wildernesscommittee.org/campaigns/historic/otherpub/reports/Vol18No05, accessed 20 August 2008).

Glavin, T. (1989) 'Indians unveil plan to save the Stein', *Vancouver Sun*, 24 June 1989: 3.

Gough, B.M. (1984) *Gunboat Frontier: British Maritime Authority and Northwest Coast Indians, 1846–90* (Vancouver: UBC Press).

Guujaaw (1996) 'The Sacred Workspaces of our Ancestors', in G. Wiggins (ed.), *Proceedings of the Cedar Symposium: Growing Western Redcedar and Yellow-cypress on the Queen Charlotte Islands/Haida Gwaii* (Queen Charlotte City, BC: Ministry of Forests, Queen Charlotte Islands Forest District): 45-48.

—— and Wanagun (1991) *Sy'aal Aboriginal Forest Utilization, Collison Point* (Report on file; Nanaimo, BC: Ministry of Forests, Vancouver Forest Region).

Ham, L., and G. Howe (1982) *Report of the 1982 Archeological Survey of the Nimpkish River Valley and Adjacent Offshore Islands* (Report on file; Victoria, BC: Archaeology Branch of British Columbia).

Harris, C. (2002) *Making Native Space: Colonialism, Resistance, and Reserves in British Columbia* (Vancouver: UBC Press).

Hebda, R., and R. Mathewes (1984) 'Holocene History of Cedar and Native Indian Cultures of the North American Pacific Coast', *Science* 225: 711-13.

Hicks, R. (1984) 'Precontact Dates Revealed by Ring Counts', *The Midden* 16.5: 11-14.

Jilek, W. (1982) *Indian Healing: Shamanic Ceremonialism in the Pacific Northwest Today* (Surrey, BC: Hancock House Publishers).

Kew, J.E.M. (1970) 'Coast Salish Ceremonial Life: Status and Identity in a Modern Village' (doctoral dissertation, Department of Anthropology, University of Washington, Seattle, WA).

Klimko, O., H. Moon and D. Glaum (1998) 'Archaeological Resource Management and Forestry in British Columbia', *Canadian Journal of Archaeology* 22: 31-42.

Lackey, S.P. (1987) *Hanson Island Culturally Modified Tree Study* (Report on file; Victoria, BC: Archaeology Branch of British Columbia).

LaDuke, W. (1984) 'Natural to Synthetic and Back', in W. Churchill (ed.), *Marxism and Native Americans* (Boston, MA: South End Press): i-viii.

Lavoie, J. (2007) 'Bands returning to agricultural roots: BC First Nations see traditional practices as road to economic health', *Times Colonist*, 8 May 2007.

Lepofsky, D. (1988) *Stein Valley Archaeological Assessment* (Report on file; Lytton, BC: Lytton Indian Band).

Lutz, J. (1992) 'After the Fur Trade: The Aboriginal Labouring Class of British Columbia, 1849–1890', *Journal of the Canadian Historical Association* 2 (n.s.): 69-93.

Matas, R. (1998) 'Courts block logging in BC: Miners, loggers fear a wave of lawsuits as Indians invoke Delgamuukw decision', *The Globe and Mail*, 17 June 1998: 1.

Mathewes, R.W. (1991) 'Connections between Palaeoenvironments and Palaeoethnobotany in Coastal British Columbia', in J. Renfrew (ed.), *New Light on Early Farming* (Edinburgh: Edinburgh University Press): 378-87.

Matson, R.G., and G. Coupland (1995) *The Prehistory of the Northwest Coast* (New York: Academic Press).

Mauss, M. (1990) *The Gift: The Form and Reason for Exchange in Archaic Societies* [originally published as *Essai sur le Don*, Presses Universitaires de France, 1950 (1925)] (New York: W.W. Norton).

McHalsie, S. (2003) 'Interview', in B. Angelbeck, *Stó:lo Forests: Traditional and Contemporary Values, Resources, and Uses of Forests* (Report on file; Sardis, BC, Stó:lo Nation): 64-69.

Menzies, C.R., and C.F. Butler (2001) 'Working in the Woods: Tsimshian Resource Workers and the Forest Industry of British Columbia', *American Indian Quarterly* 25: 409-30.

Mobley, C.M., and M. Eldridge (1992) 'Culturally Modified Trees in the Pacific Northwest', *Arctic Anthropology* 29: 91-110.

NAFA (National Aboriginal Forestry Association) (1995) *An Aboriginal Criterion for Sustainable Forest Management* (Ottawa: NAFA).

Nathan, H. (1993) 'Aboriginal Forestry: The Role of the First Nations', in K. Drushka, B. Nixon and R. Travers (eds.), *Touch Wood: BC Forests at the Crossroads* (Madeira Park, BC: Harbour Publishing): 137-70.

Nelson, J. (2003) *A Vanishing Heritage: The Loss of Ancient Red Cedar From Canada's Rainforests* (Vancouver: Western Canada Wilderness Committee/David Suzuki Foundation).

OHCHR (Office of the High Commissioner for Human Rights) (1997) *The Rights of Indigenous Peoples* (Geneva: United Nations OHCHR).

Phillip, S., R. Shintah and M. Retasket (2005) 'Respect and Reconciliation: First Nation Heritage Conservation' (Letter to the Honourable Gordon Campbell, Premier of British Columbia; Vancouver: Union of British Columbia Indian Chiefs).
Pojar, J., and A. MacKinnon (1994) *Plants of the Pacific Northwest Coast* (Vancouver: Ministry of Forests/ Lone Pine Publishing).
The Province (1998) 'Court Rules for Natives', *The Province*, 23 October 1998: A9.
Pynn, L. (1990) 'Two archeologists search for truce to end logging dispute roadblocks', *The Vancouver Sun*, 1 November 1990: A1.
Reid, W. (1971) *Out of the Silence* (New York: Outerbridge & Dienstfrey).
—— (1984) 'Foreword', in H. Stewart, *Cedar: Tree of Life to the Northwest Coast Indians* (Vancouver: Douglas & McIntyre): 8-9.
Robinson, S.A. (1963) 'Spirit Dancing among the Salish Indians, Vancouver Island, British Columbia' (doctoral dissertation, Department of Anthropology, University of Chicago).
Ross, M.L. (2005) First Nations Sacred Sites in Canada's Courts (Law and Society Series; Vancouver: UBC Press).
Rude, D., and C. Deiter (2004) *From the Fur Trade to Free Trade: Forestry and First Nations Women in Canada* (Status of Women Canada Policy Research Fund Report; Ottawa: Government of Canada).
Slocombe, D.S. (1993) 'Implementing Ecosystem-based Management: Development of Theory, Practice, and Research for Planning and Managing a Region', *Bioscience* 43.9: 612-22.
Stewart, H. (1984) 'Culturally Modified Trees', *The Midden* 16: 7-9.
Stryd, A.H., and M. Eldridge (1993) 'CMT Archaeology in British Columbia: The Meares Island Studies', *BC Studies* 99: 184-234.
Tennant, P. (1990) *Aboriginal Peoples and Politics: The Indian Land Question in British Columbia, 1849–1989* (Vancouver, BC: UBC Press).
Thom, B. (2003a) 'Aboriginal Rights and Title in Canada After Delgamuukw. Part One: Oral Traditions and Anthropological Evidence in the Courtroom', *Native Studies Review* 14: 1-26.
—— (2003b) 'Aboriginal Rights and Title in Canada After Delgamuukw. Part Two: Anthropological Perspectives on Rights, Tests, Infringement and Justification', *Native Studies Review* 14: 2-42.
Toovey, K. (2005) *Decolonizing or Recolonizing: Indigenous Peoples and the Law in Canada* (master's thesis, Department of Human and Social Development, University of Victoria, British Columbia).
Turner, N., and W. Cocksedge (2001) 'Aboriginal Use of Non-timber Forest Products in Northwestern North America: Applications and Issues', *Journal of Sustainable Forestry* 13: 31-57.
West Coast Environmental Law (2004) *'Timber Rules': Forest Regulations Lower Standards, Tie Government Hands and Reduce Accountability* (Vancouver: West Coast Environmental Law).
Wickwire, W.C. (1992) 'Ethnology and Archaeology as Ideology: The Case of the Stein River Valley', *BC Studies* 91–92: 51-78.
Wilson, I.R. (1985) *Stein River Haulroad Heritage Resources Inventory and Impact Assessment* (Report on file; Victoria, BC: Archaeology Branch of British Columbia).

8

Indigenous employment outcomes in the Australian mining industry

*Tanuja Barker**
Centre for Social Responsibility in Mining, University of Queensland, Australia

Mining has been at the forefront of colonialisation and has penetrated deeply into indigenous territories around the world. Consequently, indigenous employment also has a long history in the mining industry. In the Pilbara region of Western Australia around 300 Aboriginal people were estimated to have worked in the alluvial tin fields 24 years after its discovery in 1882 (Wilson 1980 cited in Holcombe 2004). Aboriginal labour has been an instrument of colonisation in Australia. Providing labour to resource industries may initially have been a survival strategy to enable Aboriginal people to remain on their land (Aird 2001) or to adopt to their new surrounds after their forced removal and placement in Aboriginal reserves and missions (Elder 2003; Holcombe 2004). Labour in the eyes of colonisers was considered valuable, and perhaps the only asset that Aboriginal people were able to offer that could be harnessed to meet colonisers' needs. This especially appears to be the case in northern Australia, given the scarcity of non-indigenous labour (Reynolds 2003).

Past indigenous-employment practices in the Australian mining industry were marred by discriminatory treatment and conditions. For example, according to Broome, as recently as 1969 Aboriginal workers at the Weipa mining operation in Cape York,

* This chapter is largely based on work conducted at the University of Queensland's Centre for Social Responsibility in Mining (CSRM) and is an updated and condensed version of a CSRM research report available from www.csrm.uq.edu.au/docs/t06.pdf. Apart from the literature, ideas have been derived from discussions held with numerous people at the University's Sustainable Minerals Institute (SMI) and at various mining operations and surrounding communities, especially Century Mine in the Gulf of Carpentaria. The author is grateful for these discussions and feedback received on this chapter from the book's editors and on the CSRM report from David Brereton, Tony Tiplady and David Trigger.

Queensland, were not eligible for all employment benefits provided to non-Aborigines, including bonuses, holiday pay and lodgings (Broome 1982 cited in Gientzotis and Welch 1997). These practices are reflections of such false notions as indigenous inferiority and the disregard of indigenous values, both of which have long underpinned Aboriginal employment practices in Australia (Norris 2006).

Indigenous employment levels in the Australian mining industry have increased only modestly during the past four decades, from a very low base (Cousins and Nieuwenhuysen 1984; George 2003). A 1968 survey showed that indigenous employees accounted for 1.5% of the total workforce across 23 companies (Rogers 1973). Moving forward 38 years, the 2006 National Census revealed that indigenous workers represented only 2.3% of the total mining workforce (ABS 2007). However, a 2001–2002 survey covering 112 Australian mine and petroleum operations provided a higher estimate of 4.6% indigenous employment (Tedesco *et al.* 2003). Evidently, this is not a high level of representation, given that many mining operations are located in regions where Aboriginal people comprise a substantial proportion of the population.

Nonetheless, there are several mining operations where indigenous employees now constitute a greater proportion of the mining workforce. Notable Australian examples include Century Mine in the Gulf of Carpentaria, Argyle Mine in the east Kimberley and the Newmont operation in the Tanami region (Argyle Diamonds 2006; Newmont Mining Corporation 2005; Zinifex 2005). Efforts are under way to increase indigenous involvement at other operations (Australian Government and MCA 2006; Tiplady and Barclay 2007). Several mining companies and government agencies have also developed indigenous employment policies and programmes (DITR 2005; Fowler 2005; Hall and Driver 2002; Lenegan 2005).

Underpinning these initiatives is the assumption that increased indigenous representation in the mining workforce will be mutually beneficial for mining companies and indigenous communities (Render 2005). Mining companies and operations stand to benefit in several ways. The inclusion of local Aboriginal people in the mining workforce can help to secure local support for an operation and address labour shortages in the industry (Archer 2005; Lenegan 2005). For fly-in, fly-out operations, a local Aboriginal workforce can help to alleviate the negatives associated with long-distance commuting and recruitment from a generally urban-based Australian mining workforce.

The expected benefits for indigenous peoples are assumed to be greater personal wealth, skill development and flow-on benefits to indigenous communities more generally. However, the assumption that mine work will necessarily produce such outcomes for Aboriginal peoples requires further investigation. The short- and long-term consequences of mining employment for Aboriginal employees and the communities from which they are drawn need to be understood, to assess the assumptions that are made about the benefits of indigenous employment in the industry. Gaining an appreciation of the effects of mine work on indigenous peoples will allow for the adaptation of strategies to maximise the beneficial impacts and minimise any negative outcomes.

In this chapter I will review several drivers to improve indigenous employment in the Australian mining industry. This is followed by a focus on some of the outcomes of indigenous employment at a number of Australian mining operations, leading to several questions about their consequences. The chapter concludes by highlighting the need for a culturally appropriate research framework for indigenous employment.

While the chapter has an Australian focus, the insights gained may prove useful for other countries where similar initiatives have been implemented or are contemplated.

Limitations

There are several limitations that need to be stated at the outset. First, although some valuable research reports have been written on indigenous employment, they largely exclude oral sources of knowledge (the Aboriginal tradition). By mainly drawing on written sources of information, this chapter likewise provides an incomplete picture of the situation and does not adequately convey indigenous perspectives (Nakata 2004). A non-indigenous person trained in the Western scientific paradigm has written this chapter and the interpretations presented here are inherent reflections of this.

Second, indigenous employment is a fast-moving field and current practice is likely to be ahead of what has been reported. This chapter is largely based on material published up to early 2007 and does not claim to be comprehensive. Third, mainly public sources of information have been relied on. There are other relevant documents, such as employment provisions in Indigenous Land Use Agreements (ILUAs) that remain confidential, as do internal mining company reports. Fourth, this chapter does not consider other employment opportunities up- and downstream of mining operations, including exploration and refineries. Lastly, the broad approach taken in this chapter limits the opportunity to explore the diversity involved in such terms as 'Aboriginal', 'indigenous' and 'communities' (Sarra 2000), and so, for example, the words 'Aboriginal' and 'indigenous' are used interchangeably.

Why the heightened interest in indigenous employment?

The resurgence of interest in indigenous employment within the mining industry has mainly been driven by developments in relation to the recognition of indigenous rights, coupled with increased recognition by leading companies and governments of the need to address poor socioeconomic conditions in many Aboriginal communities. While Aboriginal groups and governments help to shape the agenda of indigenous employment in the mining industry, these are beyond the scope of this chapter. The latter has been considered in further detail in Barker 2006.

Legal recognition of indigenous rights

Issues at the nexus of resource industries, their labour requirements and indigenous peoples have been at the heart of several important protests and legal developments in Australia's contact history, especially during the late 1960s and 1970s (Merlan 2005). However, it was not until the 1990s that the mining industry noticeably began to change its approach to Aboriginal relations (Howitt 2001).

This was largely in response to the recognition of indigenous rights to land within the Australian national legal framework. After a ten-year land claim struggle, Eddie Mabo made a successful High Court claim on behalf of the Meriam people in 1992. The High

Court's *Mabo* decision overturned the previous legal assumption of *terra nullius* ('vacant land'). The federal government's legislative response to the decision, the *Native Title Act 1993*, recognises that 'Indigenous people have a system of law and ownership' that can co-exist with other rights to country, as long as the loss of connection between indigenous peoples and their land, or government acts such as the grant of exclusive possession through privately owned freehold title, have not removed it (NNTT 2003). The *Native Title Act* also provided some Aboriginal parties with leverage to negotiate with mining companies about mining developments.

Several developments have occurred since, including the High Court's *Wik* decision, the *Native Title Amendment Act 1998* and numerous reviews of Indigenous Land acts in mineral-rich states and territories.[1] These developments have interacted to create an intricate web of outcomes for indigenous peoples. For instance, the 'right to negotiate' provision is a procedural right in the *Native Title Act* that allows Aboriginal parties to negotiate with mining operations about proposed developments on Native Title land. The *Wik* decision in 1996 extended the scope of the 'right to negotiate', by ruling that Native Title can co-exist with pastoral leases. Conversely, the *Native Title Amendment Act 1998* substantially restricted the scope by making several adjustments, including removing the 'right to negotiate' from certain mining lease renewals and mining infrastructure and by raising the eligibility criteria of Native Title claimants. These legal developments combined with other government policies, budgetary and legislative initiatives have effectively weakened the negotiating position of Aboriginal parties with mining companies since the advent of the *Native Title Act* (O'Faircheallaigh 2006a).

Despite these developments, numerous agreements between Native Title groups, mining companies and government authorities have been entered into over the last decade.[2] ILUAs are voluntary agreements between negotiating parties that are legally binding once registered (NNTT 2006). Employment and training provisions have been canvassed as important issues in agreement guidance documents (Quinn 2005) and have become a common feature of ILUAs (O'Faircheallaigh 2006b).

Poor socioeconomic conditions

One of the reasons why indigenous employment is often placed on the ILUA negotiation table is the hope that mining operations, which can be significant economic drivers in remote regions, can improve the socioeconomic conditions of Aboriginal peoples. More than two decades ago, Charles Perkins noted that:

> [the Aboriginal] minority is now generally recognised as the most disadvantaged and under-privileged sector of the Australian community, with the highest death rates, highest morbidity rates, the worst health and housing

1 For example, the *Aboriginal and Torres Strait Islander Land Amendment Act 2008* was proclaimed by the Queensland parliament on 17 July 2008. The *Land Right Legislation Aboriginal Land Rights (Northern Territory) Amendment Bill 2006* was passed by the Federal Government on 17 August 2006. The final report on the review of the *Land Administration Act 1997* was tabled to the Western Australian parliament on 31 August 2005.

2 Refer to the mining and minerals related agreements on the 'Agreement, Treaties and Negotiated Settlements Project' database at www.atns.net.au for further detail, accessed 15 August 2008.

> conditions and the lowest educational, occupational, economic, social and legal status of any community within the Australian society (Perkins 1982: 154).

Despite some improvements, numerous reports continue to document the high level of disadvantage experienced by Aboriginal peoples (ADCQ 2006; Baillie *et al.* 2002; Fred Hollows Foundation 2004; O'Donoghue 2003; Trudgen 2000). For instance, a socio-economic profile of the Pilbara mining region in Western Australia paints a gloomy picture for its Aboriginal inhabitants (Taylor and Scambary 2005). Table 8.1 reproduces the estimates for indigenous labour exclusion in that region for 2006.

TABLE 8.1 Indicators of the estimated scale of indigenous labour force exclusion: Pilbara region, 2006

Source: Taylor and Scambary 2005: 153

Population aged 15+	4,759
Has no post-school qualification	4,200
Has less than Year 10 schooling	1,500
Not in the labour force	2,190
Hospitalised each year (all indigenous persons)	2,800
Has diabetes (25 years and over)	1,020
Has a disability	1,020
Arrested each year	1,050
In custody/supervision at any one time	310
Achieving Year 7 benchmark literacy (current school attendees)	60%
15-year-old males surviving to age 65	<50%

Coupled with the increasing employment disparity between indigenous and non-indigenous Australians (Taylor and Hunter 1998) is the growing indigenous population. Charlie Lenegan, the former Managing Director of Rio Tinto Australia, noted that 'Our demographic studies tell us that in another decade or so, every second Australian in remote and rural regions above the tropic of Capricorn will be proud to be of Indigenous descent' (Lenegan 2005: 5).

These trends are likely to result in greater pressure for such things as education, employment, health services, housing and related infrastructure development (Taylor and Hunter 1998), which is likely to outstrip available government funds (Ridgeway 2005). According to some researchers, Aboriginal Australia is facing a looming social crisis (Behrendt 2003; Langton 2002).

Corporate social responsibility

The need to protect the rights and unique cultures of indigenous peoples has been recognised and reaffirmed by various international organisations, including the United Nations General Assembly (UN 2007)[3] and the International Labour Organisation (ILO). The ILO adopted the first international legal instrument, the Indigenous and Tribal Populations Convention (No. 107), in 1957. According to the revised ILO Convention No. 169, indigenous peoples rights include, among others, the right to participate in the benefits of mining activities (ILO 2003).

More recently, the need for mines to maximise employment benefits to communities surrounding their operations has also been recognised by industry bodies. For example, the Mining, Minerals and Sustainable Development (MMSD) project was commissioned by a consortium of companies, including nine of the world's biggest mining companies, in response to the challenges posed by sustainable development (IIED and WBCSD 2002). The MMSD project culminated in the release of the *Breaking New Ground* report in 2002 (IIED and WBCSD 2002). This report states that the employment of local people, and thus indigenous peoples for those mining operations on or near indigenous communities and land, is critical, and should be included in mining agreements and company policies (IIED and WBCSD 2002).

Following on from the MMSD project, the International Council on Mining and Metals (ICMM) released a sustainable development framework (ICMM 2003b). This framework outlines ten principles that signatories are required to uphold, including respect for fundamental human rights as well as employees' cultures and customs (ICMM 2003b). In response to these international developments, the Australian mining industry's peak body, the Minerals Council of Australia, released 'Enduring Value' in 2004, to provide guidance to companies and sites on how to implement the ICMM sustainable development principles (MCA 2004a). All signatories to this code are required to progressively implement the ICMM principles and to publicly report their performance at site level (MCA 2004b).

Company-based initiatives to increase indigenous employment, especially those of larger organisations such as Newmont, Rio Tinto and Zinifex have also developed in parallel to, or in some cases preceded, these developments at the industry level. Several companies have set indigenous employment targets and developed structured training and employment programmes to prepare and progress indigenous applicants through the mining workforce. These can include pre-employment training programmes, training programmes to develop skills, on-the-job training through traineeships and apprenticeships, career progression and tertiary education support for indigenous students.

Maintaining good Aboriginal relations continues to be in the long-term interest of mining companies. Gaining and retaining access to Aboriginal land, or land with Aboriginal interests, is required to ensure uninterrupted supply of ore to mining company customers and, ultimately, to deliver secure returns to their investors. Indigenous employment, in turn, is likely to comprise a prominent component in maintaining these relations.

3 One hundred and forty-three countries voted in favour of the UN Declaration on the Rights of Indigenous Peoples in September 2007. Eleven countries abstained and four countries, including Australia voted against the declaration (UN 2007).

These factors have intensified the pressure on mining companies to improve their indigenous employment performance, and there are several mining operations where indigenous employees now constitute a greater proportion of the workforce. This trend towards increasing indigenous involvement in the industry raises important questions about the *consequences* of employment, both for the indigenous workers themselves and the communities they come from.

Employment outcomes

Determining the consequences of mining employment for indigenous peoples is a complex task. The effects of employment span across several levels, from mine workers to their families and the communities from which they derive, as well as the communities to which they may relocate. These are neither discrete nor homogeneous categories. For instance, employees can be members of their 'home' communities and communities at remote mine sites. The effects of employment may differ substantially between various communities surrounding one mine and between local and non-local indigenous peoples. Furthermore, mining employment has indirect effects. For example, a survey respondent at the Century Mine noted that community members with no relatives employed at the mine may still be affected if mine workers reinvest their income and skills back in their communities (Barker and Brereton 2005). Similarly, they can be affected adversely if changes to community cohesion occur, owing to extended absences of miners from their home communities as a result of mine work.

There are also several impact dimensions of mining employment to consider. The effects of mine work are not solely confined to those associated with greater income levels, but can influence education, health and crime levels. The employment-related effects considered in a recent report included gender relations, income, culture, occupational skills and health-related dimensions (Barker 2006). This report showed that many questions remain regarding the employment outcomes for Aboriginal people in the Australian mining industry. For instance, further research is required to determine the post-mine employment outcomes for Aboriginal men and women. The effects of higher incomes on Aboriginal family relations and power dynamics also require further evaluation (Barker 2006). However, there is only space to consider the culture, occupational skills and health-related employment aspects in further detail here. For each of these an overview of existing research will be provided and a series of questions posed, in an effort to guide further investigations concerning the outcomes for Aboriginal peoples.

Cultural consequences

Consideration of potential cultural impacts of mining is largely confined to the start-up or expansion phase of mining operations. Cultural concerns may be raised at community forums, anti-mining protests or, more formally, in anthropological and archaeological reports invoked through the social impact and cultural heritage assessment processes. Little follow-through or monitoring of cultural impacts is apparent and, to the extent that it occurs, it is largely confined to the protection of delineated cultural sites or the removal of culturally significant artefacts and minerals[4] obtained from mining leases. Limited information is available on what effects mining employment has had on Australian indigenous cultures.

Culture comprises the value systems and beliefs that form the identity and blueprint of living for people (Rogers *et al.* 1988). Culture is therefore deeply ingrained, a lived experience and informs all aspects of employment. Determining the effects of mining employment on Aboriginal cultures is difficult, given the holistic nature of Aboriginal cultures. For the purposes of this chapter, Aboriginal cultures are interpreted as being inclusive of relations to country and family ties, core aspects of most indigenous groups. Nonetheless, there is no 'all-encompassing' indigenous culture (Smyth 1994). Each Traditional Owner group can have distinct as well as shared elements of Australian indigenous cultures. For instance, Rose (2005) notes in her work at the Comalco Weipa mining operation that indigenous employees themselves mention the importance of realising the diversity and distinctiveness of the various Traditional Owner groups present on-site.

Is mining employment culturally appropriate?

Large-scale extraction of minerals has been viewed by some as an affront to indigenous cultural values and world-views. Protests by several Aboriginal people and their supporters have centred on the incompatibility between mining and indigenous world-views (Yvonne Margarula quoted in Banerjee 2000; John Toby quoted in Dixon 1990; Ruska 1997; Wadjularbinna Nulyarimma quoted in Trebeck 2005; Helen Nunggalurr quoted in Wurm 1993; Yanner 2002). Trigger (2002) recalls the following occurrence during the development of Century Mine:

> a quietly spoken Waayni man in his fifties reflected pensively on how the proposed deep open cut pit would surely 'wake up that Rainbow Snake'—from his perspective, this was a danger wherever the mine might be dug in his country, regardless of whether the surface features of topography were particular foci of 'sacred sites' (Trigger 2002: 190).

The Aboriginal concept of 'country' denotes Traditional Owner relationships to land and waters (Smyth 1994). Country can embody spiritual significance, kinship connections and be a source of sustenance (Rose 1996; Trigger 2002). Indigenous people who uphold or espouse these world-views may therefore refrain from work within the mining industry.

This poses some serious questions for indigenous employment which have often been dismissed or relegated as a minor issue in the assessment of indigenous employment impacts, by those with a vested interest in the industry. Does work that either extracts or supports the extraction of mineral ore pose a cultural dilemma for indigenous mine workers? If it does, what is its extent, under what circumstances does it arise, and who bears the consequences? Do dire socioeconomic circumstances and lack of alternative mainstream employment opportunities in many Aboriginal communities lead some people to advocate and opt for mine work, despite possible cultural consequences? Do

4 For example, according to Century Mine the removal of hematite, a culturally significant mineral, is done under the guidance of cultural monitors representing the Waayni people, the Traditional Owners of the site (Zinifex 2007).

responses differ between members of the Traditional Owner groups on whose country mines are located and other indigenous groups? Do answers differ between mine types (underground, open-cut shallow or deep pits), ore types (for example, coal, diamonds, uranium) and job types (e.g. mining operator, cultural heritage officer, administrator)?

How does mining employment relate to indigenous cultures?

There is evidence that the desire to retain cultural practices more generally can influence indigenous attitudes to employment. Several respondents to a survey of local Aboriginal people, who voluntarily left work at Century Mine, suggested that the adoption of culturally appropriate employment practices would improve the retention of local Aboriginal employees at the mine (Barker and Brereton 2005). Some of the comments included:

> Help locals get out bush more, so they can keep their cultural ways (fishing, camping, attending ceremony).
>
> Don't get no bush tucker at the mine. The ability to go out for 4 hours or so and go fishing and live off the land is important to keep the harmony. All cooped up in one area, that's the hardest part. Can do this during shift changes; take a morning off with a group of blokes (Barker and Brereton 2005: 20).

Hall and Driver (2002) observed that the 'extended absences required for employment at Century' conflicted with the maintenance of the traditional fishing lifestyle on Mornington Island. The need for Aboriginal people in the Gulf of Carpentaria to spend time to properly care for country is documented in Memmott and Channells 2004. The need for flexible working arrangements and support by employers to enable Bininj (Aboriginal people from the Kakadu region in the Northern Territory) employees to attend ceremonial activities has similarly been raised in the Kakadu Regional Social Impact Study (Collins 2000).

Further research is required to determine the relationship between mining roster patterns, working hours and the ability to maintain indigenous cultural practices (O'Faircheallaigh 1995). To maximise production and contain labour costs, operational mine staff tend to work long hours, including 12-hour shifts and night shifts. Miners work a variety of rosters, such as nine days on, five days off, or two weeks on, one week off. To what extent do these different roster patterns constrain or enable the ability to maintain cultural activities? What work practices, leave and recreation policies can mining operations adopt that can facilitate the maintenance of such practices while also maintaining productivity? O'Faircheallaigh (1995) refers to a Canadian oil company which allowed First Nations employees to take short-term leave during the hunting and fishing season, as long as it did not interfere with operations.

Another consequence of long roster and working hours in the industry are the extended absences from family and the impact on the ability to maintain kinship relations (Lawrence 2005). The importance of extended families and the obligations that this can entail can be gleaned from a comment made by an indigenous employee at the Weipa operation: 'Funerals . . . too many in one year. The Crew Leader sees you going to too many in one year. The Crew Leader should understand that black fellas have a lot of relatives, not like white fellas' (Rose 2005: 63). Findings from a survey of local Aboriginal people who voluntarily left work at Century Mine, indicate that respondents were least happy with living away from home and roster patterns. Both were also contributing factors for leaving the mine (Barker and Brereton 2005).

Extended periodic breaks from family and social networks, especially in smaller communities, where mine workers may fulfil several vital roles, can create disruptions and voids that have ramifications for indigenous employees, their families and home communities. Some male Aboriginal workers at the Cape Flattery mine, for instance, reported a sense of loneliness and great loss for having missed out on seeing their families grow up (Holden and O'Faircheallaigh 1995). On the other hand, the extended family networks, common among many indigenous peoples, can provide a form of support for mine workers in their home communities. Knowing that family members are looked after during their absence or when issues arise can provide a sense of reassurance and offer a buffer against some of these impacts (O'Faircheallaigh 1995).

Another finding of the Century Mine survey is that employment had facilitated the movement of people out of the Gulf. Around 90% of the survey respondents resided in the Gulf when they started work at Century, but by the time the survey was undertaken 40% were living elsewhere (Barker and Brereton 2005: 26). If relocated mine workers have less frequent cultural and family interaction with their home communities, then what are the implications for the status of the workers within these networks and the functioning of these networks? What are the ramifications if customary practices and ties cannot be maintained as a result of mining employment?

Cultural adaptation does occur. The term 'circular mobility' has been used to refer to the movement of indigenous peoples to engage in mainstream employment and to maintain kin relationships and customary practices (Taylor and Bell 2004). New forms of relationships can be created as a result of mining operations. Mines that have reached a 'critical mass' of indigenous employees may provide an environment for the establishment of new networks and, depending on the composition of the indigenous workforce, allow for greater interaction between family members (Barker and Brereton 2005).

Some possible cultural impacts that indigenous employment has had, or can have, are outlined above, but this has only touched the surface. Further research is required to determine the pace and direction of change that mines bring. For instance, what are the implications of practising only those cultural activities that do not conflict with mining activities? The role and power that indigenous miners and communities have in the process of cultural adaptation also requires further investigation.

Skill development

Skill development can be a significant outcome of mining employment. However, several studies indicate that indigenous employment continues to be skewed towards entry-level and semi-skilled positions. Information about the range of skills that are developed as a result of mine work is scarce, though it is clear that the direction and degree of impact on skills development depends on numerous factors, including prior education and skills, previous work experience, opportunities for career development and duration of employment within the industry. Furthermore, much uncertainty remains regarding the impact of mine worker skills on their respective communities.

What skills can be developed at a mining operation?

Several skill sets can be developed at mining operations including occupational skills such as truck driving, trades (e.g. carpentry, electrical), engineering, administration and management, as well as general skills such as cross-cultural communication, health and safety, numeracy and literacy. These skill sets can be developed through various pathways. Mining companies can train and employ Aboriginal peoples directly. They can also facilitate the development of skills and entrepreneurial capabilities indirectly, by applying indigenous employment criteria to mining contractors and service providers, and/or facilitating the development of Aboriginal businesses to supply goods and services to mining operations (Williams 1997).

The employment of indigenous workers throughout the mine life-cycle and across all operational areas is likely to diversify the type of skills that are developed, limit disproportionate lay-offs of indigenous employees during economic downturns, and increase the transferability of skills beyond mines. The skill sets that are arguably the most transferable beyond a mine are general, trade, professional and higher management skills and those developed through indirect pathways, such as the operation of local Aboriginal businesses and joint ventures. For instance, the training and employment of local Aboriginal health and emergency workers at mines can help built local capacity to improve and maintain healthcare services in surrounding communities. Likewise, the entrepreneurial skills gained from developing and operating indigenous-owned enterprises associated with a mine could be applied to other mining operations, or be adapted to service other industries, community and public sector needs.

Mining training and employment experiences can either add to or diminish personal traits and values of workers, such as self-confidence and trustworthiness. For example, Hall and Driver (2002) observed an increase in self-confidence of successful job applicants among Aboriginal trainee graduates at Century Mine. Conversely, those who were unsuccessful experienced considerable disappointment and demotivation. Century responded by more closely aligning the training programmes with future job openings and thereby increasing the success rate of trainees in job placements (Hall and Driver 2002).

Some of these effects are summarised in a comment made by an Aboriginal participant in response to a survey question about what employees liked about work at Century Mine:

> Properly understand work and responsibility. I wasn't just given the job for being GCA [a Gulf Communities Agreement (person)], I did the job. It gave me a bit of value. I learned from other people (value and respect for different cultures). Century showed me a whole lot of job opportunities and showed me you need education to get there (Barker and Brereton 2005: 14).

What occupational skills are being developed?

Recent studies continue to demonstrate that indigenous employment remains skewed towards entry-level and semi-skilled positions, with the majority occupying truck driver or plant operator roles (Barker and Brereton 2004; O'Faircheallaigh 2002; Rose 2005; Tedesco *et al.* 2003; Tiplady and Barclay 2007). Very few indigenous employees occupy management positions (Barker and Brereton 2004; Holden and O'Faircheallaigh 1995). At the Pilbara Iron operation, close to two-thirds of the 135 indigenous workers were employed as plant operators, with only two occupying supervisory or managerial roles in mid-2004 (Taylor and Scambary 2005).

The most recent of the above studies was conducted across 12 Australian operations, representing six companies with commitments to improve indigenous employment levels. The results indicate a greater spread of indigenous employees across a range of occupations. However, out of the 1,179 indigenous workers, only 41 (or 3.5%) were employed as supervisors or superintendents, 5 (or 0.4%) as managers and none as executive managers (Tiplady and Barclay 2007).

The reasons for these continuing occupational trends require further investigation, particularly as Aboriginal managers and supervisors can act as mentors and boost the morale of other Aboriginal mine workers. The leadership qualities and skills developed in these roles are transferable beyond mines and have a greater ability to build capacity in mine-affected communities (Barker and Brereton 2004). The time it takes to progress through the various pathways of employment at a mine could be a contributing factor. If this factor is not responsible, then a lack of career progression for those with aptitude and a willingness to advance can be a major source of frustration and has been a source of criticism of the industry in the past (Banerjee 2001). These issues are being addressed by some mining operations. For example, in the case of Pilbara Iron, increased representation of indigenous employees in skilled and supervisory positions is expected as indigenous apprentices graduate (Taylor and Scambary 2005). Further investment in such initiatives, as well as the offering of career development, cadetships and scholarships, is likely to lead to improved outcomes (Tiplady and Barclay 2007).

Mining contractors and service providers have the ability to contribute substantially to the skill development of indigenous workers, especially through joint ventures and Aboriginal-owned enterprises. Gaining a clear picture of their overall contributions remains elusive, as this level of detail has not always been systematically reported across mining operations. Nonetheless, several indigenous-owned enterprises and joint ventures are now associated with the Australian mining sector and efforts are under way to increase their numbers (AEME 2005; IMETF 2003). These businesses have been established through various initiatives that may involve numerous parties, including Indigenous Business Australia, a Commonwealth Statutory Authority that promotes greater indigenous business involvement in the private sector, including mining (Indigenous Business Australia 2006).

Ngarda Construction and Civil Mining (Ngarda) in the Pilbara region of Western Australia is frequently cited as a successful business that has received support from Indigenous Business Australia. Ngarda is a joint venture, owned by Leightons Contracting, which holds half the equity, with Indigenous Business Australia and Ngarda Ngarli Yarndu Foundation each holding a quarter (Indigenous Business Australia 2007). The company was established in 2001 and offers a range of contracting services to the mining and construction industries in the Pilbara region. The company's vision statement explicitly states that 'Ngarda aspires to be the leader, and contractor of choice, in the promotion of Indigenous employment within the mining and mining services industry' (DEWR 2005: 114). The company has secured and extended numerous contracts with various operations in the Pilbara region, including a recent five-year mining contract with BHP, estimated to be worth more than A$300 million (Indigenous Business Australia 2007). Indigenous employment levels have reportedly been above 80% (DEWR 2005). However, for the reasons stated above, Ngarda's contribution to local Aboriginal skill development remains unclear.

Do communities benefit from mine skills?

It is likely that different mine phases create different impacts on the level of skill availability in affected communities. A large-scale mining operation can leave a skills vacuum in local communities, especially during the start-up phase, as skilled workers leave their jobs in the communities to pursue work at mine sites. This scenario can present an opportunity for upskilling other local Aboriginal people to fulfil these roles in the communities, if enough planning and adequate training is provided to capable candidates, or alternatively it might mean importing workers from outside to fill these roles. During the operational phase and especially on mine closure, there is also a chance that people might leave their communities and move elsewhere to pursue other opportunities.

On a larger scale, recruitment drives have been under way for some time to source indigenous professionals and other mine workers from Papua New Guinea (PNG) for the Australian resources boom.[5] This highlights the abilities of indigenous workers and the level of training and experience offered by PNG operations. However, it also highlights that the impacts of indigenous employment traverse national boundaries. On the one hand, Australian mines and communities to which PNG workers relocate receive obvious gains. On the other hand, the PNG mining and petroleum operations from which these workers are sourced are left with a 'brain drain' as the middle ranks of professionals have effectively been 'hollowed out'.[6]

Surrounding communities are likely to benefit from mine-related skills when these skills align with community needs. The ability to apply mine-related skills to the local economy is vital to sustain community benefits beyond the mine life, especially for operations with relatively short life-spans. Broadening indigenous employment mining programmes to align with local community and regional development aspirations, as well as business imperatives, could be one way to achieve sustainable benefits. Likewise, diversifying the skill sets of indigenous mineworkers can, in turn, expand their job prospects beyond a mine and minimise the work-related impacts associated with mine closures and mine-dependent communities (Brereton *et al.* 2007).

Overall, there remains much uncertainty regarding the impact of mine worker skills on communities. Further research is required to determine the range of skills obtained at mining operations and how mine work affects skill development outside mines. To what extent is it possible to converge the skill requirements and commercial imperatives of mining operation with those of indigenous individuals and the communities? Which mine skills have been the most transferable, to what contexts and whose interests do they serve? (Personal development, another mining operation, local communities and businesses?) How have the skills derived from owning and operating local Aboriginal businesses been applied beyond a mine? What have the effects been on communities with different levels of indigenous employment participation and different occupational

5 Personal communication with Greg Anderson, Executive Director, PNG Chamber of Mining & Petroleum, 22 September 2007.

6 *Ibid.*

profiles? Has mining employment affected the ability to practise the skills associated with the customary economy? What short-term employment opportunities, including consultancy work, have been provided to Aboriginal community members? More research is required to address these questions and assess the overall contribution of mining operations to skills development.

Health effects

The impact of mining employment on indigenous health status remains poorly understood and has only recently been incorporated into baseline socioeconomic assessments. Health in the broad sense of the word encompasses both mental and physical well-being. Good health is a precursor to obtaining work; it affects the ability to retain work; and it could also be an outcome of work. The relatively poor health status of some indigenous people can be a barrier to gaining work. Taylor and Scambary (2005) have demonstrated the size of the problem in the Pilbara by estimating that the proportion of indigenous people who are excluded from employment owing to health reasons could be approaching the size of the mainstream indigenous workforce in that region. If operations aim to improve local indigenous employment levels, then it is vital to gain an appreciation of the pre-existing health conditions in the communities surrounding their operations and monitor and respond to relevant issues.

What are the possible health effects of mine work?

Health can be both positively and negatively impacted by employment in the mining industry. The nature of the industry is quite hazardous. For instance, in 2003–2004, the fatal injury frequency rate for the Australian mining industry was about ten times higher than what is considered to be the 'safe' industry rate (MISHC 2005). This figure, however, represents a significant improvement in occupational health and safety performance within the last decade (MISHC 2005).

Mine emissions or discharges into the environment can also be sources of health concerns on a wider scale (Lyle *et al.* 2006; Supervising Scientist 2004). For instance, in 2005 indigenous ex-workers of the Baryulgil asbestos mine in northern New South Wales and their families were still seeking compensation for asbestos-related illnesses and deaths, thought to be induced by working and living near the mine that closed more than a quarter of a century ago (ABC 2005).

Further research is required to determine the overall net effect of employment on the health of indigenous employees, their families and communities. For example, how do the potential health hazards at mines compare to those in the communities from which Aboriginal mine workers are drawn? Do changes in the nature of food at a mine improve or exacerbate diseases such as diabetes and obesity? Is an increase in income and general socioeconomic status of indigenous employees associated with improved health status?

There is conflicting information as to the extent to which health and safety awareness at Century Mine has affected the Gulf communities. On one hand, there are indications that health and safety awareness has transferred 'beyond the mine gate' and has been applied in the wider community (Trebeck 2005), including the improvement of health and safety practices. On the other hand, Hall and Driver (2002) indicated that there was little evidence that health management skills were being transferred to the wider community. They further speculate that transfer of health education to family members improved by having more family members in the mining workforce, and that transference may be better for families of Aboriginal female trainees (Hall and Driver 2002).

Determining what methods to use to assess health outcomes and over what time period requires further consideration. Health effects may be hard to attribute to particular stints of employment and may take a long time to manifest. It will also depend on such factors as pre-employment health status, workplace health and safety standards, individual attitudes towards health, dominant attitudes towards health at mines and communities, the type of ore being mined (i.e. uranium versus mineral sands), the extraction method, job type and length of employment at the mine.

The way forward

This chapter has indicated that many questions remain about the nature of employment outcomes for Aboriginal peoples in the Australian mining industry. These outcomes cannot be considered solely in terms of the payment of salary/wages and the development of occupational skills. A common theme running through all of the employment dimensions considered is that benefits to indigenous mine workers do not necessarily accrue to their families or communities from which they are drawn. Mediating factors can include conditions prior to the development of a mine, the level and composition of the indigenous workforce, the type of indigenous employment practices being applied at a mine, the aspirations and decisions of the people affected and mobility patterns. In order to develop a broader understanding of the outcomes of indigenous employment, a comprehensive monitoring framework is required. A research approach to inform the development of such a strategy is outlined in Barker 2006. The need for a culturally appropriate framework to guide indigenous employment research is outlined below.

Indigenous mine workers and community members are the ultimate possessors of knowledge about how mine work affects them. The research agenda for indigenous employment therefore needs to be developed within an indigenous cultural framework to provide meaningful outcomes. Indigenous workers, community members and researchers are core to the process of developing conceptual frameworks that adequately convey the interlinking dimensions of indigenous employment. They need to be involved in determining what the key aspects of indigenous employment are that require research in different contexts and how they are related; what questions need to be prioritised; and which are appropriate to ask and share with wider audiences?

Strategies to incorporate Aboriginal peoples in the research process can include the use of Aboriginal perspectives to set research agendas, the involvement of indigenous researchers, and the application of indigenous concepts to develop appropriate methods to monitor outcomes. This also presents an opportunity to train indigenous peoples in social science research techniques. Aboriginal peoples who are versed in both Aboriginal world-views and social analytical techniques will be valuable contributors in mining regions.[7] In addition, decreased reliance on outside researchers and increased internal capacity to monitor long-term indigenous employment outcomes within these regions can result. Adopting a research partnership between Aboriginal mine workers, organisations and community members, experienced mine industry personnel, relevant government agencies and research institutions can provide the means to implement such an approach across mine life-cycles.

Conclusion

Indigenous employment in the Australian mining industry can potentially have wide-ranging outcomes for Aboriginal peoples. To date, little has been done to measure the benefits of mining employment for individuals or communities, or to identify and analyse possible negative impacts. Efforts to improve Aboriginal employment levels within the industry need to be coupled with a concurrent research and monitoring framework to help inform policies and practices in this area. This should be undertaken within a culturally appropriate research framework. The Australian mining industry is a significant employment contributor in mining regions and is often located near Aboriginal communities or land with Aboriginal interests. The industry therefore is ideally placed to develop an inclusive approach to evaluating the outcomes of indigenous employment initiatives and, in doing so, can provide examples of good practice for mining industries in other countries with indigenous populations.

References

ABC (Australian Broadcasting Corporation) (2005) 'Ridgeway calls for equal treatment of asbestosis sufferers', 23 March 2005; www.abc.net.au/news/stories/2005/03/23/1329595.htm, accessed 20 August 2008.

ABS (Australian Bureau of Statistics) (2007) *2006 Census of Population and Housing: 2006 Census Tables* (Cat. No. 2068.0; Canberra: Australian Bureau of Statistics; www.censusdata.abs.gov.au, accessed 26 October 2007).

ADCQ (Anti Discrimination Commission Queensland) (2006) 'Women in Prison'; www.adcq.qld.gov.au/pubs/WIP_report.pdf, accessed 20 August 2008.

AEME (Australian Enterprises in Mining and Exploration) (2005) *Inaugural Australian Aboriginal Enterprises in Mining and Exploration Conference*, Alice Springs, 4 November 2005.

7 In mining regions, Aboriginal peoples can be impacted by the presence of several mines over time. An example is the proposed development of the Aurukun bauxite resource by Chalco, the existing Comalco Weipa operation and the Alcan lease on western Cape York, Queensland. In these cases, regional assessments that take into account the cumulative effects of indigenous employment are recommended. The development of the Aurukun bauxite resource alone is estimated to employ 2,300 construction workers and 600 operational workers (Department of Infrastructure and Planning 2007). Alcan was taken over by Rio Tinto in 2007.

Aird, M. (2001) *Brisbane Blacks* (Southport, QLD: Keeaira Press).

Archer, D. (2005) 'Indigenous Employment and Labour Shortages in Mining', in *Proceedings of the Minerals Council of Australia Sustainable Development Conference*, Alice Springs, 31 October–4 November 2005; www.minerals.org.au/__data/assets/pdf_file/0020/10937/Archer_Daniel_9B2.pdf, accessed 20 August 2008.

Argyle Diamonds (2006) 'Sustainability Report 2005'; www.argylediamonds.com.au/documents/Argyle%20Sustainability.pdf, accessed 20 August 2008.

Australian Government and MCA (Minerals Council of Australia) (2006) 'Regional Partnerships Newsletter Issue 1'; www.minerals.org.au/__data/assets/pdf_file/0014/11822/RegionalPartnerships Newsletter31May.pdf, accessed 20 August 2008.

Baillie, R., F. Siciliano, G. Dane, L. Bevan, Y. Paradies and B. Carson (2002) *Atlas of Health Related Infrastructure in Discrete Aboriginal Communities* (Melbourne: Aboriginal and Torres Strait Islander Commission).

Banerjee, S.B. (2000) 'Whose Land is it Anyway? National Interest, Indigenous Stakeholders, and Colonial Discourses: The Case of the Jabiluka Uranium Mine', *Organization and Environment* 13.1: 3-38.

—— (2001) 'Corporate Citizenship and Indigenous Stakeholders: Exploring a New Dynamic of Organisational–Stakeholder Relationships', *Journal of Corporate Citizenship* 1: 39-55.

Barker, T. (2006) *Employment Outcomes for Aboriginal People: An Exploration of Experiences and Challenges in the Australian Minerals Industry* (CSRM Research Paper 6; Brisbane, QLD: Centre for Social Responsibility in Mining; www.csrm.uq.edu.au/docs/t06.pdf, accessed 20August 2008).

—— and D. Brereton (2004) *Aboriginal Employment at Century Mine* (CSRM Research Paper 3; Brisbane, QLD: Centre for Social Responsibility in Mining; www.csrm.uq.edu.au/docs/Public_Century_Report.pdf, accessed 20 August 2008).

—— and D. Brereton (2005) *Survey of Local Aboriginal People Formerly Employed at Century Mine: Identifying Factors that Contribute to Voluntary Turnover* (CSRM Research Paper 4; Brisbane: Centre for Social Responsibility in Mining; www.csrm.uq.edu.au//docs/CSRM%20Research%20Paper%204.pdf, accessed 20 August 2008).

Behrendt, L. (2003) *Achieving Social Justice: Indigenous Rights and Australia's Future* (Sydney: The Federation Press).

Brereton, D., P. Memmott, J. Reser, J. Buultjens, L. Thomson, T. Barker, T. O'Rourke and C. Chambers (2007) *Mining and Indigenous Tourism in Northern Australia* (CRC for Sustainable Tourism Technical Report; Gold Coast, QLD: Sustainable Tourism CRC; www.crctourism.com.au/CRCBookshop/Documents/Brereton_mining-indigtourism.pdf, accessed 20 August 2008).

Collins, B. (2000) 'Kakadu Region Social Impact Study Community Report'; www.deh.gov.au/ssd/publications/krsis-reports/impact-study/pubs/krsis-report.pdf, accessed 20 August 2008.

Cousins, D., and J. Nieuwenhuysen (1984) *Aboriginals and the Mining Industry: Case Studies of the Australian Experience* (Sydney: Allen & Unwin).

Department of Infrastructure and Planning, Queensland Government (2007) 'Aurukun Project'; www.dip.qld.gov.au/projects/mining-and-minerals/bauxite/aurukun-project.html, accessed 20 August 2008.

DEWR (Department of Employment and Workplace Relations) (2005) Building the Future through Enterprise: Stories of Successful Indigenous Enterprises and Entrepreneurs—Ngarda Civil and Mining; www.workplace.gov.au/workplace/Publications/Employment/CasestudiesofsuccessfulIndigenous businesses.htm, accessed 20 August 2008.

DITR (Department of Industry, Tourism and Resources, Australian Government) (2005) 'Indigenous Partnerships Program'; www.industry.gov.au/indigenouspartnerships, accessed 20 January 2006.

Dixon, R. (1990) 'Aborigines as Purposive Actors or Passive Victims: An Account of the Argyle Events by Some of the Aboriginal Participants', in R. Dixon and M. Dillon (eds.), *Aborigines and Diamond Mining: The Politics of Resource Development in the East Kimberley Western Australia* (Nedlands, WA: University of Western Australia Press): 66-94.

Elder, B. (2003) *Blood on the Wattle: Massacres and Maltreatment of Australian Aborigines Since 1788* (Sydney: New Holland).

Fowler, B. (2005) 'Employing Local Locals: The Tanami Approach to Increasing Indigenous Employment', in *Proceedings of the Minerals Council of Australia Sustainable Development Conference*, Alice Springs, 31 October–4 November 2005'; www.minerals.org.au/__data/assets/pdf_file/10215/Fowler_Brian10B1.pdf, accessed 20 August 2008.

Fred Hollows Foundation (2004) 'Indigenous Health in Australia: The Health Emergency'; www.hollows.org/upload/3381.pdf, accessed 20 February 2006.

George, K. (2003). 'Exclusive Rights: Ongoing Exclusion in Resource Rich Remote Aboriginal Australia', *Australasian Psychiatry* 11: S9-S12.

Gientzotis, J., and A. Welch (1997) 'Something to Show for Years of Work? Employment, Education and Training for Aboriginal People in Australia', *Journal of Indigenous Studies* 3.2: 1-26.

Hall, J., and M. Driver (2002) *Queensland Support for Training and Employment through the Gulf Communities Agreement and Century Mine* (Townsville, QLD: Queensland Department of Employment and Training).

Holcombe, S. (2004) *Early Indigenous Engagement with Mining in the Pilbara: Lessons from a Historical Perspective* (CAEPR Working Paper No. 24; Canberra: Centre for Aboriginal Economic Policy Research; www.anu.edu.au/caepr/Publications/WP/CAEPRWP24.pdf, accessed 7 March 2006).

Holden, A., and C. O'Faircheallaigh (1995) *The Economic and Social Impact of Silica Mining on Hope Vale* (Aboriginal Politics and Public Sector Management Monograph 1; Brisbane, QLD: Centre for Australian Public Sector Management).

Howitt, R. (2001) *Rethinking Resource Management: Justice, Sustainability and Indigenous Peoples* (London/New York: Routledge).

ICMM (International Council on Mining and Metals) (2003b) 'Ten Principles for Sustainable Development Performance'; www.icmm.com/our-work/sustainable-development-framework/10-principles, accessed 9 August 2008.

IIED (International Institute for Environment and Development) and WBCSD (World Business Council for Sustainable Development) (2002) *Breaking New Ground: Mining, Minerals and Sustainable Development Final Report of the MMSD Project* (London/Sterling, VA: Earthscan; www.iied.org/mmsd/finalreport/index.html, accessed 20 August 2008).

ILO (International Labour Organisation) (2003) *ILO Convention on Indigenous and Tribal Peoples 1989 (No. 169): A Manual* (Geneva: International Labour Office; www.ilo.org/global/What_we_do/Publications/ILOBookstore/Orderonline/Books/lang--en/docName--WCMS_PUBL_9221134679_EN/index.htm, accessed 20 August 2008).

IMETF (Indigenous Mining and Enterprise Task Force) (2003) *Indigenous Mining and Enterprise Taskforce 2002/03 Annual Report* (Darwin, NT: Department of Business, Industry and Resource Development).

Indigenous Business Australia (2006) *Indigenous Business Australia: Corporate Plan 2006–2008* (Woden, ACT: Indigenous Business Australia; www.iba.gov.au/files/IBACorporatePlan.pdf, accessed 20 August 2008).

—— (2007) 'Ngarda Civil and Mining secures 5 year contract with BHP', media release, 5 September 2007; www.iba.gov.au/files/Ngarda%20Mining%20Contract.pdf, accessed 20 August 2008.

Langton, M. (2002) 'A New Deal? Indigenous Development and the Politics of Recovery', *Dr Charles Perkins AO Memorial Oration*, University of Sydney, 4 October 2002; www.koori.usyd.edu.au/news/langton.pdf, accessed 29 May 2006.

Lawrence, R. (2005) 'Governing Warlpiri Subjects: Indigenous Employment and Training Programs in the Central Australian Mining Industry', *Geographical Research* 43.1: 40-48.

Lenegan, C. (2005) 'The Minerals Sector and Indigenous Relations', address to *Minerals Week 2005: Resourcing an Innovative Industry*, Canberra, Australia, 30 May–3 June 2005; www.minerals.org.au/__data/assets/pdf_file/0017/9062/LeneganCharlie.pdf, accessed 7 October 2008.

Lyle, D., A. Philips, W. Balding, H. Burke, D. Stokes, S. Corbett and J. Hall (2006) 'Dealing with Lead in Broken Hill: Trends in Blood Lead Levels in Young Children 1991–2003', *Science of the Total Environment* 359.1–3: 111-19.

MCA (Minerals Council of Australia) (2004a) 'Enduring Value: The Australian Minerals Industry Framework for Sustainable Development: Guidance for Implementation'; www.minerals.org.au/__data/assets/file/0020/2972/04_07_19_Consultation_Draft.pdf, accessed 10 August 2008.

—— (2004b) 'Signatory Obligations'; www.minerals.org.au/enduringvalue/sig/signatory_obligations, accessed 10 August 2008.

Memmott, P., and G. Channells (2004) *Living on Saltwater Country: Southern Gulf of Carpentaria Sea Country Management, Needs and Issues Research* (Hobart, TAS: National Oceans Office; www.oceans.gov.au/pdf/losc_capentaria.pdf, accessed 20 August 2008).

Merlan, F. (2005) 'Indigenous Movements in Australia', *Annual Review of Anthropology* 34: 473-94.

MISHC (Minerals Industry Safety and Health Centre) (2005) *Final Report to Queensland Resources Council on Underlying Causes of Fatalities and Significant Injuries in the Australian Mining Industry* (Brisbane: Minerals Industry Safety and Health Centre, University of Queensland; www.mishc.uq.edu.au/Files_for_download/QRC_report/QRC_Final_report.pdf, accessed 20 August 2008).

Nakata, M. (2004) 'Ongoing Conversations about Aboriginal and Torres Strait Islander Research Agendas and Directions', *Australian Journal of Indigenous Education* 33: 1-6; www.atsis.uq.edu.au/ajie/docs/200433001006.pdf, accessed 20 August 2008.

Newmont Mining Corporation (2005) *Now and Beyond 2005 Sustainability Report: Tanami, Australia* (Alice Springs, NT: Newmont Mining Corporation; www.newmont.com/en/pdf/nowandbeyond/NB2005-Tanami.pdf, accessed 20 August 2008).

NNTT (National Native Title Tribunal) (2003) *Short Guide to Native Title and Agreement-making* (Perth, WA: National Native Title Tribunal).

—— (2006) 'Indigenous Land Use Agreements'; www.nntt.gov.au/ilua, accessed 20 August 2008.

Norris, R. (2006) 'The More Things Change . . . Continuity in Australian Indigenous Employment Disadvantage 1788–1967' (PhD thesis, Griffith University, Brisbane; www4.gu.edu.au:8080/adt-root/uploads/approved/adt-QGU20070109.161046/public/02Whole.pdf, accessed 20 August 2008).

O'Donoghue, L. (2003) 'Third Annual Human Rights Oration: Grass Roots Human Rights: Beyond Symbolism', 10 December 2003; www.humanrightscommission.vic.gov.au/human%20rights/human%20rights%20events/2003%20human%20rights%20oration.asp?nc=3535&id=396, accessed 20 August 2008.

O'Faircheallaigh, C. (1995) 'Long Distance Commuting in Resource Industries: Implications for Native Peoples in Australia and Canada', *Human Organization* 54.2: 205-13.

—— *(2002) A New Approach to Policy Evaluation: Mining and Indigenous People* (Burlington, VT: Ashgate).

—— (2006a) 'Aborigines, Mining Companies and the State in Contemporary Australia: A New Political Economy or "Business as Usual"?', *Australian Journal of Political Science* 41.1: 1-22.

—— (2006b) 'Mining Agreements and Aboriginal Economic Development in Australia and Canada', *Journal of Aboriginal Economic Development* 5.1 (October 2006): 74-91.

Perkins, C. (1982) 'Economic Imperatives as Far as Aborigines are Concerned', in R. Berndt (ed.), *Aboriginal Sites, Rights and Resource Development* (Perth, WA: University of Western Australia Press): 153-76.

Quinn, R. (2005) *Mining Agreements: Content Ideas* (Perth, WA: National Native Title Tribunal; www.nntt.gov.au/research/files/Mining_Agreements_Content_Ideas.pdf, accessed 20 August 2008).

Render, J. (2005) *Mining and Indigenous Peoples Issues Review* (London: International Council on Mining and Metals; www.icmm.com/page/1161/mining-and-indigenous-peoples-issues-review, accessed 20 August 2008).

Reynolds, H. (2003) *North of Capricorn* (Crows Nest, NSW: Allen & Unwin).

Ridgeway, A. (2005) 'Addressing the Economic Exclusion of Indigenous Australians through Native Title', *The Mabo Lecture: National Native Title Conference*, Coffs Harbour, 3 June 2005; ntru.aiatsis.gov.au/conf2005/papers/RidgewayA.pdf, accessed 10 April 2006.

Rogers, P. (1973) *The Industrialists and the Aborigines: A Study of Aboriginal Employment in the Australian Mining Industry* (Sydney: Angus & Robertson).

Rogers, E., R. Burdge, P. Korsching and J. Donnermeyer (1988) *Social Change in Rural Societies: An Introduction to Rural Sociology* (Englewood Cliffs, NJ: Prentice Hall).

Rose, D. (1996) *Nourishing Terrains: Australian Aboriginal Views of Landscape and Wilderness* (Canberra: Australian Heritage Commission; www.ahc.gov.au/publications/generalpubs/nourishing/index.html, accessed 20 August 2008).

Rose, K. (2005) 'Managing Cultural Diversity' (Bachelor of Engineering thesis, University of Queensland).

Ruska, D. (1997) 'An Assertion of Customary Law over Invader Law: An Eco-Perspective', in R. Ganter (ed.), *Stradbroke Island: Facilitating Change. Proceedings of a Public Seminar held by the Queensland Studies Centre with Quandamooka Land Council* (Brisbane: Griffith University): 21-31.

Sarra, G. (2000) 'Linking Aboriginal Employment and Business Development Strategies to Corporate Objectives and Community Outcomes: Pinpointing Opportunities to Evolve Aboriginal Employment Strategies to Independent Business Ventures', paper presented to the *International Quality and Productivity Centre Conference*, Adelaide, SA, 21 November 2000.

Smyth, D. (1994) *Understanding Country: The Importance of Land and Sea in Aboriginal and Torres Strait Islander Societies* (Canberra: Council for Aboriginal Reconciliation, Australian Government Publishing Service; www.austlii.edu.au/au/other/IndigLRes/car/1993/1/1.html, accessed 20 August 2008).

Supervising Scientist (2004) *Investigation of the Potable Water Contamination Incident at Ranger Mine March 2004* (Supervising Scientist Report 184; Darwin, NT: Department of the Environment and Heritage, Supervising Scientist; www.environment.gov.au/ssd/publications/ssr/184.html, accessed 10 August 2008).

Taylor, J., and M. Bell (2004) *Population Mobility and Indigenous Peoples in Australasia and North America* (London: Routledge).

—— and B. Hunter (1998) *The Job Still Ahead: Economic Costs of Continuing Indigenous Employment Disparity* (Canberra: Office of Public Affairs, Aboriginal and Torres Strait Islander Commission; www.anu.edu.au/caepr/Publications/WP/jobahead.pdf, accessed 20 August 2008).

—— and B. Scambary (2005) *Indigenous People and the Pilbara Mining Boom: A Baseline for Regional Participation* (Centre for Aboriginal Economic Policy Research Research Monograph No. 25; Canberra: ANU E Press; epress.anu.edu.au/caepr_series/no_25/frames.php, accessed 20 August 2008).

Tedesco, L., M. Fainstein and L. Hogan (2003) *Indigenous People in Mining* (ABARE eReport 03.19; Canberra: Australian Bureau of Agricultural and Resource Economics; www.abareconomics.com/publications_html/energy/energy_03/er03_indigenous.pdf, accessed 2 November 2007).

Tiplady, T., and M. Barclay (2007) *Indigenous Employment in the Australian Minerals Industry* (CSRM Report; Brisbane: Centre for Social Responsibility in Mining, University of Queensland; www.csrm.uq.edu.au/docs/CSRM%20Report_FINAL%20TO%20PRINT_singles.pdf, accessed 20 August 2008).

Trebeck, K. (2005) 'Democratisation through Corporate Social Responsibility? The Case of Miners and Indigenous Australians' (PhD thesis, Australian National University, Canberra).

Trigger, D. (2002) 'Large Scale Mining in Aboriginal Australia: Cultural Dispositions and Economic Aspirations in Indigenous Communities', in *Proceedings of the Council of Mining and Metallurgical Institutions (ICMM) Congress 2002: International Codes, Technology and Sustainability for the Minerals Industry* (Carlton, VIC: Australasian Institute of Mining and Metallurgy): 189-94.

Trudgen, R. (2000) *Why Warriors Lie Down and Die* (Adelaide, SA: Aboriginal Resource and Development Services).

UN (United Nations) (2007) 'General Assembly Adopts Declaration on Rights of Indigenous Peoples' (UN General Assembly GA/10612 13 September 2007; New York: Department of Public Information; www.un.org/News/Press/docs/2007/ga10612.doc.htm, accessed 20 August 2008).

Williams, I. (1997) 'Review of Conference Discussion: Mining Industry', in *Proceedings of the Pathways to the Future: Indigenous Economic Development Conference* (Darwin, NT: Commonwealth Department of Employment, Education, Training and Youth Affairs): 65.

Wurm, J. (ed.) (1993) *People, Place, Law: Conference Report* (Sydney: Environmental Defender's Office, 8–11 September 1993).

Yanner, M. (2002) 'Murrandoo Yanner's Sorry Day Address 26 May 2002, Sydney Opera House'; www.abc.net.au/message/radio/awaye/murrandoo_trans.htm, accessed 18 June 2006.

Zinifex (2005) *Zinifex Sustainable Development Report 2005*; www.ozminerals.com/Media/docs/2005_SDR_Century-40f05706-3002-424e-ad0c-772c75145188-0.pdf, accessed 20August 2008.

—— (2007) *Sustainable Development: Community*; www.zinifex.com/sustainable_development/community.aspx, accessed 3 November 2007.

9
The fragmentation of responsibilities in the Melanesian mining sector

Colin Filer and John Burton
Research School of Pacific and Asian Studies, Australian National University

Glenn Banks
School of People, Environment and Planning, Massey University, New Zealand

Why Melanesia matters

In 1998, two multinational companies—Rio Tinto and Chevron—nominated their operations in Papua New Guinea (PNG) as examples of corporate social responsibility. Since then both companies have sold out of the operations that they nominated. Two other companies with a big stake in claims to social responsibility—British Petroleum and BHP Billiton—have also abandoned their operations in PNG since 1998. However, two of these four companies—Rio Tinto and British Petroleum—have retained an active interest in the Indonesian province of Papua. This appears to be something of a paradox, for it suggests that socially responsible companies are less comfortable with the political environment that exists in the independent Melanesian state of PNG than they are with a political environment in which the rights of an indigenous Melanesian population have been abused by the State of Indonesia.

Melanesia (or the New Guinea region) has gained a particular prominence in global debates about the corporate social and environmental responsibilities of the mining industry because it has hosted three of the world's most famously 'irresponsible' mining projects—the Panguna and Ok Tedi mines in PNG and the Freeport mine in Papua. All three thus featured as notable examples of 'unsustainability' in *Breaking New Ground*, the final report of the Mining, Minerals and Sustainable Development (MMSD) Project (IIED and WBCSD 2002). What will not be so obvious to readers of this report is the role played by the Bougainville rebellion and the forced closure of the Panguna copper mine in the global corporate policy process which led to the formation of the Global Mining Initiative (GMI) in 1998, the subsequent design of the MMSD Project, and the current charter of the International Council on Mining and Metals (ICMM).

In this chapter we investigate the ways in which different 'stakeholders' in the Melanesian mining sector have developed or applied the concept of corporate social responsibility in a social and political environment where there seems to be a significant governance deficit or a shortage of authority to determine who should be responsible to whom for what. In other words, we want to ask whether there is something about the social and political context of large-scale mining operations in Melanesia, the New Guinea region, PNG or Papua that makes it especially difficult for mining companies to demonstrate the practice of corporate social responsibility and the achievement of socially sustainable development outcomes.

Sir Robert Wilson's licence to operate

The GMI was initially conceived at a meeting held in Rio Tinto's London headquarters towards the end of 1998, and was formally established at the annual meeting of the World Economic Forum in January 1999 (Danielson 2006: 18).[1] Three of the nine mining companies that signed up to it had problems with their 'social licence to operate' in PNG. The biggest problem at that time was the charge of environmental vandalism levied against BHP Billiton as operator of the Ok Tedi mine, but similar charges had also been levied against Placer Dome as operator of the Porgera mine, because both mines were polluting the same river system with their waste materials (Banks 2006). Rio Tinto was attracting negative publicity for the practice of submarine tailings disposal from the Lihir mine, which had only recently come on stream (Bosshard 1996), but Rio Tinto also had to live with the legacy of Bougainville. Bougainville was not just an environmental disaster; it was also a social and political disaster. And PNG was beginning to look like a country in which it was very hard to distinguish between environmental and social responsibility—or the lack of it.

The leading instigator of the GMI was Sir Robert Wilson, then chairman of Rio Tinto, who went on to chair the MMSD Project Sponsors Group and the ICMM. In 1990, before the acquisition of his knighthood and chairmanships, Bob Wilson visited PNG in order to make his own assessment of what had happened on Bougainville. The reason for his visit was that he had been the Rio Tinto Zinc (RTZ) director responsible for negotiating the purchase of a bundle of mining assets from British Petroleum—a package that included the prospective gold mine on Lihir Island in New Ireland Province. At that time, Conzinc Riotinto Australia (CRA), the operator of the ill-fated Panguna mine, was a separate corporate entity, but RTZ owned 49% of it. CRA managers had provided Wilson

1 It was in effect an initiative of the nine mining companies that already belonged to the Mining and Minerals Working Group of the World Business Council on Sustainable Development.

with a copy of the paper that one of us had written about the social origins of the Bougainville rebellion (Filer 1990) because they were largely in agreement with the analysis that it contained. The focus of this analysis was the 'time bomb' created by conflict within the mine-affected communities over the distribution of money paid by the mining company as compensation for environmental damage.

The validity of the argument need not concern us here. What interested Wilson was the suggestion that this kind of cumulative social impact could eventually force the closure of big mining projects in other parts of PNG—including the prospective mine on Lihir. Indeed, Lihir was a source of particular concern because its traditional social and cultural institutions were quite similar to those of Bougainville, and because a group of Lihirian community leaders had insisted on visiting their counterparts at Panguna in the weeks immediately preceding the outbreak of hostilities. If taken to its logical conclusion, the time bomb theory would seem to suggest that a bigger compensation and benefit package would not serve to reduce the risk of violence at other mine sites, but might even serve to aggravate the causes of such violence. In which case, Wilson wondered if the best solution to the 'landowner problem' in Melanesia might be the one pursued around the Freeport mine in Irian Jaya, where the landowners got no compensation at all.[2]

We do not know what advice Bob Wilson gave to his board when he returned to St James's Square, but we do know that the Bougainville rebellion had three immediate effects on the management of PNG's mining industry. First, it raised the value of social anthropologists and other social scientists as investigators and mediators in the relationship between mining companies, government agencies and local communities (Filer 1999a). Second, it reinforced the efforts of policy-makers to change the rules of this relationship in ways that would reduce the risk of another social and political disaster (Filer 2005). But, third, it created an opportunity for politicians (including local community leaders) to engage in other games that would secure them a greater personal share of whatever the mining industry had to offer (Filer 1998).

The destabilisation of PNG's mineral policy regime was especially acute during the period of the Wingti government from 1992 to 1994. Government ministers took the view that RTZ was deliberately dithering over the development of the Lihir mine, and therefore threatened the company with a number of alternative development options that would leave it on the sidelines. Matters came to a head in 1994, when the Minister for Mining and Petroleum, Sir John Kaputin, issued a summons for the RTZ Chairman, Sir Derek Birkin, to come and explain himself to the prime minister (Filer and Imbun 2004).[3] After a heated exchange of faxes, the second knight bowed to the wishes of the first, but the visit never took place because a change of government in the middle of the year cleared the way for development of the Lihir project on terms that were more acceptable to the company.

In March 1995, Bob Wilson's earlier reflections on the relationship between Lihir and Freeport took on a new significance. The partners in the Lihir Joint Venture, including

2 These observations are based on a conversation between Colin Filer and Bob Wilson in February 1990.

3 The ministerial summons was probably intended to remind RTZ of a similar episode in 1974, when another company chairman, Sir Val Duncan, was given a rather frosty reception when he came to Port Moresby to speed up the renegotiation of the Bougainville Copper Agreement (Jackson 1982: 61). Bob Wilson, who was the company's managing director in 1994, has confirmed the extent of the irritation caused by Kaputin's gesture (personal communication, April 2007).

RTZ and the PNG government, finally signed a Mining Development Contract, but at the same time RTZ purchased a 10% stake in Freeport McMoRan and entered into a joint venture arrangement that would enable it to reap a 40% share of the profits from the expansion of the Grasberg mine in Irian Jaya (McBeth 1995). The Lihir development contract led to the public sale of more than half the equity in the new development company, Lihir Gold Ltd, which not only helped to raise the finance required for construction of the mine, but also shortened the period that would be needed for the joint venture partners to pay off their loans (Filer and Imbun 2004). The Lihir share float was followed by a public announcement that RTZ and CRA would form a single, dual-listed company with effect from January 1996. As one academic commentator remarked at the time:

> Together with its involvement in the Lihir project and in CRA's operations in PNG, RTZ is now in a unique position. While respecting the legal requirements and political sensitivities of the two host nations, RTZ needs to be capable of drawing upon experiences on either side of the border in developing a set of field practices which recognise the issues common to local communities across that border (Ballard 1995: 80).

This opportunity was diminished when the Australian branch of the company, having already been chased out of its Mt Kare prospect in 1993 (Filer 1998), decided to sell off its remaining exploration licences in PNG. However, the big event of 1996 was the disruption of Rio Tinto's Annual General Meeting in London by people protesting against (among other things) the company's perceived investment in human rights abuses at Freeport.[4] This added to Bob Wilson's conviction that something would have to be done about the industry's 'social licence to operate', and that conviction was reflected in the company's annual report for 1996, which he presented to the next AGM in his new capacity as the company's chairman.

James Wolfensohn's tri-sector partnerships

While Bob Wilson was preparing to become the champion of sustainable development in the mining industry, James Wolfensohn was asking his staff at the World Bank to construct a new kind of relationship between the private sector and what came to be known as 'civil society organisations', especially those non-governmental organisations that were deeply suspicious or highly critical of big business and the Bank itself. Since the Bank's core business was still to lend money to the governments of developing countries, this new activity came to be known as the promotion of 'tri-sector partnerships' between host governments, foreign investors and 'civil society' (Warner and Sullivan 2004).

One of the initiatives taken under this rubric was a process of engagement with some of the world's largest mining and petroleum companies to establish the extent of their interest in the practice of corporate social responsibility (McPhail and Davy 1998). Among other things, these companies were asked to indicate which of their own oper-

4 *PNG Post-Courier*, 10 May 1996.

ations might be taken as an example of 'good practice', with particular emphasis being placed on the integrated management of social and environmental issues. Rio Tinto was one of only four companies that were willing to volunteer an example, and the example that it nominated was Lihir.[5] The Bank then commissioned a study 'to obtain the views of a representative cross-section of Lihirian society on the positive and negative aspects' of that operation. The report of that study (Filer and Mandie-Filer 1998) found that Lihirians were generally rather unhappy about the social impact of the mine and were not especially impressed by the company's efforts to mitigate that impact. These findings did not impress the World Bank because they failed to discover anything that looked like an effective 'tri-sector partnership'.

The Lihir Management Company had no problem with the report apart from the appearance of 'Rio Tinto' in its title—'Rio Tinto through the Looking Glass'.[6] The puzzle that remained was why Rio decided to nominate Lihir in the first place. One obvious solution could be found in the cost of the integrated benefits package that local community leaders had negotiated as the price of their support for the Mining Development Contract. This was indeed remarkably high by anyone's standards (Filer 2005: 925), and Rio's bean counters might have figured that the amount of local community support should be proportional to the price that a company pays for it. The alternative solution was that the project's risk insurance had been covered by the Multilateral Investment Guarantee Agency, and the choice was thus a reflection of the partnership that already existed between Lihir Gold and the World Bank Group.

The plausibility of the second solution was enhanced when Lihir was subjected to further scrutiny by the Extractive Industries Review. This exercise was initiated by James Wolfensohn in response to complaints from 'civil society organisations' about the World Bank's support for irresponsible mining operations in developing countries (Bello and Guttal 2006: 77). Dr Emil Salim, the 'eminent person' appointed to lead the review, visited Lihir in 2002 and made observations similar to those made in the previous case study for the World Bank (EIR 2002).[7] Participants in a subsequent regional workshop made observations about the social impact of the Lihir project which were summarised as follows:

> This project demonstrates what happens when planning is short-sighted and far too narrow in scope. Insufficient attention was paid to the special local social and gender dynamics, the potential consequences of poor governance, anticipating the impact of cash incomes on a cashless society, involving all stakeholders from the beginning in planning discussions, having poor monitoring mechanisms, and the consequences of mine closure (EIR 2003: ix).

By this time the World Bank's involvement in PNG's mining sector reached well beyond the provision of risk insurance to the Lihir project or the occasional review of its

5 The other three companies were Placer Dome, British Petroleum and Chevron. Chevron also nominated its operation in PNG. Rio Tinto, Placer Dome and BP were represented on the steering committee established for this exercise. Rio Tinto's representative was George Littlewood, the author of the new 'communities policy' announced in Rio Tinto's 1996 Annual Report (McPhail and Davy 1998: 12).

6 The Lihir Management Company was a wholly owned subsidiary of Rio Tinto, but its managers still thought that Lihir Gold Ltd should play the part of Alice in this story.

7 His team also inspected the Kutubu project, which Chevron had nominated for the earlier study, although the World Bank Group had no financial involvement in this project.

social and environmental impact. In September 2000, when Wolfensohn initiated the Extractive Industries Review, the Bank had just agreed to provide US$10 million in technical assistance to the PNG government for the Mining Sector Institutional Strengthening Project. The president's announcement, which evidently came as a surprise to his own staff (Danielson 2006: 74), seems to have prompted the addition of a new component to this project—the design of a Sustainable Development Policy and Sustainability Planning Framework.[8] This work took place over the course of 2002 and was duly informed by the findings of the MMSD Project (Filer 2002). Extensive consultations with a wide range of stakeholders and detailed studies of the country's four major mining operations resulted in production of a Green Paper that was meant to inform revision of the mining legislation and regulatory framework (PNGDOM 2003).

That part of the Green Paper that dealt with the design of a 'Sustainability Planning Framework' placed a lot of emphasis on the need to clarify the division of responsibilities between developers, government agencies and other stakeholders in the management of mineral wealth. The Green Paper also proposed that the PNG government should

> henceforth expect any mining company proposing to invest in the development of a major mining project in PNG to include in its *Proposal for Development* a clear statement of the principles or standards of best practice for which it is prepared to be held accountable if the proposal is approved . . . In return, the investor will be expected to hold the PNG Government accountable for upholding the principles and standards of good governance to which it is committed by its own laws and policies, and by its ratification of numerous international conventions. It is proposed that an agreement by the Government and the proponent to hold each other accountable for adherence to these standards and principles should be included as an *annex to the Mining Development Contract*, and that this annex should be a *public document* (PNGDOM 2003: 19, emphasis in original).

In the event, the Department of Mining did not have the strength required to complete the policy process initiated by the Green Paper, partly because it lacked the support of other government agencies and partly because the Bank's efforts to strengthen it had only limited success (Mathrani 2003).

Retreat from the land of the unexpected

Back in 2002, it still looked as if the mining industry in PNG was a sunset industry. Placer Dome's Misima mine was about to close down, while the Porgera and Ok Tedi mines were due for closure within a decade. Furthermore, their closure was expected to coincide with a major reduction in the scale of operations at Lihir as the original ore bodies would be exhausted and the operator would begin to mine the lower-grade material contained in the waste dumps. At the same time, the level of exploration spending remained

8 The World Bank staff member sent out to negotiate this new component with the PNG Department of Mining was on secondment from Rio Tinto.

at a very low level (less than US$10 million in 2001) despite a recent overhaul of the tax regime to attract more foreign investment in the sector (Daniel *et al.* 2000). Since the hard-rock mining industry then accounted for roughly half the value of PNG's exports and 20% of its gross domestic product, there was a risk that any drastic reduction in these numbers at some future point in time would precipitate the collapse of a government already showing serious signs of weakness (PNGDOM 2003). That was the main reason why the World Bank had undertaken to strengthen the administration of the industry.

The gloomy atmosphere that surrounded the production of a new Sustainable Development Policy was compounded by the impression that those multinational companies that professed to care about high standards of corporate social responsibility now saw PNG as a very risky place in which to practise what they preached. BHP, in its capacity as operator of the Ok Tedi mine, had been right to think that its own 'landowning community' did not pose anything like the same sort of threat as the one that forced the closure of the mine at Panguna. However, the company also thought that it did not need (or could not afford) to attend to the grievances of people living downstream of the mine because these people had not been accorded the status of 'landowners' in agreements with the PNG government. It was in fact the government that (in 1989) had told BHP to do something about compensating these people for the damage being caused to their natural environment, and the company then set up a programme of research to monitor the social impact of the compensation package.[9] However, the findings and recommendations of this research were ignored by senior management until BHP was taken to court in Australia and consumed by a blaze of negative international publicity (Banks and Ballard 1997). When the case was settled out of court in 1996, BHP's managing director said that the company's great mistake was to believe that the game of generating revenues for the PNG government was worth the candle of massive environmental damage and massive damage to the company's reputation (IIED and WBCSD 2002: 348). He might also have said that the big mistake was to discount the interests and opinions of foreign journalists, lawyers, academics and environmentalists who did not count as 'stakeholders' in PNG's national policy framework.

While the World Bank was designing its institutional strengthening project in 2000, it was also responding to a request from PNG's prime minister to assess the Mine Waste Management Project Risk Assessment undertaken by Ok Tedi Mining Ltd in the wake of the Australian court action. The Bank advised that the Ok Tedi mine should be closed on environmental grounds once the company had demonstrated (to the government) that the social costs of closure would not outweigh the environmental benefits (World Bank 2000). The formation of BHP Billiton as a new corporate entity was accompanied by a new deal with the PNG government by which the former would donate its 52% share in the mine to an entity called PNG Sustainable Development Program Ltd in return for a government guarantee that it would henceforth be immune to any more 'environmental claims' (Filer and Imbun 2004; Kirsch 2007).[10] The new company was incorporated in Singapore, with a board comprising one Singaporean member, three representatives

9 The 12 reports produced by the Ok-Fly Social Monitoring Project from 1991 to 1995 are available online at: rspas.anu.edu.au/rmap/projects/Ok-Fly_social_monitoring, accessed 15 August 2008.

10 The deal was given legal effect in the Mining (Ok Tedi Mine Continuation [Ninth Supplemental] Agreement) Act of 2001.

of national institutions (the Department of Treasury, the Bank of PNG, and the PNG Chamber of Commerce), and three people (including the chairman) appointed by BHP Billiton. The new company's basic mandate was to invest two-thirds of its mining profits in a 'Long Term Fund', and to spend the balance of its income, including the interest on this long-term investment, on the implementation of 'sustainable development projects' in both Western Province and the rest of PNG, throughout and beyond the remaining life of the mine (PNGSDP 2002). This entity looks more like an aid agency than a mining company, but the fact remains that its revenues are derived from an environmental disaster, it has no economic interest in closing the mine, and its board is seemingly unaccountable to anyone else.

BHP Billiton's exit from Ok Tedi was entirely consistent with its support for the GMI and the institutions that arose from it, but this left Rio Tinto in a rather awkward position because it held a major stake in Freeport's operations on the other side of the international border. The MMSD Project had documented similarities in the scale of environmental problems created by large-volume waste disposal at both operations (van Zyl *et al.* 2002), but had also shown that human rights violations around the Freeport mine were far more serious than anything encountered at Ok Tedi (Ballard 2001). In 2002, PNG's National Executive Council agreed to prohibit the practice of riverine tailings disposal at new mine sites. In the same year, two American schoolteachers were among the victims of an armed assault on a convoy of vehicles in Freeport's mine lease area. By 2004, Western investigators were convinced that this latter action had been undertaken by elements of the Indonesian army as a way of reminding Freeport of the need to keep up payments of protection money to senior military officers (Lekic 2004). Rio Tinto's reaction was to sell its minority shareholding in Freeport-McMoRan, but to retain the joint venture arrangement through which it could profit from the continued expansion of the Grasberg mine. This act of partial disengagement was justified by Leigh Clifford, then Rio's managing director, in the following terms:

> Through our significant direct interest in Grasberg, we will continue to benefit from our relationship with Freeport, the manager of the Grasberg operations. As our preference is to invest in large, long life, low cost assets in which we have direct access to the operating cash flows, we do not generally hold long term minority positions in other listed companies (Rio Tinto 2004).

The policy declared in the second half of this quotation might explain why Rio Tinto decided to relinquish its role as operator of the Lihir gold mine and sell its minority shareholding in the operation towards the end of 2005. In that case the local and national political context would not have been a factor in the equation. But, in the previous year, Rio Tinto's Melbourne lawyers had conducted a lengthy series of interviews with people who had been involved in dealings between Lihirians and the Lihir Management Company over the 15 years since RTZ had first acquired the Lihir prospect. The apparent purpose of this exercise was to assess the possible grounds for any future legal action that might be taken against the company on behalf of the local community. We cannot say whether these interviews had any influence on the company's decision to withdraw from the Lihir project, but we do know that Rio Tinto—like BHP Billiton before it—has some reason to worry about the application of 'foreign direct liability' principles to the operators of mines in PNG.

For some years past, a court in California has been hearing the claims of a group of Bougainvillean plaintiffs seeking damages from Rio Tinto for environmental and human rights abuses on their island.[11] One of the most remarkable features of this case is a deposition from Sir Michael Somare, three times prime minister of PNG, in which he declares that the PNG government did *not* act as an independent agent in dealing with the Bougainville rebellion, but was simply acting on the instructions of Bougainville Copper Ltd (BCL) , and therefore had no responsibility for human rights abuses committed by the PNG Defence Force: 'It is my opinion that absent Rio Tinto's mining activity on Bougainville or its insistence that the Panguna mine be reopened, the government would not have engaged in hostilities or taken military action on the island' (Somare 2001: 5).

This seems rather odd coming from a man who was not only a government minister at the time when the rebellion broke out, but was also the leader of the self-governing nation during the renegotiation of the Bougainville Copper Agreement—a process long since hailed as a great assertion of national sovereignty in dealings between developing countries and foreign mining companies (O'Faircheallaigh 1984). It is also hard to imagine a current or former president of Indonesia making any such statement about Freeport's control of the Indonesian military.

What a difference the state makes

Rio Tinto's decision to detach itself from Lihir while retaining an interest in Freeport's Grasberg operation might have far less to do with questions of corporate social responsibility or legal liability than with the simple calculation of costs and profits. But it could still be argued that the weakness of the state in PNG is itself partly responsible for raising the costs and risks of doing business in that country. The game that goes on between the PNG government and major resource companies is one in which each occasionally tries to hide behind the other, but the steady diminution of the government's authority and capacity means that its own shadow keeps getting smaller and the companies feel more like shadow states. The cost of corporate social responsibility then becomes the cost of providing a range of benefits and services to mine-affected communities whose own bargaining powers have been continually enhanced by a nationwide process of political fragmentation (Filer 1998, 2001). In that case, we need to ask whether the very different balance of power between state and community in Papua has provided Freeport and Rio Tinto with cost advantages that outweigh the reputational risks associated with military malpractice.

The Indonesian state confronts centrifugal regional forces that also threaten a process of political fragmentation, albeit at a larger scale than anything observed in PNG. In East Timor, Aceh and Papua the response has been a mix of sticks and carrots to maintain the political support of ethnic minorities. In Papua, the violence directed towards the indige-

11 Two of the lawyers who acted on behalf of the Ok Tedi landowners in their suit against BHP—Peter Gordon and Nick Styant-Browne—were engaged as 'foreign consultants' to the American law firm acting on behalf of the Bougainvillean plaintiffs.

nous Melanesian population reflects the fact that they do not have the status of 'landowners' like their counterparts in PNG, but 'only' the status of citizens (Ballard 1995, 2002). This means that the land rights of local communities are less of an issue for a company such as Freeport than the maintenance of a 'partnership' with the state. By comparison with PNG, there is less of an incentive (or need) to commit resources to developing a special relationship with a small set of 'landowners', and any attempt to do so could actually raise the ire of the state.

While it is surely true that PNG has had a resource-dependent economy since the Panguna mine first came into operation (Filer and Macintyre 2006), it is also true that Freeport dominates the economic landscape of Papua to an even greater extent, since it produces more than 90% of the value of the province's (legal) exports and accounts for more than 50% of its gross domestic product (Ballard and Banks forthcoming). For this reason it is in some respects more 'state-like' than PNG's big mining projects. Not only has the company taken on social service delivery functions, it has also paid the salaries of government officials and implemented government plans for the relocation of local communities. It could even be argued that Freeport is 'above' the state, in the sense that it seems to be almost immune to criticism from national government agencies and only pays lip service to provincial and local state actors. The company's strategy of building a strong partnership with the highest levels of political and military power under the Suharto regime has continued to serve it well in the *reformasi* (reform) era (Leith 2003). At both the national and provincial level, Freeport's ability to cope with the changing political scene over this turbulent period has been based on the flexible adaptation of existing networks of patronage and influence (Ballard and Banks forthcoming).

Even in the period following the outbreak of the Bougainville rebellion, the relationship between BCL and the PNG Defence Force never bore any comparison with Freeport's version of the military–industrial complex. In response to queries from American pension funds, Freeport conceded that it had paid US$4.7 million in 2001 and US$5.6 million in 2002 for 'support costs for government-provided security' and 'infrastructure, food, travel, administrative costs and community assistance programmes run by the Indonesian military and police' (Global Witness 2005: 4). But other sources maintain that:

> from 1998 through 2004, Freeport gave military and police generals, colonels, majors and captains, and military units, nearly $20 million. Individual commanders received tens of thousands of dollars, in one case up to $150,000, according to the documents. They were provided by an individual close to Freeport and confirmed as authentic by current and former employees (Perlez and Bonner 2005).

While the company has consistently denied responsibility for the actions taken by its military partners, there is no doubt that these actions are the main obstacle to the construction of any meaningful relationship with the indigenous Amungme and Kamoro communities in the mine-affected area.

A team of social scientists engaged to investigate the company–community relationship after riots and shootings that occurred in 1996 advised the company to develop a new kind of agreement with the Amungme and Kamoro communities through a process that 'should be mediated by independent professional facilitators and must at all times remain transparent' (UNCEN/ANU 1998: 16). However, the Memorandum of Understanding put in place by 2000 was not based on any process that could be described as one of 'free, prior and informed consent' on the part of the indigenous constituency. The main point of the limited consultations that did take place was to have something on the company's website that would appease Western shareholders and stakeholders (Freeport-McMoRan 2000). Instead of coming to grips with the grievances documented by the social scientists, Freeport has recently taken to describing the agreement as a document that is 'instilling a sense of partnership and community' between itself and its 'Papuan neighbours' and 'aligning our shared interests in a sustainable and more promising future' (Freeport-McMoRan 2007: 10). Responsibility for the past is neatly sidestepped by imposing joint responsibility for the future.

Rio Tinto certainly seems to be satisfied that its own association with Freeport meets Rio's standards of corporate social responsibility:

> The Grasberg mine is committed to building and maintaining positive relationships with its Papuan neighbours, in particular the indigenous communities closest to its area of operation. It has in place a Social, Employment and Human Rights Policy, designed to provide opportunities for social, educational and economic development, including special efforts to train and hire those indigenous to the area (Rio Tinto 2007: 31).

This statement was not just based on the Memorandum of Understanding; it also anticipated the acquisition by Freeport McMoRan of another American mining company, Phelps Dodge, which was one of the nine companies to support the GMI in 1998. When the merger was completed in March 2007, the new Freeport became a member of the ICMM and acquired a bigger set of corporate policy commitments much like those of Rio Tinto. The first sentence in the ICMM's draft policy on dealing with indigenous peoples says that:

> ICMM's vision is for effective and constructive relationships between the mining and metals industry and Indigenous Peoples which are based on respect, meaningful engagement and mutual benefit with particular regard for the specific and historical situation of Indigenous Peoples (ICMM 2008: 1).

It remains to be seen whether Freeport's 'Papuan neighbours' will notice the effect of Freeport's new commitments.

A case of corporate amnesia

Placer Dome was the only member of the GMI club formed in 1998 that seems to have retained a belief in its capacity to demonstrate corporate social responsibility while operating a large-scale mine in PNG. However, in 2006 Placer Dome was taken over by another Canadian company, Barrick Gold, which was neither party to the GMI nor a founding member of the ICMM. Unlike Freeport's acquisition of Phelps Dodge, Barrick's acquisition of Placer Dome does not seem to have altered the global policy commitments of the new corporate entity. However, the story of the Porgera mine is not so much a story of how corporate policies are applied in a troublesome social environment, but another story about the gap between rhetoric and reality. Even before the takeover, the management of 'community affairs' at Porgera had been dogged by failure to maintain a proper record of who was actually entitled to receive whatever the company had agreed to distribute among the 'local landowners', let alone to understand the social impact of such allocations.

In the early years of mine construction and operation, immediately following the outbreak of the Bougainville rebellion, the company made some effort to establish an understanding of who was who and what was going on 'beyond the fence' around its operations (Burton 1992; Filer 1999b), but even that effort was repeatedly disturbed by the fractious nature of community politics in the Porgera Valley. While 'fire-fighting' was the order of the day in the Community Relations division, all forms of government were in retreat from the scene of the action.[12] The company's social monitoring programme became part of the fire-fighting apparatus, because company support ebbed and flowed with the level of community antagonism. The social monitoring reports completed in 1997 (Banks and Bonnell 1997) could find little evidence of positive social development outcomes in the mine-affected area, but no action was taken to implement the new form of 'tri-lateral partnership' which they recommended. Instead of persisting with a programme of action to address the local governance deficit, the mine managers were diverted by Placer Dome's new enthusiasm for 'sustainability reporting'—an enthusiasm that sprang directly from its participation in the GMI. In the six Sustainability Reports that Porgera sent to Vancouver (PDAP 1999–2004), the section on 'social progress' became a four-page litany with little evidence of practical remedies being applied to specific local problems or the achievement of measurable outcomes on the ground. As in the case of Freeport's Memorandum of Understanding with local communities, these reports illustrate the way in which the concept of corporate social responsibility was then being applied to the task of reassuring an audience of Western shareholders, stakeholders and regulators, not the task of dealing more effectively with people in the mine-affected area.

Then the reporting stopped. The last year for which Placer Dome reported on the sustainability of any of its mining operations was 2003.[13] It is not clear whether this break in 'normal service' was due to the rapid growth in the number of international standards and guidelines to which the mining industry might wish to hold itself accountable or to Placer's anticipation of the forthcoming merger with Barrick Gold. In any event, Barrick has developed its own line in 'responsibility reporting' against a number of these international standards,[14] and these should now be the yardsticks against which it measures its performance at Porgera. However, while the introduction to the 2006 Responsibility Report declares that it 'will include information on EHSS [environmental, health, safety and social] performance at the former Placer Dome properties for 2006', the only mention of Porgera is in a statement about the control of trespassing by artisanal miners in the mining lease (Barrick Gold 2007: 16). In sum, the reporting gap in respect of sus-

12 It was the Porgera mine managers who persuaded the PNG government to introduce a tax credit scheme in 1992 so that the company could then charge the government for the cost of operating as a shadow state.

13 This is evident from the most recent cache of the Placer Dome website at the Internet Archive (web.archive.org), dated 22 August 2005.

14 At the last count these included the UN Global Compact, the Extractive Industries Transparency Initiative, the Voluntary Principles on Security and Human Rights, and 'Towards Sustainable Mining' (Canada). Barrick Gold joined the ICMM at the start of 2008, after the period we describe.

tainable development issues at Porgera has now lasted a full decade. While this does not necessarily mean there is a lack of practical initiatives to improve the lives of people in the mine-affected communities, it does suggest a lack of transparency and commitment.

If the time bomb theory were correct, then Porgera should be close to the point of detonation by now. The level of mutual misunderstanding—or distorted communication—between company managers and local landowners is certainly as great as anything observed in Bougainville in 1988, when the Panguna mine had been operational for roughly the same period of time. In April 2007 local landowners closed down the Porgera mine for several days because of grievances over the company's plans to resettle all those still living within the boundaries of the Special Mining Lease.[15,16] Many local families had been relocated during the early years of the mining operation, and the social impact of relocation had been one of the topics covered in the original social monitoring programme (Banks 1999; Bonnell 1999). The need for a long-term resettlement plan aligned to the company's long-term mining (and mine closure) plans was the subject of active discussion when the social monitoring programme was abandoned in 1997 (Banks and Bonnell 1997). For several years the issue was placed in the '"too hard" basket', and, while it remained there, the staff turnover in Community Relations rapidly depleted the institutional memory required to do anything more about it. But then the rising price of gold persuaded Placer management to extend the mining operation, both in time and space, and that entailed another round of relocation. To comply with the World Bank Group's safeguard policies on involuntary resettlement, the company would first have to figure out the criteria that it had previously used to decide who was entitled to what kind of resettlement package, but even before the Barrick takeover it had lost the capacity to do this. The takeover only made matters even more bewildering.

Conclusion

PNG and Papua both exhibit symptoms of the so-called resource curse, not only in the sense of suffering the negative economic effects of a dominant extractive industry sector, but also because of the environmental damage caused by mining operations, the corrupting and sometimes violent contestation over access to resource rents, and the promotion of a local or national culture of 'resource dependency' (Filer and Macintyre 2006). From the industry's point of view, the curse takes the form of a responsibility vacuum, in which partnerships with host governments or local communities are difficult, dangerous or downright impossible because they will not take responsibility for their own actions. In the words of one former manager of 'community affairs' in PNG's extractive industry sector:

> Everyone expects development, but everyone expects someone else to make it happen. There is a very poor understanding of the degree of commitment required to make the desired state attainable. The Government just sits back

15 *The National*, 25–27 and 30 April 2007.

16 Landowner relocation is not the only bone of contention in the history of company–community relationships, but happened to be the bone of choice in 2007.

> and expects the developer to make things happen. The landowners expect everything to be done for them because the developer is on their land. The developer is reluctant to take over what they see as the role of the Government (Power 2000: 86).

If we consider only the case of PNG, we might be tempted to argue that there is a specific kind of 'irresponsibility' that is characteristic of Melanesian political culture, both at the level of the village and the level of the state, which has something to do with the traditional cultural diversity and political fragmentation of the region (Filer 1998). Should we then conclude that Rio Tinto's willingness to retain a stake in Freeport reflects the higher level of 'responsibility' exhibited by the Indonesian army?

The responsibility vacuum is not just a feature of the triangular relationship between company, state and community in some particular region or location. Where mining companies behave like shadow states, their performance is not just a function of what the 'real' state is or is not doing. Corporate behaviour is also influenced by variations in the physical and economic quality of mineral deposits, the economic cost of compliance with particular standards of responsible management, and the legal, political or reputational risks of failing to meet those standards. The Grasberg deposit is a bonanza whose grades clearly offset the reputational risk of a partnership with an army whose own behaviour falls well short of compliance with international standards pertaining to the rights of indigenous peoples. The cost of compliance with standards of best practice in the mitigation of social and environmental impacts is generally very high in the New Guinea region. This is not just because people and politicians are fickle and unpredictable, but also because these are qualities of the physical environment itself. If a big company can safely store its waste materials and still make a reasonable profit, then it will generally hesitate to engineer an environmental disaster such as Ok Tedi. The reasons why New Guinea is a hard place in which to practise the art of corporate environmental responsibility are also the reasons why it is a region of such enormous biological and cultural diversity. Mining companies could still live with the creation of social and environmental disasters if there were some way of hiding them from public scrutiny by influential stakeholders. But herein lies a paradox. For while the cost of transport and communication in this region is higher than it is in most other parts of the world, the world is still populated with an awful lot of public information about the social and environmental problems which its mining industry has created—especially the mining industry in PNG.

PNG is not a good place for the mining industry to hide its dirty laundry. Its indigenous political and intellectual leaders are very adept at complaining to a global English-speaking audience about anything that they regard as a violation of their rights and customs. It is a happy hunting ground for environmentalists who want to maintain its status at the 'last great place' for the conservation of biodiversity. It is also infested with social scientists such as ourselves who have made it their business to publish numerous books and articles about the relationship between mining companies and local communities. However, if this makes PNG a place in which corporate social responsibility has to be something more than a set of optical illusions, it also makes it a place in which the mining industry is more likely to be taken over by companies operating in regimes of accountability where the optics do not even matter.

References

Ballard, C. (1995) 'Citizens and Landowners: The Contest over Land and Mineral Resources in Eastern Indonesia and Papua New Guinea', in D. Denoon, C. Ballard, G. Banks and P. Hancock (eds.), *Mining and Mineral Resource Policy Issues in Asia-Pacific: Prospects for the 21st Century* (Canberra: Australian National University, Research School of Pacific and Asian Studies, Division of Pacific and Asian History): 76-81.

—— (2001) 'Human Rights and the Mining Sector in Indonesia: A Baseline Study' (MMSD Working Paper 182; London: Mining, Minerals and Sustainable Development Project of the International Institute for Environment and Development).

—— (2002) 'The Signature of Terror: Violence, Memory and Landscape at Freeport', in B. David and M. Wilson (eds.), *Inscribed Landscapes: Marking and Making Place* (Honolulu, HI: University of Hawai'i Press): 13-26.

—— and G. Banks (forthcoming) 'Between a Rock and a Hard Place: Corporate Strategy at the Freeport Mine in Papua, 2001–2006', in B. Resosudarmo and F. Jotzo (eds.), *Development and Environment in Eastern Indonesia: Papua, Maluku and East Nusa Tenggara* (Singapore: Institute of Southeast Asia Studies).

Banks, G. (1999) 'The Next Round of Relocation', in C. Filer (ed.), *Dilemmas of Development: The Social and Economic Impact of the Porgera Gold Mine 1989–94* (Pacific Policy Paper 34; Canberra: Asia Pacific Press): 191-221.

—— (2006) 'Mining, Social Change and Corporate Social Responsibility: Drawing Lines in the Papua New Guinea Mud', in S. Firth (ed.), *Globalization and Governance in the Pacific Islands* (Canberra: ANU E Press): 259-74.

—— and C. Ballard (eds.) (1997) *The Ok Tedi Settlement: Issues, Outcomes and Implications* (Pacific Policy Paper 27; Canberra: Australian National University, National Centre for Development Studies and Resource Management in the Asia Pacific).

—— and S. Bonnell (1997) 'Porgera Social Monitoring Programme: Annual Report for 1996 and 1997 Action Plan' (unpublished report to Porgera Joint Venture).

Barrick Gold (2007) 'Barrick Responsibility: 2006 Environmental, Health, Safety and Social Performance' (Toronto: Barrick Gold Corporation; www.barrick.com, accessed 20 August 2008).

Bello, W., and S. Guttal (2006) 'The Limits of Reform: The Wolfensohn Era at the World Bank', *Race and Class* 47: 68-81.

Bonnell, S. (1999a) 'The Landowner Relocation Programme', in C. Filer (ed.), *Dilemmas of Development: The Social and Economic Impact of the Porgera Gold Mine 1989–1994* (Pacific Policy Paper 34; Canberra: Asia Pacific Press): 128-59.

Bosshard, P. (1996) 'Tainted Gold from the Pacific: A Case Study about MIGA's Lihir Island Goldmine Project in Papua New Guinea' (Zurich: Berne Declaration).

Burton, J.E. (1992) 'The Porgera Census Project', *Research in Melanesia* 16: 129-56.

Daniel, P., K. Palmer, A. Watson and R. Brown (2000) 'Review of the Fiscal Regimes for Mining and Hydrocarbons', unpublished report to the Independent State of Papua New Guinea Tax Review.

Danielson, L., with C. Digby (2006) *Architecture for Change: An Account of the Mining Minerals and Sustainable Development Project* (Berlin: Global Public Policy Institute; www.iied.org/mmsd).

EIR (Extractive Industries Review) (2002) 'Extractive Industries Review Project Visit to Papua New Guinea, 2–11 August 2002'; www.worldbank.org.

—— (2003) 'Asia and Pacific Regional Workshop, Bali, Indonesia, 26–30 April 2003: Testimonials and Consultation Report'; www.worldbank.org.

Filer, C. (1990) 'The Bougainville Rebellion, the Mining Industry and the Process of Social Disintegration in Papua New Guinea', in R.J. May and M. Spriggs (eds.), *The Bougainville Crisis* (Bathurst, NSW: Crawford House Press): 73-127.

—— (1998) 'The Melanesian Way of Menacing the Mining Industry', in L. Zimmer-Tamakoshi (ed.), *Modern Papua New Guinea* (Kirksville, MO: Thomas Jefferson University Press): 143-73.

—— (1999a) 'The Dialectics of Negation and Negotiation in the Anthropology of Mineral Resource Development in Papua New Guinea', in A. Cheater (ed.), *The Anthropology of Power: Empowerment and Disempowerment in Changing Structures* (London: Routledge): 88-102.

—— (ed.) (1999b) *Dilemmas of Development: The Social and Economic Impact of the Porgera Gold Mine, 1989–1994* (Pacific Policy Paper 34; Canberra: Asia-Pacific Press).

—— (2001) 'Between a Rock and a Hard Place: Mining Projects, "Indigenous Communities", and Melanesian States', in B.Y. Imbun and P.A. McGavin (eds.), *Mining in Papua New Guinea: Analysis and Policy Implications* (Waigani, Papua New Guinea: University of Papua New Guinea Press): 7-24.

—— (2002) 'Implications of the "Mining, Minerals and Sustainable Development" Project' (Working Paper 1; Port Moresby: PNG Department of Mining, Sustainable Development Policy and Sustainability Planning Framework).

—— (2005) 'The Role of Land-owning Communities in Papua New Guinea's Mineral Policy Framework', in E. Bastida, T. Wälde and J. Warden-Fernández (eds.), *International and Comparative Mineral Law and Policy: Trends and Prospects* (The Hague: Kluwer Law International): 903-32.

—— and B.Y. Imbun (2004) 'A Short History of Mineral Development Policies in Papua New Guinea' (Working Paper 55; Canberra: Australian National University, Resource Management in Asia-Pacific Program).

—— and M. Macintyre (2006) 'Grass Roots and Deep Holes: Community Responses to Mining in Melanesia', *Contemporary Pacific* 18.2: 215-31.

—— and A. Mandie-Filer (1998) 'Rio Tinto through the Looking Glass: Lihirian Perspectives on the Social and Environmental Aspects of the Lihir Gold Mine' (unpublished report to the World Bank).

Freeport-McMoRan Copper and Gold Inc. (2000) 'Amungme, Kamoro and Freeport Indonesia announce agreement; initial cooperative projects launched', news release, 18 August 2000.

—— (2007) 'Underlying Values: 2006 Working towards Sustainable Development Report'; www.fcx.com/envir/wtsd/pdf-wtsd/2006/WTSD.pdf, accessed 1 October 2008.

Global Witness (2005) *Paying for Protection: The Freeport Mine and the Indonesian Security Forces* (Washington, DC: Global Witness; www.globalwitness.org/media_library_detail.php/139/en, accessed 20 August 2008).

ICMM (International Council on Mining and Metals) (2008) 'Position Statement: Mining and Indigenous Peoples'; www.icmm.com/document/293, accessed 1 October 2008.

IIED (International Institute for Environment and Development) and WBCSD (World Business Council for Sustainable Development) (2002) *Breaking New Ground: Mining, Minerals and Sustainable Development Final Report of the MMSD Project* (London/Sterling, VA: Earthscan; www.iied.org/mmsd/finalreport/index.html, accessed 20 August 2008).

Jackson, R. (1982) *Ok Tedi: The Pot of Gold* (Port Moresby, Papua New Guinea: University of Papua New Guinea).

Kirsch, S. (2007) 'Indigenous Movements and the Risks of Counterglobalization: Tracking the Campaign against Papua New Guinea's Ok Tedi Mine', *American Ethnologist* 34.2: 303-21.

Leith, D. (2003) *The Politics of Power: Freeport in Suharto's Indonesia* (Honolulu: University of Hawai'i Press).

Lekic, S. (2004) 'Indonesian Army ordered deadly ambush', Associated Press, 3 March 2004; www.minesandcommunities.org/Action/press272.htm, accessed 20 August 2008.

Mathrani, S. (2003) 'Evaluation of the World Bank Group's Activities in the Extractive Industries: Papua New Guinea Country Case Study', report to World Bank Operations Evaluation Department; www.ifc.org/ifcext/oeg.nsf/AttachmentsByTitle/oed_ccs_papua/$FILE/oed_ccs_papua.pdf, accessed 20 August 2008.

McBeth, J. (1995) 'Yours, Mine, Ours: Freeport, RTZ Lock Drills on Irian Jaya', *Far Eastern Economic Review*, 23 March 1995: 48-50.

McPhail, K., and A. Davy (1998) 'Integrating Social Concerns into Private Sector Decision-making: A Review of Corporate Practices in the Mining, Oil, and Gas Sectors' (Discussion Paper 384; Washington, DC: World Bank).

O'Faircheallaigh, C. (1984) *Mining and Development: Foreign Financed Mines in Australia, Ireland, Papua New Guinea and Zambia* (London: Croom Helm).

PDAP (Placer Dome Asia Pacific) (1999–2004) *Porgera Mine: Sustainability Reports 1998–2003* (Port Moresby, PNG: Placer Dome Asia Pacific).

Perlez, J., and R. Bonner (2005) 'Below a Mountain of Wealth, a River of Waste', *New York Times*, 27 December 2005.

PNGDOM (Papua New Guinea Department of Mining) (2003) *Sustainable Development Policy and Sustainability Planning Framework for the Mining Sector in Papua New Guinea: Green Paper* (Port Moresby: PNGDOM).

PNGSDP (Papua New Guinea Sustainable Development Program Ltd) (2002) *Annual Report 2002* (Port Moresby: PNGSDPL; www.pngsdp.com).

Power, T. (2000) *Community Relations Manual: Resource Industries. Volume 1* (Port Moresby: PNG Chamber of Mines and Petroleum).

Rio Tinto (2004) 'Rio Tinto reaches agreement to sell shares in FCX', media release, 22 March 2004; www.riotinto.com/media/5157_3383.asp, accessed 20 August 2004.

—— (2007) '2006 Sustainable Development Review: Access to Resources: People, Land, Capital'; www.riotinto.com.

Somare, M. (2001) 'Declaration of Sir Michael Somare, Former Prime Minister of Papua New Guinea' (dated 23 November 2001), Exhibit 1 in *Alexis Holyweek Sarei et al. v. Rio Tinto plc et al.* (United States District Court, Central District of California, Western Division).

UNCEN/ANU (Universitas Cenderawasih and Australian National University) (1998) 'UNCEN-ANU Baseline Studies Report No. 6: Draft Final Report' (unpublished report to PT Freeport Indonesia).

Van Zyl, D., M. Sassoon, A.-M. Fleury and S. Kyeyune (2002) 'Mining for the Future' (Working Paper 68; London: Mining, Minerals and Sustainable Development Project; www.iied.org/mmsd/activities/mine_closure_policy.html, accessed 20 August 2008).

Warner, M., and R. Sullivan (eds.) (2004) *Putting Partnerships to Work: Strategic Alliances for Development between Government, the Private Sector and Civil Society* (Sheffield, UK: Greenleaf Publishing).

World Bank (2000) 'Ok Tedi Mining Ltd. Mine Waste Management Project: Risk Assessment and Supporting Documents'; www.abc.net.au/4corners/content/2000/worldbankreview.doc, accessed 20 August 2008, with covering letter from Klaus Rohland, World Bank Country Director, Papua New Guinea to Mekere Morauta, Prime Minister, Government of Papua New Guinea, 20 January 2000; www.abc.net.au/4corners/archives/2000a_Monday10April2000.htm, accessed 20 August 2008.

10

Shareholder activism and corporate behaviour in Ecuador

A comparative study of two oil ventures

Emily McAteer
RiskMetrics Group, USA

Jamie Cerretti
Environment America, USA

Saleem H. Ali
University of Vermont, USA

The corporate structure of extractive industries is highly variable and many indigenous communities have to contend with complex ownership structures and evasive management practices. Historically, these complex corporate structures have presented significant challenges to indigenous groups, often impeding their efforts to effectively exert influence over a company operating in their communities. Recently, however, indigenous communities have begun to identify certain opportunities in the extractive industry that offer new grounds for strategising. Publicly traded companies in particular provide an opportunity for indigenous communities to exert influence through transnational activist networks. This chapter examines these new mechanisms of influence and the ways in which strategies for leveraging their negotiating power have helped indigenous communities reach agreements with oil companies as well as resist particular aspects of development. We use two case studies of oil extraction in Ecuador to understand the nuances of power relationships between corporations, communities, activist groups and the government. Our fundamental research question is: what are the key characteristics of shareholder activism that can make it effective within existing corporate structures and governance mechanisms in a diverse, pluralistic society?

Ecuador, like many other South American countries, is highly diverse and characterised by stark divisions between regions, socioeconomic classes and ethnic groups. These regions provide a useful locale for empirical research about cross-cultural conflicts between indigenous and settler populations, particularly in the context of environmental social movements. The Andes mountain range separates the plains of the Pacific coast from the astoundingly biodiverse Amazon rainforest, known as the Oriente, or East. The nine indigenous groups in the vast Amazon, the Cofán, Secoya, Siona, Shiwiar, Huaorani, Achuar, Zápara, Shuar and Oriente Kichwa, are both culturally and geographically isolated from the highland region, which is home to the majority of Ecuador's population and 96% of its indigenous peoples (Gerlach 2003).

Land use before the oil boom

The inhabitants of the Oriente traditionally met their basic needs by use of the wide array of food items and building materials provided by the biodiversity of the Amazon. Hunting, fishing, gathering and small-scale shifting agriculture provided kin groups with their livelihoods (Korovkin 2002). Large amounts of land were required for these traditional activities, which were unlike the settled agriculture suited to the richer soils of the highlands and coastal areas. Territories were held communally and individual use of the land for hunting, gathering and travel did not prevent other individuals in the community from engaging in the same activities (Gerlach 2003). The idea of exclusion was related more to concern for tribal sovereignty than for personal ownership (Cronon 1983). As Pawson and Cant (1992) note in their analysis of indigenous land rights, 'land is the source of communal identity, the place of belonging, the link between present, past, and future generations'.

Historically, the Amazon's indigenous people expressed a relative lack of desire for production and accumulation of material goods owing to their experiences with the bounty of materials offered by the surrounding forest (Korovkin 2002). This clashed with the values of Europeans who felt that the abundance of raw materials in the equatorial region promoted laziness, which was the 'greatest of all obstacles to labor and industry' (Curtin 1964: 61-62, cited in Arnold 1996). The Spanish conquerors defined the 'best use' of the land differently from its traditional inhabitants, being concerned with economic productivity and its resultant accumulation of individual wealth. After the Spanish rose to power in Ecuador, land itself became a commodity, and the right of indigenous tribes to subsist on the land came into conflict with the right of others to profit from the land. In the highlands, this resulted in a concentration of land into large haciendas held by whites and worked by an indigenous underclass (Gerlach 2003).

The relative remoteness and impenetrability of the Amazon rainforest made colonisation more difficult, but contact with Spanish settlers and Catholic missionaries did result in several disease epidemics that ravaged the Oriente's indigenous population beginning in the late 17th century. The Amazon rubber boom of the late 1800s through the 1920s brought an influx of tappers and traders to the area (Korovkin 2002). The Ecuadorian government gave Standard Oil the first concession for petroleum exploration in the region in 1921, but the endeavour proved largely unsuccessful and was followed by several equally lacklustre attempts by the Leonard Exploration Company from 1923 to 1931, the Anglo Saxon Petroleum Company in 1937, and lastly the Shell Company Ecuador in 1939 (Gedicks 2001). Shell eventually abandoned its efforts in the Oriente in 1950, but not before bringing in the Summer Linguistic Institute/Wycliffe Bible Translator Inc., an organisation comprising Protestant missionaries who were charged with translating the Bible into various indigenous languages and converting the population to Christianity (Martin 2003). These evangelical missionaries had a long-term impact on the indigenous community, with future alliances between the church and petroleum companies facilitating resource extraction in indigenous territory. In 1942, approximately one-half of the Oriente was annexed by Peru, resulting in a loss of close to one-third of the Ecuador's total original territory. Ecuador refuses to acknowledge this defeat, and has designated the Oriente a 'national security area', resulting in a strong military presence throughout the region (Kimerling 1991).

Texaco and the oil era

In 1949, President Galo Plaza Lasso dismissed oil prospects in the Oriente as 'a myth', asserting that there was 'simply no oil to be found in the Amazon' (Gerlach 2003: 33). There was subsequently little interest in the region's resource potential until 1967 when Texaco-Gulf struck oil at Lago Agrio, slightly north of the traditional lands of the Huaorani indigenous group (Gedicks 2001). This discovery launched a new era of Ecuadorian history in which the economy became almost completely dependent on petroleum. In 1972, Ecuador passed the Hydrocarbon Law, which declared all oil reserves as property of the state and created the Ecuadorian State Petroleum Corporation (CEPE) to absorb 25% of the rights and profits of Texaco-Gulf. That same year, a 313-mile pipeline was built to connect Lago Agrio to the Pacific Ocean at Esmeraldas, traversing the Andes (Korovkin 2002; Kane 1995).

Texaco played a significant role in Ecuadorian policy-making, mostly through funding of various presidential campaigns, as petroleum production quadrupled and Ecuador's GNP rose from $2.2 billion in 1971 to $5.9 billion in 1977. Between 1984 and 1993, 12 more companies, most of them based in the USA, were given petroleum concessions in the Amazon over six rounds of international bidding (Martin 2003). By the time Texaco pulled out of Ecuador in 1992, ceding full control to the state oil company, the Oriente had been divided into over a dozen 202,350 ha (500,000 acre) drilling blocks. Almost 1.5 billion barrels of oil had been extracted from the region between 1971 and 1991 (Kimerling 1991), leaving behind hundreds of oil wells, roads and pumping stations crisscrossing over 1 million ha (2.5 million acres) of forest (Gedicks 2001). The government embraced the petroleum economy as a means of modernisation, development and poverty alleviation, and by 2001 46% of the country's income was based on oil revenues (Gerlach 2003). An Ecuadorian environmental agency was not established until 1984, and, until then, Texaco extracted and exported oil in collaboration with Petroecuador (formerly CEPE) with no oversight or impact reports. Thus, with essentially no intervention by outside actors and minimal control over its activities, the company 'acquired a private, authoritative role within the Ecuadorian government' (Martin 2003: 75).

In most cases, the government did not recognise any form of indigenous land ownership in the Oriente throughout the petroleum boom. As described in Sabin (1998: 150-51), the state pushed soldiers and former construction workers to settle near the Texaco production area, even flying in 'vagrants and delinquents from the cities' to increase a non-indigenous presence in the region. Oil extraction required road building, which in turn provided settlers with a way to travel into the formerly isolated Oriente. The government promoted colonisation in the region as a method of relieving population pressures in the highlands and along the coast, citing the existence of empty tracts of land expansive enough to support more than double the current population of the country. In order to legitimise land claims under the agrarian reform laws of the 1960s and 1970s, indigenous inhabitants and newly arrived colonists were required to develop at least half of their land, resulting in an increase in cattle ranching that was supported by government subsidies.

Differences in the rights conferred on indigenous people in the Oriente to use and own land have had a significant effect on the distribution of both income and power in this region. In 1994, Luis Macas, former president of CONAIE (Confederation of Indigenous Nationalities of Ecuador), described the relationship between the fight for land, the environment and human rights:

> The problems facing the indigenous peoples are deeply connected to the issue of land ownership. When the colonizers arrived, they cleared out the Indians. Today, land is concentrated in the hands of the few, and many of our people don't have any land. In the Amazon region, there is a crisis caused by the presence of oil and mining companies and their violations of indigenous peoples' rights. The displacement of people from their homes has made it impossible for indigenous people to meet basic living conditions. The oil companies have not only caused the decomposition of our communities and the decomposition of our culture but also the destruction of the ecology. The fight for land is thus extended to the struggle for maintaining the ecology (Gerlach 2003: 66).

Oil, structural adjustment and the indigenous population

Oil revenues were used to finance a variety of public initiatives, including electrification programmes, universities, ambitious infrastructure and social service programmes, large energy and food subsidies, and tax breaks (Sawyer 2004). A significant amount of funds were spent to heavily subsidise domestic petroleum products at half the cost of production, dramatically increasing domestic consumption of petroleum; the number of automobiles in the country, for example, rose from 80,000 in 1970 to 223,000 just seven years later (Gerlach 2003). This export-led development path also included government incentives to support manufacturing and industry, and Ecuador's agricultural economy quickly turned into an economy based on services, manufacturing, industry and mining. The country's cities—Quito and Guayaquil—expanded at unprecedented rates, and a new urban middle class quickly grew as large populations from neighbouring provinces moved into the growing cities. By the time President Jaime Roldós was democratically elected in 1979, ending the military regime's seven-year rule, the state budget had expanded by 540% (Gerlach 2003).

When oil revenue was insufficient to maintain this high level of spending on the ambitious modernisation plans, the military regimes turned to foreign borrowing, relying on future oil reserves as loan guarantees. Foreign debt rose just as quickly as state revenue; between 1974 and 1982, Ecuador's foreign debt had climbed from 18% to 60% of the country's GDP (Sawyer 2004).

Unfortunately, this high dependence on oil revenues put Ecuador's economy in a vulnerable position, and as world oil prices quickly dropped in the 1980s, falling from $34.48/barrel in 1981 to $12.70/barrel in 1986, the country began to realise the full impact of this vulnerability (Gerlach 2003). Ecuador's debt continued to climb throughout the 1980s; by 1994, the debt had nearly doubled from its 1982 level, and Ecuador had the highest per capita foreign debt in all of Latin America (Sawyer 2004). Due to the price drops, oil revenues were no longer sufficient to service Ecuador's accruing debt, and the state turned to structural adjustment loans to repay its mounting debts. Multilateral lending institutions, including the World Bank, the International Monetary Fund and the Inter-American Development Bank, collaborated to impose conditions on the state to institutionalise neoliberal policies that cut public spending. Throughout the 1980s and 1990s, Ecuador was forced to implement austerity programmes that included unpopular restrictions on federal expenditures, such as cutting state subsidies for consumer goods and education/health services.

The austerity programmes implemented by Ecuador in an effort to balance its books after massive debt accruement proved to be particularly hard on the country's indigenous populations, as the state's vulnerability to the demands of foreign institutions such as multilateral development banks only further undermined the state's ability to compensate indigenous groups for the detrimental social and environmental impacts of oil development in the Amazon. As anthropologist Suzana Sawyer observes, 'in the wake of the growing social, economic, political, and environmental ills caused by unchecked capitalist activities, *indigenas* found the state unresponsive to the tragedies tormenting their lives' (Sawyer 2004).

A member of a Quito-based environmental NGO working on oil issues summarised the effects of decades of oil on Ecuador's population stated in January 2007:

> The reality of Ecuador is that there have been all these years of oil production and really it hasn't helped the Ecuadorian people live better. In fact, it is completely the contrary; before oil production the Ecuadorian population lived better . . . there was not as much debt as there is today—the huge foreign debt that we have—nor the poverty indexes as high as they are today. Before there were limits to the poverty . . . now we are poor.[1]

Advocacy networks and their influence on corporations

Years of oil extraction, an embedded racial hierarchy, a consistent lack of indigenous representation in the political sphere, a markedly intimate and dependent relationship between the state and multinational oil corporations, a large vulnerability to the demands

1 Interview with author, January 2007.

of multilateral lending institutions, and an infamously volatile government have certainly done much to close the political opportunity structure for advocates of indigenous rights in Ecuador's Amazon rainforest. However, beyond offering an emblematic story of indigenous marginalisation from state-led development, Ecuador's long history of oil development in fact illuminates an intricate mosaic of both opportunities and limitations for the country's indigenous peoples. Sawyer (2004: 18) argues that 'neoliberal maneuvers and oppositional tactics are mutually constitutive', and 'the very policies that nurtured transnational capital undermined people's rights and produced disruptive subjects'. In other words, although oil development in the Amazon has been accompanied by severe social and environmental impacts that have taken a heavy toll on the indigenous peoples of Ecuador, these effects have served to spur a strong and unique indigenous resistance. Indigenous opposition in Ecuador has expanded from grassroots resistance tactics to a myriad of national and international strategies, most notably the use of shareholder pressure aimed directly at the top management of oil corporations in their home country. As an article in *Business Week* explained, 'The plight of indigenous groups is [now] penetrating the boardrooms of multinationals' (quoted in Gedicks 2001: 2).

Concepts such as the transnational advocacy network (TAN) and the related boomerang model put forth by Margaret Keck and Kathryn Sikkink (1998) offer a useful framework for analysing the indigenous resistance strategies to oil development in the Amazon. TANs, according to Keck and Sikkink (1998: 9), are 'dense webs of connections' between various domestic and international actors that 'share values and frequently exchange information and services'. These networks are 'organized to promote causes, principled ideas, and norms' often focusing on a specific target (such as a government) (Keck and Sikkink 1998: 8). When channels between domestic actors and states are blocked, TANs can set into motion the 'boomerang pattern of influence', in which domestic groups ally with international actors to apply pressure on the state from the outside. This strategy is particularly important in the developing world, where domestic actors often lack power or influence over state activities, and therefore must rely on the leverage and information offered by international allies.

While Keck and Sikkink do acknowledge that states are not always the primary target of advocacy networks, traditional TAN theory has historically been applied to networks targeted at state institutions. Largely absent from this literature is a non-state-centric analysis of TANs that reflects the widespread recent shift from states to corporations as the targets of TAN activities, and changes in network strategies that have accompanied this shift. Newell describes this new focus on corporations as a 'strategic turn' by members of advocacy networks (primarily NGOs) 'towards the private sector' (Newell 2000: 34). While this shift in NGO practice is well documented and has attracted considerable attention in the past decade, these changes have yet to be integrated into TAN theory. Developments in Ecuador's indigenous rights movement in the Amazon rainforest during the past decade offer a basis on which to consider a new form of TAN and a modified boomerang model.

The TAN centred on indigenous rights and oil development in Ecuador demonstrates a new type of transnational advocacy network, referred to here as the shareholder transnational advocacy network (STAN), which links indigenous communities with NGOs and corporate shareholders to influence the policies and practice of multinational corporations. STANs operate according to a modified version of Keck and Sikkink's boomerang model, called the corporate boomerang model, which assumes that indigenous communities, with few avenues to influence corporate practice from the local level, align with domestic and international NGOs, who in turn provide connections to corporate shareholder groups, including socially responsible investing firms, religious investors and other institutional investors. These shareholder groups implement certain mechanisms, such as shareholder resolutions and dialogues with the company, to pressure corporations in their home country. The leverage strategies employed by corporate shareholders are legally grounded in Section 14(a) of the Securities Exchange Act of 1934, which establishes the shareholder proposal process and outlines the official Securities and Exchange Commission (SEC) rules on shareholder resolutions.[2] The SEC shareholder proposal rules establish the right for shareholders with holdings of $2,000 or more in a public company to file formal resolutions with the company's board of directors. If granted approval by the SEC, these proposals are then voted on by the company's shareholders at its Annual General Meeting.

Shareholder resolutions can serve as an effective means for activists to communicate their concerns directly to a company's executive management, thereby forcing certain issues onto the corporate agenda. Annual shareholder meetings provide an official forum for shareholders to frame issues of concern in a manner that will be most appealing to the company. For example, resolutions calling on an oil company to improve its relations with indigenous communities translate a network value—such as indigenous rights and environmental protection—into a 'business case' framework, presenting the issue in terms of a manager's fiduciary duty to maximise shareholders' interests (Hetherington 1969). While social and environmental shareholder resolutions typically receive low votes at the annual meeting, the power to file resolutions with the company's board of directors provides shareholders with a unique opportunity to exert pressure directly on the company's top management, giving indigenous communities a new, and often more effective, route of access to the company.

STAN members can be categorised into four levels (moving from the most grassroots level to the international/corporate actors): the first level is composed of domestic indigenous federations; the second, domestic NGOs; the third, international NGOs; the fourth, shareholders. Indigenous federations provide necessary testimony and local knowledge of the issue, which serves to legitimise and strengthen NGO and shareholder action. Domestic NGOs serve primarily to empower indigenous communities, providing necessary resources and appropriate connections to more powerful allies. International NGOs play a similar role as their domestic counterparts, but they are also instrumental in both informing shareholders of relevant local issues and facilitating contact between shareholders and indigenous communities. Lastly, shareholders, empowered by their support and testimony from local organisations, exert pressure on corporate senior management to change the company's policy and practice. This type of advocacy network clearly demonstrates a boomerang pattern of influence as indigenous communities, supported by domestic and international NGOs, align with more powerful actors to pressure the corporation from the outside, rather than at the domestic level.

Beginning in 2002, two distinct cases of shareholder transnational advocacy networks emerged in Ecuador's Amazon rainforest. Both networks are centred on changing the policies and practices of a specific oil company with operations in the Amazon. The first

2 For further information, refer to Investor Responsibility Research Center 1994.

case study examines a STAN operating in the southern Amazon, where indigenous resistance successfully blocked all attempts by Burlington Resources to initiate oil production in the region. The second case study evaluates a STAN in the northern Amazon that has been less successful in influencing the practices of Chevron, which it holds responsible for the operations of Texaco, now a Chevron subsidiary.

As both case studies demonstrate, the threat of oil extraction in the Ecuador's Oriente has prompted anti-oil mobilisation by indigenous groups in both the southern and northern Amazonian communities. Shareholders, in turn, have become key players in this indigenous resistance, as they have increasingly recognised the positive correlation between shareholder value and the improved environmental and social performance of a company. Interestingly, however, the outcomes of the two case studies differ widely in the networks' success in influencing their targeted company. The two case studies offer an illuminating comparison for consideration of the factors that determine the success of shareholder transnational advocacy networks.

Burlington Resources

Burlington Resource's concessions in Ecuador are located in Blocks 23 and 24 of the Pastaza and Morona Santiago provinces, a region known more commonly as the Oriente's *centro-sur*. Both blocks were first included in oil negotiations in 1995, during the eighth round of concession leasing (Acción Ecológica 2006). In 1996, Argentine company Compañia General de Combustibles CGC was awarded 100% of the drilling rights to Block 23, serving the block's operator. Three years later, after purchasing CGC's partner company, Chevron acquired a 50% share of the concession, which it sold to Burlington Resources in March 2003. CGC continues to be the operator of the block.

Drilling rights to Block 24 were granted to Atlantic Richfield Company (ARCO) in 1996. Following its acquisition by British Petroleum in 1999, ARCO sold its share to Burlington Resources, granting the company 100% of the licence to the concession (Acción Ecológica 2006). In March 2006, ConocoPhillips acquired Burlington Resources, becoming the official owner of Burlington's 50% stake in Block 23 and its 100% operating share in Block 24. Burlington Resources did not succeed in conducting any oil drilling in its duration as owner of the two concessions, and ConocoPhillips has not proceeded further to develop oil in these two blocks.

The shareholder transnational advocacy network organised around Burlington's oil activities in the southern Amazon follows the four-level structure outlined above. The first tier of the network includes the southern Amazon's five indigenous groups, represented through a network of federations with varying degrees of collaboration. Four principal national NGOs make up the second level of the network. These organisations, based in Quito, offer various forms of support to indigenous federations and communities, including legal assistance, capacity-building, facilitation of dialogues, and financial support. Amazon Watch, an international NGO headquartered in San Francisco, plays a pivotal role in the third tier of the STAN, orchestrating shareholder–indigenous links in the transnational advocacy network targeted at Burlington Resources. In September 2003, Amazon Watch recruited Boston Common Asset Management, a Boston-based socially responsible investing firm, to put pressure on Burlington Resources by filing a shareholder resolution related to the company's indigenous rights policy. Complementing an array of other resistance strategies, including legal action and public demonstration, the Burlington STAN's shareholder advocacy included the filing of four resolutions with Burlington Resources via Boston Common Asset Management; the arrangement of several indigenous delegates from the Shuar and Achuar federations and the Sarayaku community to present their position at the company's annual general meetings; and a 'fact-finding mission' to Ecuador taken by several Burlington shareholders. By the time ConocoPhillips took over the company in 2006, Burlington Resources had held several dialogues with STAN members, issued an indigenous rights policy and pledged not to proceed into oil concessions by force. In addition, three of the four resolutions filed with Burlington were withdrawn after negotiations between the shareholders and the company. A chronology of activities in this case is provided in Box 10.1.

Box 10.1 Time-line of shareholder activity, Burlington Resources/ConocoPhillips

- **April/May 2003.** Amazon Watch organises indigenous leaders to speak in Houston and New York City (NYC)
- **May 2003.** Conference call with Amazon Watch, Earthrights International, Boston Common and other socially responsible investors
- **September 2003.** Amazon Watch contacts some of the conference call participants to solicit interest in filing resolution with Burlington. Boston Common gets approval to file from its client Brethren Benefit Trust (BBT), who held shares in Burlington
- **November 2003.** Boston Common—on behalf of BBT—files first resolution asking Burlington to adopt an indigenous peoples rights policy
- **Early December 2003.** Burlington contacts Boston Common, indicating interest to hold dialogue; said that it already had some policies on indigenous rights
- **Mid-January 2004.** Burlington produces a one-page policy on indigenous peoples rights
- **Early April 2004.** Boston Common sends letter to Burlington, signed by Boston Common, BBT and several other institutional investors

  Burlington states in annual report that it would only enter into Block 24 where it had majority approval from block's indigenous residents
- **21 April 2004.** Steven Heim and Pablo Tseres, President of FICSH (Federación Interprovincial de Centros Shuars) and coordinator of Comité Interfederacional, present at shareholder meeting
- **June/July 2004.** Boston Common begins fact-finding. Writes to presidents of FIPSE (Federación Independiente del Pueblo Shuar del Ecuador), FICSH, FINAE (now NAE [Nacionalidad Achuar de Ecuador]) and Sarayaku asking for clarification of indigenous positions. Burlington's website claimed 95% of indigenous peoples of Blocks 23 and 24 were receptive to petroleum activities

- **Summer 2004.** Boston Common sends email to Burlington requesting meeting with shareholders and asking for details regarding its consultation process with indigenous communities in Ecuador. Burlington does not respond to questions
- **October 2004.** Boston Common arranges shareholder meeting with Burlington in New York City, hosted by New York State Common Retirement Fund, one of the largest US public pension funds and a large Burlington investor
- **November 2004.** Boston Common and several others file second resolution, asking company to publish sustainability report and noting in resolution concerns regarding indigenous rights
- **December 2004.** Burlington challenges second resolution on grounds that (a) resolution contained false and misleading statements, and (b) the company had plans to publish sustainability report in 2005
- **February 2005.** SEC rules in favour of Boston Common. Burlington agrees to detail its consultation process and its indigenous policy. Second resolution withdrawn
- **March 2005.** Shareholder delegation to Ecuador
- **April 2005.** Burlington meets several shareholders in New York City after FIPSE and FICSH presidents agree secretly to negotiate with Burlington and government. Heim challenges Burlington on several issues, including territory under jurisdiction of Sarayaku, FINAE, FICSH; 2002 Civic Anti-Corruption Commission ruling to revoke Burlington's oil concessions in Blocks 23 and 24; 2002 lawsuit by FIPSE, FICSH and FINAE to block Burlington from negotiations with individual communities; and assembly authorisation to presidents of FICSH and FIPSE for negotiations. Heim presents at Burlington's annual general meeting; chairman of board of directors confirms that Burlington will respect the governance processes of the indigenous communities
- **November 2005.** Shareholders file two different resolutions with Burlington: indigenous rights report and sustainability report
- **December 2005.** ConocoPhillips announces acquisition of Burlington. NYC Pension Funds is lead filer of resolution requesting a sustainability report. Withdraws resolution after Burlington agrees to provide information to ConocoPhillips for ConocoPhillips' next sustainability report
- **May 2006.** Indigenous leader presents at ConocoPhillips' annual meeting
- **December 2006.** Indigenous leaders from Peru travel to ConocoPhillips' Houston headquarters after no response to letters and phone calls requesting a meeting. Successful in meeting briefly with vice president, health, safety and environment
- **January 2007.** ConocoPhillips does not challenge resolution with SEC
- **May 2007.** ConocoPhillips' annual meeting held in Houston. First shareholder vote on the issue after three years of shareholder engagement

ChevronTexaco

In 1964, Texaco was granted rights to the oil concession in the northern Amazonian region surrounding Lago Agrio. In its 28 years of operation, the company extracted roughly 1.5 billion barrels of crude, netting an estimated $30 billion in profits (Kimerling 2006; Amazon Defense Coalition 2006; Maass 2007). Throughout the 1970s, Nueva Loja, the town located closest to Texaco's first commercial oil field, rapidly developed into a 'boom town', as settlers poured into the Amazon on newly developed roads that opened up previously inaccessible areas of the Oriente (Kimerling 2006). Lago Agrio (or 'Sour Lake', the name of Texaco's founding town in Texas) as the city is nicknamed, now has a population of over 35,000 (Hearn 2006; Maass 2007).

Texaco's operations spread throughout the northern Amazonian provinces of Orellana and Sucumbíos and were conducted from the initiation of its contract in 1964 until 6 June 1992, when the company's contract expired and all of Texaco operations were turned over to Petroecuador (Kimerling 2006). As noted above, Texaco operated under very little environmental regulation by the state, which expected the company to take responsibility for implementing proper drilling technology. According to Judith Kimerling, author of a study that first revealed the detrimental environmental and health impacts of Texaco's operation, Texaco operated 'in [an] environmental law vacuum . . . Texaco set its own standards and policed itself' (Kimerling 2006: 436). Texaco reportedly constructed 350 wells and 1,000 open-air waste pits, allegedly discharging 18 billion gallons of toxic waste-water into the Amazon's surface and sub-surface waters and soils in the duration of its period in Ecuador (Amazon Defense Coalition 2006). In 2001, Chevron and Texaco merged to form ChevronTexaco.

Network activities in the northern provinces of Amazon are centred primarily around the landmark class-action lawsuit *Aguinda* v. *Texaco, Inc.*, a case originally filed against Texaco in a federal court in White Plains, NY, by 74 plaintiffs representing 30,000 indigenous and *colono* (non-indigenous settler) inhabitants of Amazonian areas affected by Texaco's operations (Kimerling 2006). The lawsuit, which was grounded on common law claims such as negligence, public and private nuisance, and strict liability, sought environmental remediation and compensation for the injuries incurred by Texaco's alleged sub-standard practices (Kimerling 2006). Although plaintiffs originally sought to pursue the case in the US court system, the case was dismissed from the New York court by Judge Jed Rakoff, who argued that a US court was not the proper forum for the case to be argued. In October 2003, the case reopened in Lago Agrio's superior court, filed by 46 of the original Aguinda plaintiffs and two additional plaintiffs (Kimerling 2006). In 1994 and 1995, Texaco signed several agreements with Ecuador, agreeing to conduct limited environmental remedial work and make payments for socioeconomic compensation projects. The remediation implemented, however, is alleged to be unsubstantial; a 2003 study conducted by Global Environmental Operations estimated a complete clean-up to cost roughly $6.14 billion (Amazon Defense Coalition 2006). On 19 March 2007, the Ecuador judge hearing the class-action environmental lawsuit ordered that the final phase of the trial—including a damage assessment—be completed in 120 days (Amazon Watch and Frente de la Defensa de la Amazonia 2007). The case is currently still under review; most recently, a court-appointed expert in Ecuador recommended an $8–16 billion settlement. The chronology of this case is provided in Box 10.2.

Box 10.2 Time-line of shareholder activity, Chevron

- **Fall 2002.** Trillium Asset Management and Amnesty International co-file a shareholder resolution with Chevron requesting that the company adopt a human rights policy
- **Spring 2003.** Company agrees to develop a corporate-wide human rights policy and resolution is withdrawn
- **November 2003.** Trillium files second resolution with Chevron requesting that the company 'prepare a report on new initiatives instituted by management to address the specific health and environmental concerns of villagers living near unremediated waste pits and other sources of oil-related contamination in the area where Texaco operated in Ecuador'
- **March 2004.** Several shareholders of Chevron participate in a delegation to Texaco-affected areas of Ecuador's northern Amazon
- **April 2004.** Chevron annual shareholder meeting. Attended by two Ecuadorian representatives from Texaco-affected areas and human rights activist Bianca Jagger. Voting results: 9.5%
- **Fall 2004.** Trillium files third resolution with Chevron with the same request as the previous year. Co-filers include New York City State Common Retirement Fund, Amnesty International, and the Sisters of Mercy (Burlingame, CA)
- **April 2005.** Chevron annual shareholder meeting. Attended by two Ecuadorian representatives from Texaco-affected areas. CEO David Reilly turns off microphone during presentation by healthcare worker from a northern Amazon community. Voting results: 9.2%
- **Fall 2005.** Trillium files fourth resolution requesting a report on Chevron's '(a) annual expenditures by category for each year from 1993 to 2005, for attorneys' fees, expert fees, lobbying, and public relations/media expenses, relating in any way to the health and environmental consequences of hydrocarbon exposures and Chevron's remediation of Texaco drilling sites in Ecuador and (b) expenditures on the remediation of the Ecuador sites'. Co-filers include Boston Common Asset Management, New York State Common Retirement Fund, and other filers. A second resolution, filed by Society of Jesus, Wisconsin Province and 21 co-filers affiliated with the Interfaith Center on Corporate Responsibility (ICCR), called on the company to develop a comprehensive human rights policy by October 2006. A third resolution related to Chevron's environmental standards was also filed by ICCR members
- **April 2006.** Chevron annual shareholder meeting. Attended by two Ecuadorian representatives from Texaco-affected areas. Voting results: 8.4%
- **Fall 2006.** Trillium files fifth resolution with Chevron, along with New York City Employees Retirement System, Amnesty International and Catholic Healthcare partners, calling on the company to prepare a report 'on the policies and procedures that guide Chevron's assessment of the adequacy of host country laws and

regulations with respect to their adequacy to protect human health, the environment and our company's reputation'

- **April 2007.** Chevron annual shareholder meeting. Voting results: 8.6%
- **Fall 2007.** New York City Employees Retirement System, along with Trillium and the School Sisters of Notre Dame Cooperative Investment Fund, re-files resolution with Chevron
- **May 2008.** Chevron annual shareholder meeting. Voting results: 8.3%

Comparative analysis

A network comprising local actors, domestic NGOs, international NGOs, and shareholders formed to employ a 'corporate boomerang mechanism' targeted at Burlington and Chevron, using shareholder advocacy to put pressure directly on the company to address the environmental and social impacts of Texaco's legacy. Local nodes of the network include five indigenous groups of the northern Amazon as well as a non-indigenous organisation formed to coordinate local legal strategies related to the *Aguinda* lawsuit. Amazon Watch's role in the Chevron STAN is nearly identical to its position in the Burlington Resources network: the organisation acts as the primary bridge between shareholders and national/local actors in Ecuador. Trillium Asset Management, a social investment firm founded in 1982 and located in four offices across the USA, first filed a resolution with Chevron related to the Ecuador issue in 2004 (as of 2005, the firm held 3,000 shares of common stock in Chevron). Institutional investors such as the New York City Employees' Retirement System, continue to file resolutions each year with the company. Several shareholders participated in a delegation trip to Ecuador, one year prior to the Burlington 'fact-finding mission'.

Unlike the Burlington STAN, however, Chevron network members found the company largely unresponsive to the shareholder tactics. Shareholders were greatly frustrated at the outcomes of shareholder activity with the company, explaining that shareholder advocacy has produced little change in the company's position regarding the environmental and health concerns of communities living in Texaco-affected regions of the Amazon.

Despite their similarities in structure and composition, the outcomes of the two STANs of Ecuador's Amazon rainforest differ widely. The relative success of the Burlington STAN in comparison to the Chevron network can be explained by significant differences in both the density/strength of the network and the vulnerabilities of the two targeted companies. The regional context of the Burlington network—namely, the lack of oil development in the southern provinces of the Amazon—has afforded this network certain advantages in maintaining a cohesive identity and strong communication channels among domestic network members. Lack of oil development in the southern Amazon has allowed for the preservation of traditional indigenous advocacy culture, in turn promoting a more cohesive local-level identity among first-level network participants. In addition, traditional indigenous governance structures form the basis of local representation in the southern Amazon, whereas non-indigenous representation of indigenous interests has caused tensions among the northern network's local nodes. Dynamics between indigenous groups and domestic NGOs also partly explain differences in the outcomes of the two networks: the goals of principal NGOs in the southern network were more closely aligned with their local-level partners than those of the north, weakening intra-level relations between network members.

In addition, owing to the different timings of the arrival of oil conflict in the different regions, network members of the Burlington case study have had the opportunity to learn from the experiences of oil-affected northern Amazonian regions, whereas domestic actors of the northern network have had more limited access to information regarding the impacts of oil development. Analysis of different company characteristics has also revealed several notable explanations of Burlington's relative receptiveness to shareholders in comparison with Chevron. Once again, the different network contexts—particularly the fact that shareholder demands of Chevron concern past actions of a subsidiary company, whereas pressure from Burlington's shareholders are focused on the company's current operations—has moulded company reactions to shareholders. Chevron has avoided reputational damage by distancing itself from the former Texaco, and, because it no longer is invested in Ecuador, the company has less incentive to accommodate shareholder and community demands. In addition, the different corporate cultures of the company, largely shaped by Burlington's relative inexperience of shareholder advocacy, also helps to explain the different outcomes experienced by shareholder advocates. Lastly, the fact that the Chevron STAN concerns pending litigation has also afforded the company certain advantages in terms of leveraging power.

Broader applicability of findings

Ecuador is only one of many countries in which natural resource exploitation has affected vulnerable populations. Petroleum operations have come under particular scrutiny, as evidenced in the World Bank Extractive Industries Review which resulted in a recommendation for the lending institution to phase out investments in oil production by 2008 (World Bank 2003). The review highlighted many of the problems common to countries reliant on the extractive industries to finance economic growth, including a lack of government capacity to transparently manage projects and revenues, a skewed distribution of benefits in which local communities are forced to bear an inordinate cost, and legal systems that do not ensure resource and land tenure rights for indigenous peoples. In countries such as Colombia and Nigeria, these problems have led to violent conflict. Despite many fierce clashes between indigenous communities and the state, the situation in Ecuador's Amazon region has not yet disintegrated into widespread violence. Oil extraction may be acceptable to some communities with appropriate safeguards, while for others it may be an untenable proposition because of value-based concerns.

However, the corporate boomerang model suggests that modern companies, with appropriate regulatory mechanisms, can make oil exploration socially responsible by internalising many of the otherwise ignored economic 'externalities' of business. Shareholder participation devolves power in a constructive way for corporations to allow for self-critique and a longer-term planning horizon for projects to be considered within the business model for corporate growth and development. While such efforts may seem to be a hindrance for companies in the short term, they are likely to mitigate conflict and lead to a more robust strategy for corporate behaviour. Thus, the following lessons drawn from the Ecuadorian case studies are particularly useful to other areas where resource extraction is happening on indigenous territory.

Preventative measures are more successful than reactionary strategies

As evidenced in the case of Burlington Resources, indigenous groups hold an advantage when resource extraction has not yet begun. Corporations are more easily influenced when little investment has been made in a particular drilling block, in contrast to the case of Texaco where the damage had already been done and the company had no financial incentive to remedy it. The current ideology of 'participation' often results in indigenous communities being forced to respond to a process that has already been set in motion instead of playing a role in the construction of overall development planning.

Best practices exist that lend legitimacy to the demands of indigenous communities who want to ensure resource extraction occurs only after their input and consent. The World Bank Extractive Industries Review, for example, produced a set of recommendations designed to make certain that the involvement of lending institutions in extractive industries projects is compatible with the mission to alleviate poverty and contribute to sustainable development. These recommendations include the promotion of economic diversification and building government capacity to handle social and environmental challenges and manage revenues responsibly. There are also provisions for free, prior and informed consent on the part of affected communities and a suggestion to require integrated environmental and social impact assessments. The review argues for the establishment of 'no-go' zones for oil, gas and mining projects, particularly in protected areas and biological hotspots. It includes the important idea that the International Bank for Reconstruction and Development and the International Development Association work with governments to 'clarify and strengthen' the legal framework for resource and tenure rights of indigenous peoples and recommends a 'rebalancing of institutional priorities' of the entire World Bank Group if extractive industries projects are to proceed. While many of these recommendations have been deemed too extreme by many in development circles, they have made some impact on governments in negotiating contracts.

The petroleum industry has also developed operating guidelines for extraction projects. The Oil Industry International Exploration and Production Forum, now known as the International Association of Oil and Gas Producers, published several reports including the *Oil Industry Operating Guidelines for Tropical Rainforests* (Exploration and Production Forum 1991) and 'Principles for Impact Assessment: The Environmental and Social Dimension' (International Association of Oil and Gas Producers 1997). These reports recognise that local governments do not always work in the best interest of affected peoples and that the company has a responsibility to ensure appropriate, 'two-way' consultation. They also note that environmental monitoring and impact assessment and mitigation should be ongoing processes that last through the life of the project.

While these best practices and recommendations are a start, they remain as voluntary provisions that companies and lending institutions can choose to adopt. A monitoring mechanism needs to be put in place that ensures these practices are actually being followed in cases where companies claim they are being employed. Environmental and social justice groups such as the Rainforest Action Network have pointed out that these 'best practices' often serve to simply justify the presence of petroleum companies in sensitive ecosystems (RAN 1998). There is often no consideration of the cumulative impacts of several projects in one area, as can be seen in the Amazon basin. Finally, while companies have become adept at developing and utilising technology that causes less environmental impact, the guidelines for assessing and mitigating social and cultural impacts remain vague and ineffective.

Cohesive indigenous networks are more effective than fragmented groups

The communities organising to prevent Burlington Resources from drilling have been more effective in part because they maintain open lines of communication with each other and the higher levels of their network. They rely on traditional indigenous knowledge and governance structures to ensure that their interests are represented in all phases of resource extraction. Indigenous leaders are aware of the full range of impacts petroleum extraction can have on local populations and realise that they have legal rights that their government has not guaranteed. They have developed proposals that take the interests of other stakeholders into account, and have begun to create projects that contribute to the economic growth of their communities while maintaining cultural and environmental integrity. They have recognised that alternative development schemes offer a sense of empowerment and contribute to the diversification of the Ecuadorian economy, something that is greatly needed if the country is to survive in the long term. The missing link evident in the Chevron case study is a lack of focus on building and strengthening the relationship between different groups in the northern Amazon. Leaders of the formally recognised indigenous organisations should be responsible for organising their respective communities in order to reach consensus on overall strategy and messaging before they can effectively partner with international groups. In order to prevent divisiveness within a network, a formal process in which each party plays a specified role needs to be developed. As can be seen in the Chevron example, local communities continue to be conceived of as 'just another stakeholder' when in reality they have a much larger stake in the outcome of extractive projects than multinational petroleum companies, the Ecuadorian bureaucracy and international environmental organisations.

Despite a clear lack of negotiating power in many areas, indigenous peoples in Ecuador and other countries wrestling with conflict over resource extraction have created an international movement with some influence over government and corporations. Because of the work of this advocacy movement, the standards of behaviour for petroleum corporations have improved since Texaco struck oil in Ecuador in 1967. A university professor in Quito explained the importance that community pressure has had in holding corporations accountable for their actions in the Amazon in 2006:

> I don't think any enterprise could actually act, let's say, publicly, like Texaco 30 years ago or 20 years ago. But this does not result from some kind of a philanthropy or sudden change to generosity or to social conscience from the market. This results from the pressure of civil society and from the organisation of these communities and their ability to demand for the respect of their rights.[3]

As T. Macdonald (2004) points out, indigenous peoples in Latin America are 'now a political force, not simply a category of marginal victims'. Conflicts and fractures have formed within the indigenous movement as they attempt to harness this force and define an agenda that represents the diversity of peoples it includes (Ali 2000). Yet real gains have been made in Ecuador towards creating the multi-ethnic and multicultural state espoused in the 1998 constitution. Petroleum corporations and the government have long held the other points in the triangle that comprises Ecuadorian society, and it is only by recognising and strengthening their negotiating power that indigenous communities and the rest of civil society can play an equal role in the country's future.

References

Acción Ecológica (2006) *Atlas Amazónico del Ecuador* (Quito, Ecuador: Acción Ecológica, CONAIE).

Ali, S.H. (2000) 'Shades of Green: NGO Coalitions, Mining Companies, and the Pursuit of Negotiating Power', in J Bendell (ed.), *Terms for Endearment: Business, NGOs, and Sustainable Development* (Sheffield, UK: Greenleaf Publishing): 79-95.

Amazon Defense Coalition (2006) 'Rainforest Catastrophe: Chevron's Fraud and Deceit in Ecuador'; www.texacotoxico.com, accessed 20 August 2008.

Amazon Watch and Frente de la Defensa de la Amazonia (2007) 'Texaco in Ecuador'; www.chevrontoxico.com, accessed 19 March 2007.

Arnold, D. (1996) *The Problem of Nature: Environment and Culture in Historical Perspective* (Oxford, UK: Blackwell).

Cronon, W. (1983) *Changes in the Land: Indians, Colonists, and the Ecology of New England* (New York: Hill & Wang).

Exploration and Production Forum (1991) *Oil Industry Operating Guidelines for Tropical Rainforests* (Report No. 2.49/170; London: Exploration and Production Forum).

Gedicks, A. (2001) *Resource Rebels: Native Challenges to Mining and Oil Corporations* (Cambridge, MA: South End Press).

Gerlach, A. (2003) *Indians, Oil, and Politics: A Recent History of Ecuador* (Wilmington, DE: Scholarly Resources Inc.).

Hearn, K. (2006) 'Big Oil On Trial', *E Magazine*, March/April 2006: 36-39.

Hetherington, J.A.C. (1969) 'Fact and Legal Theory: Shareholders, Managers, and Corporate Social Responsibility', *Stanford Law Review* 21.2: 248-92.

International Association of Oil and Gas Producers (1997) *Principles for Impact Assessment: The Environmental and Social Dimension* (Report No. 2.74/265; London: International Association of Oil and Gas Producers; www.commdev.org/content/document/detail/709, accessed 23 October 2008).

Investor Responsibility Research Center (1994) *The SEC and Social Policy Shareholder Resolutions in the 1990s* (Washington, DC: Investor Responsibility Research Center).

3 Interview with author, January 2007.

Kane, J. (1995) *Savages* (New York: Alfred A. Knopf).

Keck, M.E., and K. Sikkink (1998) *Activists Beyond Borders* (Ithaca, NY: Cornell University Press).

Kimerling, J. (1991) *Amazon Crude* (Washington, DC: Natural Resources Defense Council).

—— (2006) 'Indigenous Peoples and the Oil Frontier in Amazonia: The Case of Ecuador, ChevronTexaco, and Aguinda v. Texaco', *NYU Journal of International Law and Politics* 38: 413-664.

Korovkin, T. (2002) 'In Search of Dialogue? Oil Companies and Indigenous Peoples of the Ecuadorian Amazon', *Canadian Journal of Development Studies* 23.4: 633-63.

Maass, P. (2007) 'Slick', *Outside*, March 2007: 100-19.

Macdonald, T. (2004) 'Ecuador's Past Offers Direction for the Future', *Cultural Survival Quarterly* 28.3: 39-42.

Martin, P.L. (2003) *The Globalization of Contentious Politics: The Amazonian Indigenous Rights Movement* (New York: Routledge).

Newell, P. (2000) 'Globalisation and the New Politics of Sustainable Development', in J. Bendell (ed.), *Terms for Endearment: Business, NGOs, and Sustainable Development* (Sheffield, UK: Greenleaf Publishing): 31-39.

Pawson, E., and G. Cant (1992) 'Land Rights in Historical and Contemporary Context', *Applied Geography* 12: 95-108.

RAN (Rainforest Action Network) (1999) 'Ecuador Declares Two Sensitive Rainforest Parks Off Limits to Oil Drilling'; www.ran.org/home/victories.html, accessed 20 August 2008.

Sabin, P. (1998) 'Searching for Middle Ground: Native Communities and Oil Extraction in the Northern and Central Ecuadorian Amazon, 1967–1993', *Environmental History* 3.2: 144-68.

Sawyer, S. (2004) *Crude Chronicles: Indigenous Politics, Multinational Oil, and Neoliberalism in Ecuador* (Durham, NC: Duke University Press).

World Bank (2003) *Striking a Better Balance: The Final Report of the Extractive Industries Review* (Washington, DC: World Bank).

11

Environmental justice concerns with transnational mining operations

Exploring the limitations of post-crisis community dialogues in Peru

Isabelle Anguelovski

Department of Urban Studies and Planning, Massachusetts Institute of Technology, USA

In South America, questions of environmental injustice and inequity are increasingly linked to the impact of extractive industries in low-income and minority communities, in particular indigenous communities. In the Andes and Amazon regions, indigenous peoples are disproportionately affected by the destructive impacts of oil, gas and mining operations on their health and their subsistence food sources, leading to the disruption of their lifestyle and eating patterns (Christian Aid 2005; Wheatley and Wheatley 2000). For instance, studies in the Bolivian Amazon have shown that the mercury released by gold mines in the upper Beni River basin has dramatic health impacts on indigenous people who have a regular fish diet,[1] especially young children being breast-fed, as compared to non-indigenous residents who have different dietary habits (Maurice-Bourgoin *et al.* 2000).

As a result, mining operations have often been associated with conflicts and with resistance from host communities, as a response to loss of land and natural resources, forced relocation, environmental impact and lack of respect for human rights and community

1 The mercury concentrations found in piscivorous fish from the Beni River are of great concern since they can exceed by four times the safety limits established by the World Health Organisation.

consultation. The delicate position of mining companies operating in Latin America is mainly due to four contributing factors: the legacy of conflict in the region, struggles over the distribution of mining benefits, legislative inconsistencies between reform processes, and a perceived lack of legitimacy in laws and regulations on which foreign companies rely (Thompson and Joyce 2002). Consequently, communities such as Yanacocha, Choropampa or La Oraya in Peru regularly make the headlines of the national press for showing resistance against transnational mining operations on their land.[2]

Commonly referred to as corporate social responsibility (CSR), companies have resorted to a range of responses to answer communities' concerns, mitigate the large impacts of their mining and oil operations and benefit societies in ways that go above and beyond what companies are legally required to do (Vogel 2005). At present, one of the most promising and sought-after approaches has been the creation of multi-party deliberative spaces engaging community members, NGOs, state representatives and corporate executives in dialogue processes. This evolution is linked to multiple interrelated factors inside and outside corporations, such as:

- The pressures exerted by national or international NGOs and shareholders
- The recognition by company executives of the need to gain a social licence to operate and their increased commitment to social justice for local host communities
- New decrees or executive orders obliging companies to improve their practices and offer social and economic benefits to communities
- The political ascendancy of the indigenous movement and the organised resistance of indigenous and *mestizo* (mixed-ancestry) communities near the mining operations (Ballard and Banks 2003; Imbun 2007)

Mining companies, in particular, have come under great public scrutiny in recent years for their controversial environmental and social practices in the developing world, and have thus initiated a variety of corporate social responsibility strategies to build better relations with local communities to minimise conflicts (Humphreys 2000; Kapelus 2002). In Peru, several companies, such as BHP Billiton at the Tintaya copper mine and Newmont at the Yanacocha gold mine, have agreed to participate in multi-party deliberative spaces in an attempt to improve their practices and build trust with local people. However, such processes have been challenged by continued community protests and some have subsequently been abandoned.

This chapter attempts to offer a comparative view on two such dialogues, using the theoretical framework of environmental justice and data from semi-structured interviews I conducted in 2006 and 2007 with environmental activists, NGOs, indigenous representatives and company executives in the USA and Peru, and reports and original documents published on the Yanacocha and Tintaya dialogue tables. I attempt to show that transnational mining companies operations in Peru have violated the three environmental justice principles defined by Schlosberg (2003): (a) equitable distribution, (b) recognition of the views of local populations, and (c) community members' participation. Even in situations in which companies commit to improving the CSR standards

2 See www.oxfamamerica.org/whatwedo/where_we_work/south_america, accessed 15 August 2008.

these principles are often ignored. A violation of these principles creates negative precedents and might explain the resistance of communities to extractive industries on their land, their mistrust of corporations and the continued disappointments of dialogue processes.

Lack of equitable distribution of environmental burden and risk

Baseline extraction activities and contamination and expropriation issues

In Peru, indigenous communities have historically been more exposed than other communities and geographic areas to environmentally hazardous activities on their territory, to contamination of their forests, land and rivers, and to land expropriation due to extraction activities. Today, 39% of mining companies in Peru operate on indigenous lands and 3,200 of Peru's 5,660 indigenous communities are affected by mining activities (Oxfam America 2006). The reality is that the large majority of native Andean Aymara and Quechua people live in the Andes, which are very rich in ore deposits and have seen important mining activity develop along the mountain range (Cardozo and Cedillo 1990). Today, more than 40 different metallic and non metallic resources are exploited in the country (see Table 11.1). Since 1992, mining claims in Peru have increased from 4 million to 22 million ha, mostly in Andean departments, where large-scale mining operations have been privatised and where new mineral resources are likely to be exploited in the next few years (see Fig. 11.1).

TABLE 11.1 Peruvian mineral production and exports

Source: adapted from Bury 2005

Mineral	Production 2000 (metric tons)	World ranking (reserves)	Percentage of world production (2000)	Percentage increase (1990–1999)	Export value (US$ million) 2000
Copper	567,751	6	4.5	65	931
Gold	135	8	5.8	534	1,145
Lead	252,257	4	9.1	30	190
Silver	2,353	2	13.0	15	180
Zinc	773,757	4	10.5	67	496

To attract foreign investment and boost the country's economic growth, President Fujimori revised land-tenure rights in 1996 through the new National Mining Cadastre Law (Law 26615), which has eliminated many previous mining-claim procedures and

FIGURE 11.1 Map of national mining claims distribution by department, 2000

Source: Peruvian Ministry of Energy and Mines

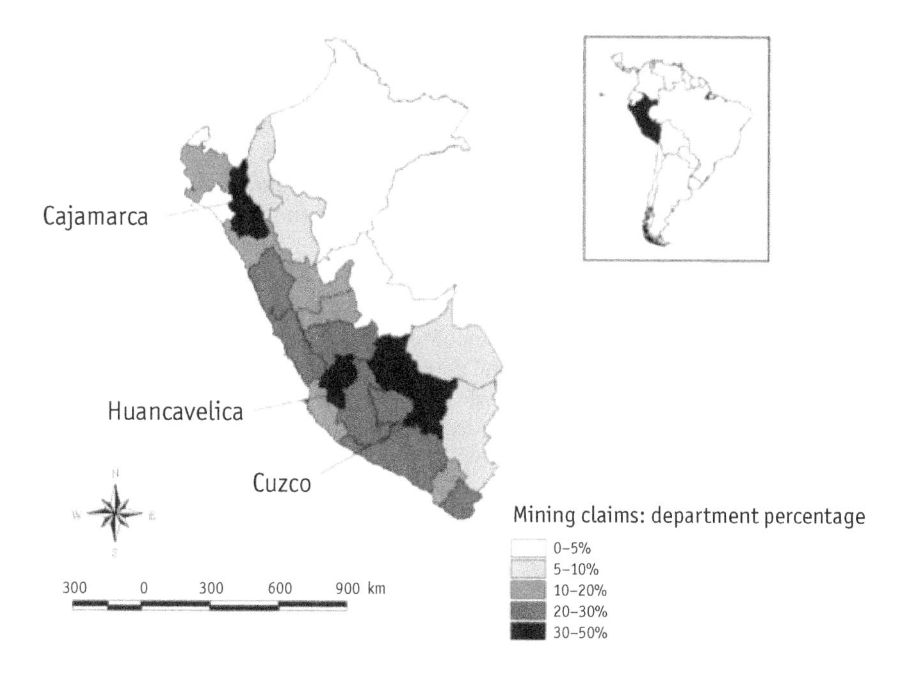

centralised and unified concessions under a new geographic reference system (Ministry of Energy and Mines 2000). This law has effectively granted national and foreign mining firms exclusive control of all the necessary land resources to implement their extraction operations, and made indigenous territory more vulnerable to mining projects. Since then, the development and expansion of copper, gold and zinc mines have led to skyrocketing export earnings, especially during the presidency of Alejandro Toledo, from $7 billion dollars in 2001 to $17 billion in 2005, and to the strongest economic growth rates (5% per year from 2002 to 2006) since 1962 (McClintock 2006).

However, these extraction projects place host indigenous communities in delicate positions, because their traditional lifestyle is more affected by mining activities than urban areas or rural communities living in the plains, and because the by-products of mines are contaminating rivers and large areas of land that indigenous people use daily for their food resources. Several studies, such as that conducted in northern Peru by Bech and his colleagues, have shown that plants and soil around copper mines are contaminated by heavy metals such as mercury and arsenic (Bech *et al.* 1997). Schlosberg (2003) argues that the first principle of environmental injustice is that low-income communities and racial and ethnic minorities tend to face more environmental risks than richer and white communities, violating the idea of justice as the appropriate division of social advantages (Rawls 1971). The following two case studies of the Tintaya copper mine and the Yanacocha gold mine illustrate this principle in operation and exemplify the environmental burden suffered by indigenous communities owing to mining extraction in Peru.

The Tintaya copper mine is located in the southern Andes of Peru in the Espinar province. It was first established in 1985 as a state enterprise following an expropriation from 125 indigenous farmer families in the community of Marquiri. In return for the land, the Peruvian government offered farmers about $3 for 2.45 acres and the promise of mining jobs (Echave *et al.* 2005: 63). However, these promises did not avert widespread poverty in the region. After its privatisation in 1994, Australia-based BHP Billiton bought the Tintaya mine in 1996 and rapidly expanded it through mining and milling process improvements (Oxfam Community Aid Abroad 2003: 32). In 2001, BHP built a copper oxide plant and a new tailings dam in the community of Huinipampa containing waste ore left over from the refining process, which is one of the most pressing concerns of communities around Tintaya.

In 2000, to contest land losses, extensive mining-related environmental degradation and allegations of human rights abuses, a coalition of five affected communities forged an alliance with national and international NGOs. They claimed that BHP had 'illegally and unethically' purchased an additional 2,386 ha of land from them in the 1990s, since many inhabitants did not have full knowledge of their legal rights and of the true value of their land when they agreed to sell it to BHP Billiton (Echave *et al.* 2005: 9). In this particular case, environmental injustices arise from the lack of access to transparent information on land tenure issues and land prices among indigenous communities and, following the expropriation of these lands, the loss of livelihoods from agriculture. Other concerns were raised regarding waste-water from the mine leaking into local rivers, contaminating pasture and making the water unsafe for humans and animals (Oxfam Community Aid Abroad 2001). This pollution has threatened the rights of indigenous peoples to health and to secure livelihoods.

Despite the pressure and public denunciation initiated by the Peruvian NGOs Cooper-Acción and CONACAMI (National Coordinating Body of Communities Affected by Mining in Peru) and by Oxfam Australia's Mining Ombudsman in 2000 and 2001, BHP Billiton's corporate headquarters or Peruvian authorities did not take subsequent actions to address the problems. Environmental injustices are linked to the lack of reaction of the state to obvious violations of mining standards and human rights and to the general lack of attention by public authorities and international corporations to the threats affecting the sustainable development of indigenous communities.[3] Injustices also stem from the fact that the historic exclusion of indigenous peoples has left native tribes with the feeling that the government does not protect them. In fact, it was only in December 2001 that the Mining Ombudsman and BHP Billiton's executives agreed to meet in Peru to start addressing community concerns through a dialogue process (Kasturi *et al.* 2006).

The second case studied is the Yanacocha open-pit gold mine in the Northern Cajamarca Province. This project is one of the most influential and largest transnational mining operations in Peru. Minera Yanacocha has been operating since 1993 as a partnership between Newmont Gold Corporation (US-based, 51.3%), Buenaventura (Peru-based, 43%), and the International Finance Corporation (IFC) of the World Bank Group (5%).

3 The definition of sustainable development accepted here is the Brundtland Commission's definition: a 'development that meets the needs of the present without compromising the ability of future generations to meet their own needs' (see WCED 1987).

Today, gold represents the largest export of the country, partly due to the rapid expansion of Yanacocha.[4] Since operations started, communities living in the surroundings of the mine and those displaced by it have expressed serious grievances about the effect of mining activities on the quality of their water sources, which apparently changed in colour and smell, lost fish and frog stocks, and no longer provide potable water (Project Underground 2000). The Yanacocha mine's cyanide heap-leaching operations led to a dramatic shift of land-cover patterns and a widespread alteration of environmental processes in the region. Across more than 10,000 ha, the mine has altered watercourses, shifted millions of tons of earth, decreased community access to irrigation water supplies and to land resources for livestock and agriculture, and lowered water and land quality (Bury 2005). Between 1991 and 1999, Minera Yanacocha violated the standards of the Ministry of Energy related to watershed and air contamination by as much as 93% for solids in suspension (Bury 2004).[5] In short, as in Tintaya, environmental injustices arise because community livelihoods and culture have been deeply affected by mining operations.

Communities displaced by the mine complain that promises of return to their land after the closure of the mine have not been met, despite the agreement they reached with the mine when they sold their land for cash, which included a right to return (Project Underground 2000). Walzer (1983) considers that the criteria for distributive justice will differ according to the different values people assign to things; being 'just' will mean helping different people have access to things that they value more because of their identity and choices. In Yanacocha, resettled families report that the low quality and value of their new land has not provided them with enough productivity to feed their whole household, or with enough chances to sell it at decent prices. Faced with this hardship, families had to move to urban areas such as Cajamarca, where they continue to face poverty because of lack of qualifications for the types of job offered there, and where they have lost the social networks from their former communities. Here, environmental injustices appear because local peoples do not have the capacity to live according to their own choices, forced to relocate to a setting incompatible with their traditional way of life.

In June 2000, a major environmental and health incident occurred to confirm the concerns of the communities about contamination when, in June 2000, 150 kg (330 lb) of mercury were spilled from a mine contractor's truck on a road in the town of Choropampa near Yanacocha. It is widely recognised that high exposure to mercury leads to a higher prevalence of stillbirths and deformities in children, neurological disorders, as well as depletion of fish when in contact with water resources.[6] However, according to a report published after the spill, 'local inhabitants collected most of the spilled mercury, either to play with (in the case of children) or to extract gold and other metals they believed were associated with the mercury' (Vega *et al.* 2001). Between 200 and 300 peo-

4 In 2001, Peru's gold was valued at $1.2 billion for a production of 4.3 million ounces (12,1903 kg) of gold (see Webber 2002).

5 For example, average dissolved and suspended solids, as well as concentrations of copper, iron, zinc, manganese and sulphate, have exceeded many times the mine's standards on average and several times at extremely high concentrations.

6 Cyanide and mercury are both integral in processing gold ore, used to separate gold from crushed rock, but they are highly toxic for humans and the environment around them. See Global Response 2005.

ple were subsequently hospitalised with mercury poisoning. The lack of access to environmental education and to transparent information prevented indigenous peoples from understanding the risks associated with exposure to toxic compounds.

Interviews with Newmont staff reveal that the company believed its response was exceptional from a technical perspective through the quick mobilisation of technical experts and remediation efforts. However, the IFC's independent social and environmental auditor, the Compliance Advisory Ombudsman (CAO), who conducted an evaluation of the spill, released a report stating that the mining company had not followed sufficient safety procedures in transporting the mercury. It also accused Newmont of depicting the incident inaccurately to the communities in terms of the potential severity of the mercury spill (Environmental Media Services 2001). Local staff members have recognised that doubts were raised within Newmont about the company's capacity to understand community perspectives related to the remediation and to include them in the post-spill phase. Claims of environmental injustices are linked to Newmont's failure to protect poor, disempowered and isolated communities from spills and to inform them about the danger of exposure to toxic metals. This confirms the results of other studies that have shown that land-use conflicts in mining areas between indigenous communities and corporations are directly caused by preventable environmental accidents (Hilson 2002).

Continued environmental concerns during conflict resolution processes

To resolve conflicts with local communities, Newmont and BHP Billiton have attempted to address several of the four main foundations of corporate social responsibility (CSR) in the extractive industries sectors:

- Human rights principles (including the right to a clean environment and the principle of indigenous free, prior and informed consent for extractive development on indigenous lands)
- Sustainability (including refraining from mining in protected and environmentally sensitive areas
- Economic efficiency
- A social licence to operate (Ali and O'Faircheallaigh 2007)

To respond to community concerns, the two companies have agreed to participate in multi-party dialogue tables (*mesas de diálogo*) in Tintaya and Yanacocha.

In Yanacocha, since communities believed they had not been fairly compensated by Newmont, the CAO decided in October 2001 to initiate a round-table process, the Mesa de Dialogo y Consenso (the Mesa), to hear the perspectives of community members on the effects of the spill, as well as suggestions for addressing immediate and long-term needs. However, during the Mesa, farmers reported to the CAO that their health was still impacted by the spill declaring:

> We don't have any confidence in the diagnosis of the other doctors. They told me in the hospital, contracted by the mine, that my children are suffering from malnutrition, but they never had malnutrition . . . My children, like other children here, suffer from shaking, dizziness and others have rashes. Before the spill they never had these problems (Rey-Sanchez 2002).

The details and sources of the medical problems are still in dispute and the issue of trust among the stakeholders continues to be a central concern because of several recent accidents around the mine. In January 2004, 26,500 litres (7,000 gallons) of petroleum were spilled by DAMARSH, the contractor of Mobil in the proximity of Choropampa, contaminating the River Jequepeque. In May 2002, 8,000 trout were killed at the Granja Porcón farm, 2,200 in August 2002, and more than 26,500 in October and November 2002 (FOEI *et al.* 2006). Lastly, during the Mesa meetings themselves, Newmont apparently intimidated citizens who filed charges against land expropriation and denied that there had been negative social and environmental impacts (FOEI *et al.* 2006).[7] Problematically, this position was imposed on the members of the Mesa (Friends of the Earth International [FOEI], Amigos de la Tierra and Grupo de Formación e Intervención para el Desarrollo Sostenible [GRUFIDES]). From these different events, it becomes clear that environmental injustices have not been redressed through official claims and dialogue processes and that contamination in Yanacocha has not been stopped to the community's satisfaction.

In Tintaya, the Dialogue Table participants agreed in 2002 to form four working commissions to investigate grievances of communities, formulate recommendations and implement changes dealing with: (1) loss of land, (2) environmental impacts, (3) human rights violations, and (4) sustainable development. However, despite the apparent goodwill of NGOs and corporations to build a sustainable future for Tintaya, the process was challenged twice by protests of local inhabitants. First, in May 2003, 1,000 Espinar inhabitants, including many from the Ccañipia River basin, stormed the Tintaya mine site, claiming that they had not been adequately consulted about the company's construction of a new tailings dam close to their lands, which might affect their farming practices and livestock. To respond to these concerns, BHP Billiton signed the Framework Agreement in September 2003, which also includes a development fund for the province.[8] In parallel, despite the signature of an agreement in December 2004 between BHP, NGOs and communities called the Tintaya Agreement, a second protest occurred in May 2005 when 2,000 people attacked BHP's facilities. According to interviews with the activists, protesters demanded a faster implementation of the 2003 Framework

7 In October 2005, the CAO published the results of water monitoring through April, May, June and July in canals providing water to various communities. The CAO detected the presence of heavy metals in the water that exceeded the allowed maximum limit by Peruvian standards. These data confirmed results found both in the Environmental Audit by INGETEC and in the study by Stratus Consulting. However, Yanacocha mine company, while being part of this initiative, denied the results, arguing that measurements were taken incorrectly by calling Class III waters Class I. In response, the CAO organised workshops and meetings, arguing that the waters in which the metals were found were used only for irrigation and that the population did not run any risks, contradicting its own findings.

8 Through this agreement, 3% of the mine's income (before interest and taxes) would go to the fund or, alternatively, when profits were low, a minimum of $1.5 million. The fund was to be managed through a Management Committee (Comité de Gestión) composed of local government officials and community organisations.

Agreement[9] and the 2004 Tintaya Agreement, a revision of certain clauses of the Framework Agreement related to job creation and development projects in communities, and closer environmental monitoring.

In both Tintaya and Yanacocha, communities have been concerned by the ongoing contamination and the lack of improvements of practices and clean-up of environmental damage, despite the creation of dialogue processes. The lack of progress in such ostensibly corrective gestures illustrates their failure to address distributive aspects of environmental injustices.

Lack of recognition of the cosmology and way of life of indigenous peoples

A second aspect of environmental injustice in mineral extraction is related to the special relations of indigenous peoples to their land, which public authorities and companies in Peru have failed to take into account when they developed mining operations. As Iris Young (1990) points out, injustices also stem from a lack of recognition of identity and difference. In Peru, indigenous peoples are not recognised for their particular distinctiveness and are not seen as deserving the same dignity as other members of the Peruvian society. Disrespect is shown when people or institutions violate the bodies of others, deny their rights and denigrate their way of life (Honneth 1992). In both Tintaya and Yanacocha, indigenous peoples have suffered from multiple forms of disrespect. These include expropriation of their land, the use of violence against local people by the mine workers and the denial by companies of prior and informed consent before the beginning of mining operations by BHP and Newmont, violating Article 15.2 of the ILO Convention 169 on indigenous peoples. The corporations have not respected native peoples' own cultural traditions related to the use of their land for their livelihoods and religious practices.

The core problem here is that, until recently, companies have not been obliged by the Peruvian government to improve their practices and compensate communities fairly. For example, in Yanacocha since 1992 the mine has initiated a land-acquisition programme and resorted to state-sanctioned expropriation procedures and forced land evictions, while compensating landholders at what the company determined to be the estimated market value of the land (Bury 2005). Schlosberg (2003) considers that this lack of recognition of people's values and traditions impairs them in achieving a positive understanding of themselves, which is usually acquired by inter-subjective means. Mis-recognition and lack of protection of people's vision for their own development are cultural and institutional forms of injustice and lead to the oppression of those particular groups.

9 In the past year the Espinar Framework Agreement's Comité de Gestión had struggled to identify good projects to fund. A variety of legal complications and a slow approval process linked to extensive government bureaucracy meant that the committee had yet to disburse the $1.9 million BHP Billiton had committed to the province for 2004.

Indigenous peoples' unique relation to land and territory

In Peru, many aspects of indigenous peoples' way of life and social relations are based on the land, its products and resources. However, mining extraction entails the risk of destroying the ties between indigenous peoples and their land. In the Andean region of South America, the Aymara notion of *chacha-warmi* (or *hari-warmi* in Quechua) is an essential part of the cosmology of Aymara and Quechua people. According to this cosmology, a human being (*jaqi*) is complete and an integral part of the community when he or she has formed a couple (man–woman = *chacha-warmi*).[10] In a political vision, this means that it is not only about 'man' and 'woman', it is also about the fact that everyone together needs to be part of constructing the future of the indigenous people in unity with nature. For indigenous people, 'everything is *chacha-warmi'* (Aymara), 'everything is *hari-warmi*' (Quechua). '*Pitaj ima niwasun kariwarmita, warmiltarita*' means that everything in nature has a masculine or feminine connotation with natural elements being complementary of one another, such as stones, mountains and trees.[11] Indigenous peoples consider that the spirituality and rituals they practise are also tied to the land they inhabit.[12] For these reasons, contamination from mining extraction will separate and isolate human beings and threaten the special relation of indigenous peoples with the natural elements. According to native tribes, mining will destroy communities, with no more family unity and life continuity.[13]

My interviews with the NGO CONACAMI show that indigenous communities are not systematically opposed to mining activities. But their traditional way of life and community relations have been affected by the presence of mining companies and, in most cases, by the contamination of by-products from the mine and the extension of the land needed for extraction, which has determined their position against mining. Mining extraction, in its current state, is incompatible with their vision of indigeneity, but this could possibly change with a different approach to social engagement by mining companies.

Beyond this crucial bond to the land, indigenous people aim for the control of a territory, as indigeneity is both a function of land and territory. Among indigenous peoples, the term 'territory' (*territorio*) is used to refer to an ancestral space, and has a more holistic meaning than 'land' (*tierra*). *Tierra* refers to a space with resources that indigenous people currently occupy. *Territorio* encompasses the historical belonging of a people within a lived landscape. It cannot be sold or bought, since it is communally owned and has a complex significance associating an identity with a locality, not an economic item that shows wealth or possession. Therefore, expropriation of indigenous land as at Yanacocha and Tintaya compromises this historical notion of territory. As Sawyer (2004: 83) explains, 'much more than signifying the physical and material contours of a region,

10 The concept of *chacha-warmi* (the Andean couple) expresses the duality and complementary nature of the relationship between men and women in a harmonious relation with the surrounding land and natural resources.

11 Information obtained through interviews conducted by Isabelle Anguelovski in Peru and Bolivia, May–June 2002 (see also Stephenson 1999).

12 Several times per year, both men and women travel within their land to rivers or mountains to honour deities and express their thanks for the past harvests or for the help received to maintain strong relations within the community.

13 Information obtained through interviews conducted by Isabelle Anguelovski in Peru and Bolivia, May–June 2001 (see also Stephenson 1999).

"*territorio*" encompasses moral-cosmological and political-economic complexes that shape identity and social relations'. It is important to note that 'territory' is a recent political construction and has emerged from the indigenous leadership in the 1970s to clearly mark a difference between the construction of rationally calculated mining concessions that companies bid on and the practical and conceptual indigenous cosmology, extending territory over river basins in which multiple systems of property control can co-exist. However, this vision is not recognised by either the state or companies, as proved by the mineral extraction in Tintaya and Yanacocha, with companies extracting copper and gold through wide parts of indigenous territory.

Precisely, for indigenous peoples, the *Sumak Allpa* (territory in Quechua language) is made of three parts, or spaces: *Jahuapacha* (the sky), *Caipacha* (the earth) and *Ukupacha* (the subsoil). Indigenous peoples administer these three spaces and the resources according to their traditions.[14] The rituals that indigenous people practise set clear sociocultural and symbolic borders between them and the dominant culture, creating two separate 'worlds of meaning' (Wilson and Donnan 1998). Indigenous peoples believe that, if they control the use of natural resources (rivers, forests, oil and minerals) on their territory, they will maintain their traditional way of life, inter-community relations, socialisation and remain true to their fundamental belief system, or cosmology.

Weak protection of indigenous peoples' cosmology through laws and regulations

The main setback for indigenous claims to their traditional land is that the Peruvian state does not recognise any special relation between indigenous peoples and their land and territory. Peruvian legislation has not established clear rules for prior and informed consent of communities, and companies can conduct extraction on traditional community land without offering fair compensation. The new 1995 Land Law established a process to transfer land to third parties called *servidumbre* (easement) through which the Ministry of Energy and Mines can grant permits to access the property of a landowner for mineral extraction. Despite the fact that the Land Law eliminates the right of the state to expropriation, indigenous communities living on land with minerals have lost almost all negotiation power, as they face the risk of losing their land without compensation if they refuse to sell it to companies (Christian Aid 2005: 7-8). The 1995 Land Law seems to be in contradiction with Article 89 of the 1993 Constitution, which provides guarantees for communally held property, as is the case for many indigenous communities. With few exceptions, farmers' (*campesino*) and indigenous land are considered 'inalienable' and cannot be transferred to third parties.[15]

This lack of respect for indigenous peoples' values and way of life has been translated into weak mining regulations, which do not adequately protect indigenous peoples' cos-

14 Information gathered during an interview between Isabelle Anguelovski and Inés Shiguango in March 2006. According to Shiguango, the territory is an essential element in people's cosmology: 'mountains, rivers, cascades and plains are strength, wisdom, protection, and nutrition given to mankind'.

15 According to Article 89 of the Constitution, the farmers and native communities have legal existence, are juridical persons, and are recognised to have autonomy in their organisation and communal work and in the use and free disposal of their lands. The Peruvian state also respects the cultural identity of the farmers and native communities.

mology. In the early 1990s, President Alberto Fujimori undertook a series of reforms as part of a structural adjustment plan supported by the World Bank, which included laws to promote foreign investment in mining operations through advantageous taxation, repatriation, foreign exchange policies for companies, and less stringent environmental regulations. For example, through the 1992 New Mining Law, the Ministry of Energy and Mines took over virtually all regulatory functions that were under the responsibility of other ministries and, as a result, environmental authorities exercise little influence over the mining sector. Ironically, Peru has no Ministry of the Environment (Christian Aid 2005: 5). In addition, several laws have modified the 1990 Environment Code.[16] For example, Article 56, which permitted the creation of protected areas by national, regional and local governments, was suppressed, attributing this authority exclusively to the federal level. Consequently, areas that indigenous peoples might farm or use for traditional ceremonies might be unilaterally transferred to companies, without the possibility for local governments to protect them. This change has major implications, since governors of provinces or mayors are more likely to be of indigenous origin than political leaders at the federal level. As a result of these recent laws, environmental injustices are quite flagrant; indigenous communities can be virtually expropriated from land they rely on for the sustainability of their livelihoods.

In summary, the Peruvian government has failed to establish a fair process that takes into account the cosmology of indigenous peoples and their traditional way of life, when environmental justice requires that their rights, culture, and decision-making processes over community development be respected. As this section has attempted to show, when faced with virtual expropriation, indigenous peoples have no sustainable alternatives, which is contrary to environmental justice principles and to their traditional cosmology.

Lack of participation of indigenous communities

The last aspect of environmental justice as defined by Schlosberg (2003) relates to the question of voice and participation. Schlosberg considers that communities unfairly affected by environmentally hazardous activities need to be seen and heard by mainstream environmental movements and politicians, who have generally ignored them. Environmental policies should develop ways to initiate community participation in both the design and the ongoing monitoring of environmental risks of such activities. Young (1990) goes even further by arguing that increased justice will be achieved only through the elimination of institutionalised domination and oppression in favour of more democratic decision-making procedures. However, as I describe in this section, the failures to allow native peoples' adequate participation in decision-making processes over mining development, redistribution of mining revenues and resolution of conflicts boils down to a violation of the notion of environmental justice as a participative process.

16 Legislative Decree No. 613. Environment and Natural Resources Code (07/09/90).

Lack of participation in decisions over mining development and revenue redistribution.

First, indigenous peoples consider that it is essential for them to participate in decisions related to the development (or non-development) of mining operations. In Peru, indigenous people are citizens of a nation-state in which a dominant culture seems to impose values, norms and institutions that are central in decisions related to mining development and the redistribution of mining revenues. If indigenous peoples can achieve some control over a border between themselves and the dominant culture, and some control over what happens at and within this border in regards to resource extraction, they will increase their chances of establishing a more balanced relationship with companies on their land.

Three main types of 'border' determine the relations in or among societies: (1) social and symbolic boundaries that define memberships in collectivities and that people might choose to use differently according to different pressures, (2) cultural boundaries that separate different worlds of meaning among people, and (3) geopolitical boundaries that define visible and tangible territories (Wilson and Donnan 1998). The presence of a border is needed to control indigenous land and territory and manage relations at this border. However, in Peru, the status of indigenous peoples prevents them from being able to control natural resources on their land and the borders of their territory against private industries. Currently, despite the promises of Article 89 of the 1993 Constitution, which recognises the legal existence of farming and indigenous communities and the inalienability of their land, the absence of true autonomy or self-determination status for indigenous peoples weakens their bargaining power with companies and an adequate participation in the decisions affecting their land. The procedural notion of justice, as defined by Schlosberg (2003), is thus violated.

Moreover, the capacity of mining activities to contribute to poverty reduction for local communities such as Tintaya and Yanacocha has not been demonstrated. The final report of the Extractive Industries Review published in 2003 looks at specific case studies of mining operations around the world and suggests that in Peru revenues have not been transferred to affected communities and that poverty reduction has not been significant in these places (World Bank 2003). From 1991 to 2002, the reduction that was noticed was only short-term and limited to Lima and the surrounding urban Sierra. Rural highlands in the Andes, where Tintaya and Yanacocha are located, did not benefit significantly from this poverty alleviation (World Bank 2003: 29). On the contrary, according to the Peruvian Compensation Fund for Social Development (FONCODES), Cajamarca, near Yanacocha, has moved from being the fourth poorest region in the country in the 1980s to the second poorest in 1993 (cited in Bosshard *et al.* 2003). In the Tintaya case, Kasturi *et al.* (2006) estimate that in 2004 the copper mine provided employment to 615 direct employees and 236 contractors and that 65% of the mine's employees and contractors were from Espinar. However, poverty reduction cannot be measured only in terms of jobs provided to local inhabitants. In both 2003 and 2005, one of the protestors' complaints was that the development fund promised by BHP Billiton was not being implemented.

The main issue here is that CSR initiatives are not intended to tackle structural dimensions of poverty and social exclusion, but rather performance enhancement and image (Blowfield and Frynas 2005; Frynas 2005). Furthermore, policies to help communities participate in the financial benefits of mining operations and redistribute government mining revenues have failed to target poor communities. In Peru, the main redistribution fund of federal revenues from mining, the Mineral Rent (Canón Minero) is criticised for being inefficient, poorly supervised and slow in payments. According to Ciudadanos al Día, a public interest group, the Canón redistributes only 10% of its resources to district governments and fails to address the development needs of communities that bear mining extraction. Mechanisms to increase citizens' access to transparent information and adequately monitor the use of funds at the local and district level are also very limited (Ciudadanos al Día 2005). Interviews with the two mining companies, Newmont and BHP Billiton, show that large sums of money contributed by companies to the state are sitting in bank accounts, unutilised, without the capacity of governmental structures at the central and regional levels to transfer this money to development programmes. According to the companies, another major obstacle in the quick implementation of development projects is the National System of Public Investment (SNIP), which obliges parties to respect complicated guidelines and standards and go through a long series of approval before being able to execute development projects.

Lack of culturally appropriate participation frameworks in dialogue processes

The most flagrant violation of the procedural aspect of environmental justice is related to the difficult and inequitable participation of indigenous peoples in dialogue processes, undermining corporations' efforts at community engagement. In both Tintaya and Yanacocha, challenges related to broad community engagement in the dialogue tables were raised (see Table 11.2).

In Yanacocha, on learning about the June 2000 spill, Newmont mobilised analysts to identify the effects of the spill and design a strategy for rapid response. However, this process did not involve discussions with community members or local representatives, and those affected by the spill were not part of defining the clean-up strategy, when achieving environmental justice would require the transparent, open and independent participation of communities. Later, in September 2001, the CAO of the IFC decided to create three parallel dialogue processes in three different communities and, in November 2001, another round-table dialogue was created by the executive powers, the congressmen of Cajamarca, the CTAR (the regional government), the Provincial Municipality of Cajamarca, the Frente de Defensa (Defence Front), various NGOs and the mining company. Interviews with Newmont CSR staff reveal that the company's goal was to 'attempt to build a mutually beneficial relationship together [with communities] by meeting each other's needs and a higher way of life for communities in direct areas of influence'. However, immediately, the dialogue table of the CAO lost credibility and legitimacy, as a large part of the population perceived it as an attempt to serve as a parallel non-independent process, which would impair the efforts of the community and its institutions to obtain an institutionalised way of managing the conflicts (FOEI *et al.* 2006). In contrast, for Newmont the CAO dialogue table was 'very meaningful and successful for the people who chose to participate in the dialogue process', with good outcomes in terms of 'relationship building and understanding of the issues, concerns, and risks that parties are taking'.

Table 11.2 Dialogue tables in Tintaya and Yanacocha (continued opposite)

Evaluation criteria	Tintaya	Yanacocha
Objectives	*Dialogue table:* • To investigate community grievances and improve corporate practices in four areas (sustainable development, land loss, human rights violations, and environmental impact) *Framework Agreement dialogue:* • To implement the clauses of the 2003 Framework Agreement in the province	To broach and resolve conflicts between the Yanacocha Mine and the Cajamarca community, with the participation of public and private institutions, in a transparent, open, independent and involved way
Structure	*Two dialogue processes:* • A dialogue table at the local level to address the concerns of the five most affected communities • Framework Agreement dialogue at the provincial level	*Two dialogue tables:* • The Mesa supported by the CAO (a member of the World Bank) • The CTAR (the regional government) dialogue table
Duration	*Dialogue table:* February 2001–ongoing *Framework Agreement signed in 2003:* September 2003–ongoing	*CAO dialogue table:* October 2001–ongoing, but effectively deactivated* *CTAR dialogue table:* November 2001–2003
Composition	*Dialogue table:* • Five closest and most affected communities around the Tintaya Copper Mine • Provincial mayor • National NGOs • Oxfam America • The mining company *Framework Agreement dialogue:* • Provincial authorities • Representatives from communities • Representatives from the mining company • Oxfam America as observer	*CAO table:* • Group of experts of the CAO • Representatives of public and private institutions of Cajamarca • Community members *CTAR dialogue table:* • Peruvian Ministry of Energy and Mines • Members of the executive power (Ministry of Internal Affairs, Ministry of Energy and Mining, INRENA and CONAM) • Peruvian congressmen representing Cajamarca • CTAR and Provincial Municipality of Cajamarca • Frente de Defensa • Various NGOs • Mining company

* Although the CTAR round-table dialogue was deactivated after two years whereas the CAO round-table dialogue still exists today, at present the latter has no credibility or legitimacy in the eyes of the population. Because the CTAR dialogue existed at the same time, members of the CAO dialogue did not do what was necessary to build a single, strong, shared and transparent dialogue process.

TABLE 11.2 (from previous page)

Evaluation criteria	Tintaya	Yanacocha
Rules	• Broad and active participation • Consensus-based decisions to ensure trust-building and shared interpretation • Joint fact-finding • Confidentiality of meetings to increase trust among parties	• Mutual listening • Respectful behaviour, especially to women • Skills development • Respect for identity of participants • Division of the discussion and matrix of concerns into two broad categories: environmental management and social responsibility

Furthermore, the members of the CAO dialogue table did not operate on a truly equal basis with the company, which did not help develop negotiation and conflict management skills for local people. As a result, trust-building was a major challenge. In September 2005, local peoples, concerned with the potential contamination of two of the main water sources of the region, the rivers Porcón and Grande, camped on a mountain to prevent the expansion of Yanacocha into a protected area Cerro Quilish, paralysing mining activities in the region (Riley and Griffin 2004). For Newmont, this mobilisation represented evidence that (1) changes were occurring too fast for people to be able to cope with them, and (2) the uncertainties about the future expansion of the mine had undermined people's trust.

In Tintaya, the dialogue table included elected leaders and interested citizens from five communities around Tintaya, and each of the four commissions formed during the dialogue process was composed of the communities' elected leaders and interested residents, municipal and NGO representatives and BHP Billiton staff. Most important, several principles, ground rules and procedures were outlined and implemented (See Table 11.2). However, recurring obstacles impaired effective participation:

1. Technical and legal aspects of the discussions and reports which impeded adequate dissemination of information by community members
2. Lack of effective communication between community members at the dialogue table and non-participating members
3. Lack of negotiating and dispute resolution experience

According to the activists who participated in the May 2005 protest at the Tintaya mine, NGOs did not strengthen sufficiently the organisational capacity of local social organisations and help them communicate information on the dialogue table to their wider constituencies. Interviewed activists considered that the dialogue table was a secretive mechanism with meetings occurring at the mine offices, without being opened to members of the communities outside the dialogue table. As a result, the population did not seem informed about the design, objectives, mechanisms and progresses of the dialogue table: 'My community was very much part of the Dialogue Table. I want that the reports of the Table reach us. I also want micro-indicators of air, water, and soil . . . But, there is now no true information.'[17]

Other challenges related to the participation of communities in dialogue tables are the perceived lack of capacity of NGOs as adequate intermediaries for the representation of groups' concerns in front of companies (Newell 2005). In Yanacocha, according to the NGOs FOEI *et al.* (2006), the mining company entered into institutional cooperation agreements and working contracts with the main organisations that participated in the CAO's round-table dialogue. These organisations were first in opposition to Newmont—for example, Federación de Rondas Campesinas Femeninas del Norte Del Peru (FEROCAFENOP)—but then benefited personally and economically from agreements with Newmont, impeding their independence. Shortly after, they started attacking leaders of independent NGOs in Yanacocha.[18]

The dialogue tables and the national NGOs in Tintaya and Yanacocha were partially funded by Oxfam America and the World Bank, which might have induced scepticism in some communities regarding the capacity of donor-induced participation processes to challenge the mechanisms at the root cause of their problems. In Yanacocha, the people who knew about the existence of the CAO dialogue table immediately related it to the IFC, which was a partner of Yanacocha Mine, and therefore doubted its independence as an Ombudsman (FOEI *et al.* 2006). In Tintaya, some activists participating in the May 2005 protest resented Oxfam for failing to build the capacity of social organisations to resist co-option, train communities in strategic planning, facilitate a permanent joint monitoring of the environmental contamination of the Tintaya mine, and diffuse accessible technical information to communities on the dialogue process in Tintaya and the impact of mining activities.

Lastly, when NGOs and trade unions play the role of mediators between companies and communities, issues arise as to who speaks for whom, on what basis, and whether the interests of the most vulnerable communities are well represented (Blowfield and Frynas 2005; Newell 2005; Prieto-Carrón *et al.* 2005). In mining areas, women are the ones who work in the fields, but generally mining companies negotiate only with men for access to land and resources. When communities move, the various survival networks that women have built tear, and women have to live with the physical and mental traumas due to the responsibility of family survival (Anguelovski 2002).

The impact of mining on women is exacerbated by a failure to identify them as a distinct group of stakeholders in the planning and operation of mine sites and establish trusted means of communication (Gibson and Kemp, Chapter 6 this volume; Musvoto 2001, cited in Walker and Howard 2002). In dialogue processes, although channels of

17 Interview data, 2007.

18 According to FOEI *et al.* (2006), the majority of the members of the round-table dialogue of the CAO became involved in institutional or personal agreements with Yanacocha. In the media they attacked leaders and organisations denouncing human rights violations and contamination. The members of FEROCAFENOP, which initially had been in open opposition to the mining activities, quickly changed their position when they started to profit from agreements with Yanacocha Mine. In the media they started to openly attack leaders of the Defence Front, of peasant organisations and of independent NGOs. This was why the FEROCAFENOP eventually broke ties with Project Underground, an NGO that had helped them file complaints with the CAO and the Multilateral Investment Guarantee Agency (MIGA) in 2001.

communication exist, women feel threatened by the presence of men throughout the hierarchy and are reluctant to speak up and raise concerns (Macdonald and Rowland 2002). In Tintaya, the low participation of uneducated citizens and women, despite the availability of Quechua–Spanish translation, was a recurrent concern of the members of the dialogue table (Barton 2005). One notable measure was taken: the requirement that each working commission have a least one female member. However, most of the time women did not speak in the dialogue table meetings and several participants reported that the quality of translation was often poor and incomplete. Interviews reveal that members of the five communities and from the rest of the province felt marginalised and showed mistrust towards the capacity of the dialogue table to address their concerns. One of the women activists explained: 'People of the Table are only doing meetings and the Presidents tell us after during the assemblies. They don't let anybody go in . . . Members of the Table have credentials so this is the way they enter.' Another activist declared: 'Some go and participate in the Dialogue Table but the rest is on the floor, like us the women.'

Last, several protesters in the 2005 mobilisation, especially social leaders and women, resented the company staff—both executives and workers—for furthering the symbolic and physical distance between the population and the mine. Activists criticise promises of open doors to executives' offices when apparently their demands are not heard and they are labelled as agitators:

> The mine buys the leaders of communities. 'I'll give you a pack of milk, a pack of sugar or temporary work.' And then the leader will think in favour of mine while hundreds of community members will never benefit. The mine is abominable. They buy leaders, any class of leaders who are opposing.[19]

Faced with this criticism, BHP Billiton's community relations staff responds that it is in fact constantly going to rural communities and participating in the committees created to implement the 2003 Framework Agreement discussed earlier. Employees listen to people's concerns and proposals for projects and implement them, such as the construction of water systems, the electrification of remote areas, or deliveries of tractors and other machinery for agriculture.

As this last section suggests, Schlosberg's (2003) demands for broader and authentic public participation to address issues of distribution and cultural mis-recognition are challenged by the difficulties of empowering communities and engaging them democratically in dialogue processes. In Yanacocha, it seems that the mining company eventually subdued and manipulated the institutions that civil society sought to use to solve social and environmental conflicts. As a result, NGOs and local communities demanded in October 2004 that the CAO Dialogue Table be deactivated (FOEI *et al.* 2006). In Tintaya, interviews with activists of the May 2005 mobilisation reveal that protesters attempted to voice and promote their own model of dialogue, with large popular participation and the views of everybody present at the mine. On the other side, BHP Billiton had its own definition of dialogue—discussions should be in restricted selective settings with elected leaders. The 2005 mobilisation was the latest of a series of historical social movements in the province and served to re-legitimate and strengthen dia-

19 Interview data, January 2007.

logue processes in Tintaya, as well as the population's and social organisations' own voices and participation in them.

Conclusion

The cases of Tintaya and Yanacocha suggest that notions of equity, recognition and participation closely interplay in claims of environmental injustices related to mining operations in indigenous communities. As Schlosberg (2003) points out, these three notions must be interrelated to address environmental injustices, as the vision and cosmology of indigenous peoples needs to be recognised by companies and states, and participation needs to be culturally adequate to redress inequities in the distribution of the environmental burden caused by mineral extraction. Tintaya and Yanacocha demonstrate the importance of new standards set by the state and by companies in the relationship between companies and local communities—most importantly their inclusion in the planning, design, definition and implementation of mining projects. BHP has already taken very encouraging measures in this direction and has been elected by the Roberts Environmental Center in 2006 as the best company in the metals, mining and crude oil sectors in terms of environmental and sustainability reporting and quantitative social performance. (For more information see ICMM 2006: 3.) Newmont is also putting greater emphasis on transparency and inclusion in processes and decisions to monitor the environmental impact of mining activities.

In the future, the dialogue processes in Tintaya and Yanacocha should work to incorporate both small and large group discussions to maintain the viability and survival of deliberative democracy practices. Discussions could sometimes focus on technical and environmental issues and at other times allow space for multicultural collaboration. This would require the support of a professional facilitator and a widely accepted community leader to help people move ideas into achievable projects and change grievances into proposals. Careful attention should also be devoted to ensuring community-building and broad participation in both planning and implementation of agreements designed to improve the sustainable development of resource extraction areas. Other planning projects have even used task forces to conduct outreach to the wider community, discuss decision-making and review draft plans for implementation (Briggs *et al.* 1996). In Tintaya, for example, a broad variety of stakeholders could be identified to participate in a task force, based on their knowledge of the community, their history of community involvement and their affiliation. This would ensure the involvement of different constituencies (women, youth, farmers, business owners and social leaders).

These ideas emphasise a greater but far more focused role for NGOs, which have been less active in Tintaya, in improving dialogue mechanisms. NGOs should strengthen local social organisations, diffuse accessible information on mining extraction and help to ensure the quality of discussion in meetings. NGOs eager to promote deliberation for conflict resolution will need to ensure adequate representation of indigenous interests, deliberative engagement, application of different kinds of knowledge appropriate for every circumstance and actor, and social learning. This follows the model of empowered participatory governance—a focus on tangible problems, the involvement of ordinary people affected by problems and officials close to them, and the use of deliberation to create solutions—to make decisions more effective and accepted by a wide number of community members (Fung 2004; Fung *et al.* 2003).

The companies themselves need to better understand community dynamics and go through a significant behavioural change in the way they frame their daily relations with communities. Assessing the state of stakeholder relationships and sociopolitical networks of relationships among the company's stakeholders seems highly important in Tintaya and Yanacocha (On Common Ground Inc. 2003).[20] A community relations strategy is required that is culturally sensitive, relevant to the community and matched to the scale and intensity of the operation. For example, in Tintaya, the community relations officers are not accepted by villagers, who see them as 'spies' who report their criticisms of the mine to company executives. These employees have been in place for several years and are seen as an old guard of paternalist executives. It is also essential for mining companies to take relevant procedures to enable ample citizen participation and ensure the timely and culturally appropriate delivery of information to the population—in particular the disempowered members of communities—in order to avoid fragmentation in communities.

Furthermore, voluntary initiatives in the mineral sector should include objectives that go beyond legal requirements and should be complemented by other instruments, such as international cooperation and national policy and regulations. Encouraging examples of these voluntary initiatives are three programmes started by mining companies in Peru, including Newmont and BHP Billiton, aimed at improving emergency response planning and implementation and preventing mining accidents, such as the Safe Transportation Initiative and the creation of Civil Defence Committees.[21] With improved planning, implementation of sound environmental management tools and cleaner technologies, extended social responsibility to stakeholder groups, the formation of sustainability partnerships, and improved training, a mine can improve performance in both the environmental and socioeconomic arenas, and thus contribute to sustainable development (Hilson and Murck 2000; Labonne 1999).

At the national and global level, it is important to put in place stricter mechanisms towards sustainable mining through the certification of mining activities or the external verification of the mining industry's environmental and social performance around systems and outcomes, such as the Australian Minerals Industry Code for Environmental Management (Solomon 2000) or WWF Australia's Mining Certification Evaluation Project (MCEP). The MCEP investigates the feasibility of third-party certification of

20 The Centre for Innovation and Management at Simon Fraser University in Vancouver recently developed a sociopolitical measurement system known as the Stakeholder 360 to help managers gain a better understanding of power structures, networks and relations among stakeholders in communities. The methodology recommended for managers consists of collecting perceptions from representatives of organisations and assessing the state of social capacity in the community through a rapid survey of their leaders. Later, the company convenes other dialogue meetings to track the sharpening and clarification of the economic vision of the community. The main problem of this Stakeholder 360 model, however, is that it excludes unorganised aggregates of people. It is important for managers to understand that spontaneous revolts can actually arise from disaggregate marginalised people and that they cannot just limit their consultation work to official representatives of structured organisations. For more information, see On Common Ground Inc. 2003.

21 For more information, see www.icmm.com/page/784/losses-in-transporting-mercury, accessed 1 October 2008.

environmental and social performance of mine sites to establish a knowledge platform for broader international debate and future effort.[22] In the USA, an interesting mechanism, which does not exist in Peru, is the Good Neighbor Agreement. The major strength of such agreements is that they are legally binding for the parties engaged and for future owners and managers.[23] In Peru, if agreements following dialogue processes were legally binding and underlined clear sanctions in case of non-implementation, then local populations might actively support them.

In summary, environmental justice and sustainable development cannot be separated in the mining industry. To be just, mining should maximise human well-being and identify and internalise environmental and social costs; ensure a fair and efficient distribution of the costs and benefits of development; and respect and reinforce the fundamental rights of the communities in terms of cultural land-use practices, despite their incongruence with dominant norms. Respecting the traditional organisations of indigenous peoples and their communication channels is an essential prerequisite to getting the 'social licence to operate' extractive ventures.[24] Promoting responsible stewardship of natural resources requires a clear support of participatory decision-making, no matter how long and arduous the process may initially seem.

22 See wwf.org.au/ourwork/industry/mining, accessed 15 August 2008. Experiences in certification in other sectors show that the credibility and effectiveness of certification programmes is challenged by three main issues: setting standards for technologies and working practices with particular groups, assessment and assurance, and governance.

23 In the Stillwater Good Neighbor Agreement, Stillwater and Sweet Grass Counties citizens and the Stillwater Company, which operates two platinum/palladium mines in the area, have signed an agreement to protect the local quality of life and provide sustainable and responsible economic development. For more information, see www.northernplains.org/ourwork/goodneighbor/files/GNA_Citizens_Guide.pdf, accessed 15 August 2008; www.northernplains.org/ourwork/goodneighbor, accessed 15 August 2008.

24 For example, communication barriers between indigenous peoples and companies need to be decreased. Indigenous and non-indigenous people may have trouble communicating because they have a vastly different fundamental understanding of the universe and because indigenous peoples rely on the words of their elders. For instance, some indigenous people pay strict attention to their elders, who have intimate knowledge of traditions and local events and tend to explain facts through story-telling and parables. As a result, non-indigenous listeners may become frustrated and even angry when they try to get straightforward information from an indigenous person. See Centre for Traditional Knowledge 1997.

References

Ali, S.H., and C. O'Faircheallaigh (2007) 'Introduction', *Greener Management International* 52 (theme issue on 'CSR, Extractive Industries and Environment: Values and Principles'): 5-16.

Anguelovski, I. (2002) *Lineamientos de política de género para la Oficina Regional de Oxfam América en América del Sur* (Lima, Peru: Oxfam America).

Ballard, C., and G. Banks (2003) 'Resource Wars: The Anthropology of Mining', *Annual Review of Anthropology* 32 (October 2003): 287-313.

Barton, B. (2005) 'A Global/Local Approach to Conflict Resolution in the Mining Sector: The Case of the Tintaya Dialogue Table' (MA thesis, Fletcher School of Law and Diplomacy, Tufts University, Medford, MA; fletcher.tufts.edu).

Bech, J., C. Poschenrieder, M. Llugany, J. Barcelo, P. Tume, F. Tobias, J.L. Barranzuela and E.R. Vasquez (1997) 'Arsenic and Heavy Metal Contamination of Soil and Vegetation around a Copper Mine in Northern Peru', *Science of the Total Environment* 203: 83-91.

Blowfield, M., and J.G. Frynas (2005) 'Setting New Agendas: Critical Perspectives on Corporate Social Responsibility in the Developing World', *International Affairs* 81.3 (May 2005): 499-513.

Bosshard, P., J. Bruil, K. Horta, S. Lawrence, C. Welch (2003) *Gambling with People's Lives: How Does the World Bank's 'High-Risk/High-Reward' Funding Strategy Threaten People and the Environment?* (New York: Environmental Defense, Friends of the Earth and International Rivers Network).

Briggs, X. de S., A. Miller and J. Shapiro (1996) *Planning for Community Building: CCRP in the South Bronx* (Washington, DC: American Institute of Certified Planners).

Bury, J. (2004) 'Livelihoods in Transition: Transnational Gold Mining Operations and Local Change in Cajamarca, Peru', *Geographical Journal* 170.1: 178-91.

—— (2005) 'Mining Mountains: Neoliberalism, Land Tenure, Livelihoods, and the New Peruvian Mining Industry in Cajamarca', *Environment and Planning* 37: 221-39.

Cardozo, M., and E. Cedillo (1990) 'Geologic-metallogenetic Evolution of the Peruvian Andes', in L. Fontbotd, G.C. Amstutz, M. Cardozo, E. Cedilla and J. Frutos (eds.), *Stratabound Ore Deposits in the Andes* (Berlin: Springer-Verlag): 37-60.

Centre for Traditional Knowledge (1997) *Prototype Guidelines for Environmental Assessments and Traditional Knowledge* (Ottawa: Centre for Traditional Knowledge).

Christian Aid (2005) *Unearthing the Truth: Mining in Peru* (London: Christian Aid).

Ciudadanos al Día (2005) *El canón minero en el Perú: Transparencia fiscal* (Lima, Peru: Ciudadanos al Día).

Echave, J., K. Keenan, M.K. Romero and Á. Tapia (2005) *Los procesos de diálogo y la administración de conflictos en territorios de comunidades: El caso de la Mina de Tintaya en el Perú* (Lima, Peru: CooperAcción).

Environmental Media Services (2001) *Background and Summary: CAO Independent Commission Report* (Washington, DC: Environmental Media Services).

FOEI (Friends of the Earth International), Amigos de la Tierra and GRUFIDES (2006) *The CAO in Peru, Lessons Learned from Dialogue as a Strategy to Reduce Conflict* (Amsterdam: FOEI).

Frynas, J.G. (2005) 'The False Developmental Promise of Corporate Social Responsibility: Evidence from Multinational Oil Companies', *International Affairs* 81.3 (May 2005): 581-98.

Fung, A. (2004) *Empowered Participation : Reinventing Urban Democracy* (Princeton, NJ: Princeton University Press).

——, E.O. Wright and R. Abers (eds.) (2003) *Deepening Democracy: Institutional Innovations in Empowered Participatory Governance* (London/New York: Verso).

Global Response (2005) *Stop Gold Mine Expansion: Peru* (Boulder, CO: Global Response).

Hilson, G. (2002) 'An Overview of Land Use Conflicts in Mining Communities', *Land Use Policy* 19: 65-73.

—— and B. Murck (2000) 'Sustainable Development in the Mining Industry: Clarifying the Corporate Perspective', *Resources Policy* 26: 227-38.

Honneth, A. (1992) 'Integrity and Disrespect: Principles of a Conception of Morality Based on the Theory of Recognition', *Political Theory* 20.2 (May 1992): 187-201.

Humphreys, D. (2000) 'A Business Perspective on Community Relations in Mining', *Resources Policy* 26: 127-31.

ICMM (International Council on Mining and Metals) (2006b) 'Biodiversity Survey Highlights Mining Improvement', *Good Practice: The Newsletter of the International Council on Mining and Metals,* 2006 5.2: 3.

Imbun, B.Y. (2007) 'Cannot Manage without the "Significant Other": Mining, Corporate Social Responsibility and Local Communities in Papua New Guinea', *Journal of Business Ethics* 73: 177-92.

Kapelus, P. (2002) 'Mining, Corporate Social Responsibility and the "Community": The Case of Rio Tinto, Richards Bay Minerals and the Mbonambi', *Journal of Business Ethics* 39.3: 275-96.

Kasturi, R.V., B. Barton and R. Reficco (2006) *Corporate Responsibility and Community Engagement at the Tintaya Copper Mine (A and B)* (Harvard Business School Case 506-023; Boston, MA: Harvard Business School Publishing).

Labonne, B. (1999) 'The Mining Industry and the Community: Joining Forces for Sustainable Social Development', *Natural Resources Forum* 23: 315-22.

Macdonald, I., and C. Rowland (eds.) (2002) *Tunnel Vision: Women, Mining and Communities* (Melbourne: Oxfam Community Aid Abroad).

Maurice-Bourgoin, L., I. Quiroga, J. Chincheros and P. Courau (2000) 'Mercury Distribution in Waters and Fishes of the Upper Madeira Rivers and Mercury Exposure in Riparian Amazonian Populations', *Science of the Total Environment* 260: 73-86.

McClintock, C. (2006) 'A Left Turn in Latin America? An Unlikely Comeback in Peru', *Journal of Democracy* 17.4 (October 2006): 95-109.

Ministry of Energy and Mines (2000) *Reference Plan for Mining 2000–2009* (Lima, Peru: Ministry of Energy and Mines).

Newell, P. (2005) 'Citizenship, Accountability and Community: The Limits of the CSR Agenda', *International Affairs* 81.3 (May 2005): 541-57.

On Common Ground Inc. (2003) *Assessing the State of Stakeholder Relationships: The Stakeholder 360* (Vancouver, BC: On Common Ground Inc.).

Oxfam America (2006) 'Tintaya Copper Mine', Oxfam America; www.oxfamamerica.org/whatwedo/where_we_work/south_america/news_publications/tintaya/art6242.html, accessed 20 August 2008.

Oxfam Community Aid Abroad (2001) *Mining Ombudsman Annual Report 2001* (Melbourne: Oxfam Community Aid Abroad).

—— (2003) *Mining Ombudsman Annual Report 2003* (Melbourne: Oxfam Community Aid Abroad).

Prieto-Carron, M., P. Lund-Thomsen, A. Chan, A. Muro and C. Bhushan (2005) 'Critical Perspectives on CSR and Development: What We Know, What We Don't Know, and What We Need to Know', *International Affairs* 82.5 (September 2005): 977-87.

Project Underground (2000) *Andean Odyssey: Campesinos Challenge Newmont's Conquistador-Like Return to Cajamarca* (Berkeley, CA: Project Underground).

Rawls, J. (1971) *A Theory of Justice* (Cambridge, MA: Belknap Press of Harvard University Press).

Rey-Sanchez, V. (2002) 'Peruvian Gold Company mercury spill impact to be studied', *Dow Jones Newswires*, February 2002.

Riley, M., and G. Griffin (2004) 'Fighting Back', *Denver Post,* 13 December 2004; www.wman-info.org/news/newmont%20part%202/newsitem_view, accessed 23 October 2008.

Sawyer, S. (2004) *Crude Chronicles: Indigenous Politics, Multinational Oil, and Neoliberalism in Ecuador* (Durham, NC: Duke University Press).

Schlosberg, D. (2003) 'The Justice of Environmental Justice: Reconciling Equity, Recognition, and Participation in a Political Movement', in A. Light and A. De-Shalit (eds.), *Moral and Political Reasoning in Environmental Practice* (Cambridge, MA: MIT Press): 77-106.

Solomon, F. (2000) 'External Verification of the Australian Minerals Industry Code for Environmental Management: A Case Study', *Australian Journal of Environmental Management* 7.2 (June 2000): 91-98.

Stephenson, M. (1999) *Gender and Modernity in Andean Bolivia* (Austin, TX: University of Texas Press).

Thompson, I., and S. Joyce (2002) *Changing Expectations: Future Social and Economic Realities for Mineral Exploration* (Vancouver: On Common Grounds Consultants).

Vega, M., M. Scobble and M.L. McAllister (2001) 'Mining with Communities', *Natural Resources Forum* 25.3: 191-202.

Vogel, D. (2005) *The Market for Virtue: The Potential and Limits of Corporate Social* (Washington, DC: Brookings Institution Press).

Walker, J., and S. Howard (2002) *Finding the Way Forward: How Could Voluntary Action Move Mining towards Sustainable Development?* (London: Environmental Resources Management and International Institute for Environment and Development).

Walzer, M. (1983) *Spheres of Justice: A Defense of Pluralism and Equality* (New York: Basic Books).

WCED (World Commission on Environment and Development) (1987) *Our Common Future* ('The Brundtland Report'; Oxford, UK: Oxford University Press).

Webber, J. (2002) 'Impoverished Peru is a Gold Prospect for Miners', *Forbes*, 21 May 2002.

Wheatley, B., and M.A. Wheatley (2000) 'Methylmercury and the Health of Indigenous Peoples: A Risk Management Challenge for Physical and Social Sciences and for Public Health Policy', *Science of the Total Environment* 259: 23-29.

Wilson, T.M., and H. Donnan (1998) *Border Identities: Nation and State at International Frontiers* (Cambridge, UK/New York: Cambridge University Press).

World Bank (2003) *Striking a Better Balance: The Final Report of the Extractive Industries Review* (Washington, DC: World Bank).

Young, I.M. (1990) *Justice and the Politics of Difference* (Princeton, NJ: Princeton University Press).

12

Indigenous people and mineral resource extraction in Russia

The case of diamonds

Susan A. Crate
Department of Environmental Science and Policy, George Mason University, USA

Natalia Yakovleva
BRASS Research Centre, Cardiff University, UK

This chapter presents a case study of indigenous[1] people and diamond mining in contemporary Russia. Although there can be some fairly consistent and predictable environmental and social effects from extractive industries the world over, indigenous responses vary widely, depending on specific histories and level of rights exercised by different groups. Mining and indigenous peoples of Russia need to be understood within the context of the circumpolar north.[2]

1 One important point of clarification for our discussion of Sakha is the classification of indigenous people within the former Soviet Union versus on a global scale. The 1925 Soviet classification of 'small-numbered peoples' or 'indigenous' included 26 peoples (now numbering 42, due to the fact that Soviet authorities grouped many ethnicities together and in the post-Soviet period they have been able to re-establish themselves) who practised hunting, gathering and reindeer herding and whose populations did not exceed 50,000 (Slezkine 1994: 2). This category did not include the 'numerically large' peoples including Komi, Yakut (Sakha), Tuvan and Buriat (Shnirelman 1999: 119). However, these same 'large-numbered peoples' also fit the global classification of 'indigenous'. Because in this volume we are discussing the issues of indigenous peoples and mining on a global scale, it is this latter classification, or the global definition of indigenous, that we use in our analysis of Sakha (see Article 1 of the ILO 'Convention on Indigenous and Tribal Peoples' [ILO 2003]). Similarly, we use the terms 'indigenous', 'Aboriginal' and 'native' interchangeably (Brown 2003: xiii).

2 The circumpolar north includes all or part of eight nations: Alaska, Canada, Greenland, Iceland, Norway, Sweden, Finland and Russia. All except Iceland are home to indigenous peoples.

Since World War II governments have increasingly moved into territorial northlands for resources, transportation corridors, education, healthcare and administration facilities, and national defence. Concomitantly, Arctic indigenous peoples have defended their homelands and cultures (Jull 2003: 23), making the Arctic a proving ground for localising indigenous human rights (AHDR 2004; Caulfield 1997; Habeck 2003; Nuttall 1998; Sejersen 2002; Sirina 2005; Wilson 1999). Arctic indigenous groups have made claims to their resources, knowledge and rights with some success (Berkes 1999; Dahl *et al.* 2000), showing that northern indigenous peoples can achieve levels of social and economic equity, cultural survival and political devolution. Self-government and self-determination are the central foci of contemporary indigenous concerns in the Arctic. The establishment of the Alaska Native Claims Settlement Act (1971), Greenland Home Rule (1979) and the first indigenous territory of Nunavut (1999) are three such successes and serve as examples for other Arctic and global contexts.

Comparative analysis between northern Russia's plights and of other circumpolar cases has been ongoing for at least a decade (Chance and Andreeva 1995) and reveals that Russia's indigenous peoples are severely behind. They lack land claims, resource rights, political experience, strong visionary leadership and an overall mind-set of self-determination and self-government. Similarly, they suffer the consequences of the almost complete absence, until the 1990s, of any mechanism through which social pressure might be applied for diamond-mining corporations to behave in a socially responsible way. All these factors are largely a result of colonial history, and especially of the Soviet period or the last century of *perestroikas* (Grant 1995: xiii).

With these similarities and constraints in mind, we move to our case study to explore the specific chronological, environmental and cultural history of exploiting diamonds in the western part of the Republic of Sakha (Yakutia), actions by both the populace and the diamond company in response to mining's environmental and social issues, and a discussion of the human rights and environmental justice implications of contemporary extractive industries.

A case study of diamond mining

Russia is the world leader in proven reserves of diamonds and annually produces 20–25% of world production (Matveev 1998; Goskomekologia 2000). Russia's deposits are situated in three regions: the Republic of Sakha (Yakutia) which accounts for 82% of reserves, Arkhangelsk *oblast* (region) with 18% and less than 1% in Perm *oblast* (Matveev 1998). Almazy Rossii-Sakha Company Ltd. (ALROSA) carries out the major diamond operations in the Republic of Sakha (Yakutia), where 99.8% of Russian diamonds are mined (Matveev and Cherny 2000) (see Fig. 12.1).

The Republic of Sakha (Yakutia) covers 3 million square kilometres and makes up 18% of Russia's territory, with a small population of 1 million. The territory of the Republic consists of 33 *ulus* (administrative districts) and the two administrative territories of Yakutsk and Neryungri (see Fig. 12.2). According to the 2002 Russian census, ethnic Sakha make up 46% of the Republic's population, Russians 44%, minority indigenous

FIGURE 12.1 Map of the Russian Federation

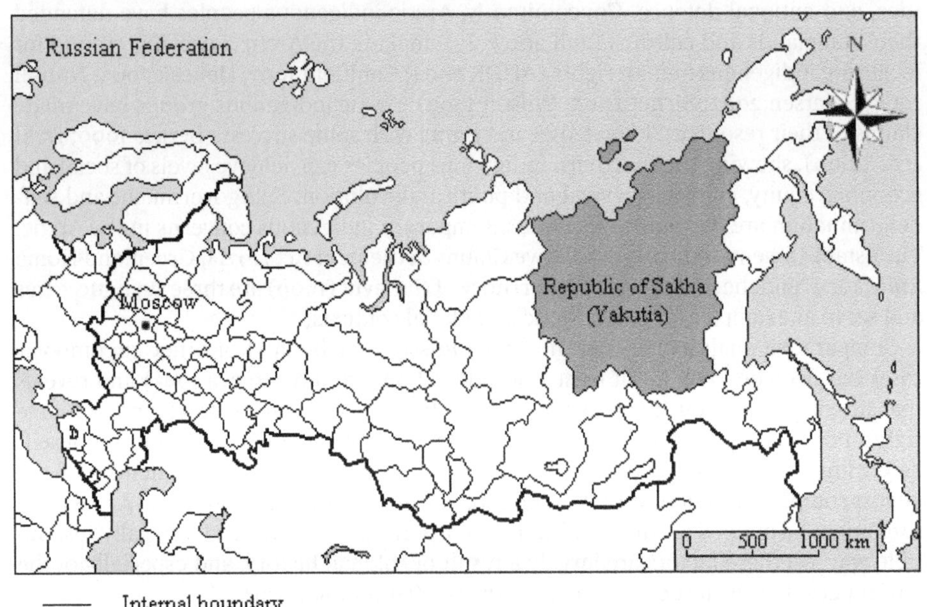

peoples including Evens, Evenks, Yukagirs and Chukchis, 3%, with the remainder including Ukrainians, Tatars and Belorussians (FSGS 2005).

Sakha are the main non-Russian group inhabiting the Republic.[3] Most contemporary rural Sakha are agropastoralists practising horse and cattle breeding. Sakha's Turkic ancestors transmigrated from Central Asia to the shores of Lake Baikal in the 800s and then travelled north, beginning in the 13th century, following the Lena River to their present home (Ksenofontov 1937). Vilyuy Sakha are a Sakha group inhabiting the Vilyuy watershed regions in the Republic's western area. Russians began colonising Sakha in the mid-1600s and demanded *yasak* (a fur tax). For many centuries the Vilyuy regions were occupied by Russians and were a fairly stable colonial society in the Tsarist period. During the Soviet period Sakha herding, hunting and fishing activities were gradually collectivised then consolidated in the late 1950s into agro-industrial state farms. With the 1991 fall of the USSR, local state farm authorities disbanded their operations on the Vilyuy. Since that time rural Vilyuy Sakha depend on mixed cash economies based on state transfer payments, supplemented with household-level food production centred on breeding horses and cattle (Crate 2006a, 2003c). Urban Vilyuy Sakha depend on salaries and wage labour.

3 Sakha is the self-designation of Yakuts, the titular nation of the region, while Yakut is the Russian designation for Sakha.

FIGURE 12.2 Map of the Republic of Sakha (Yakutia)

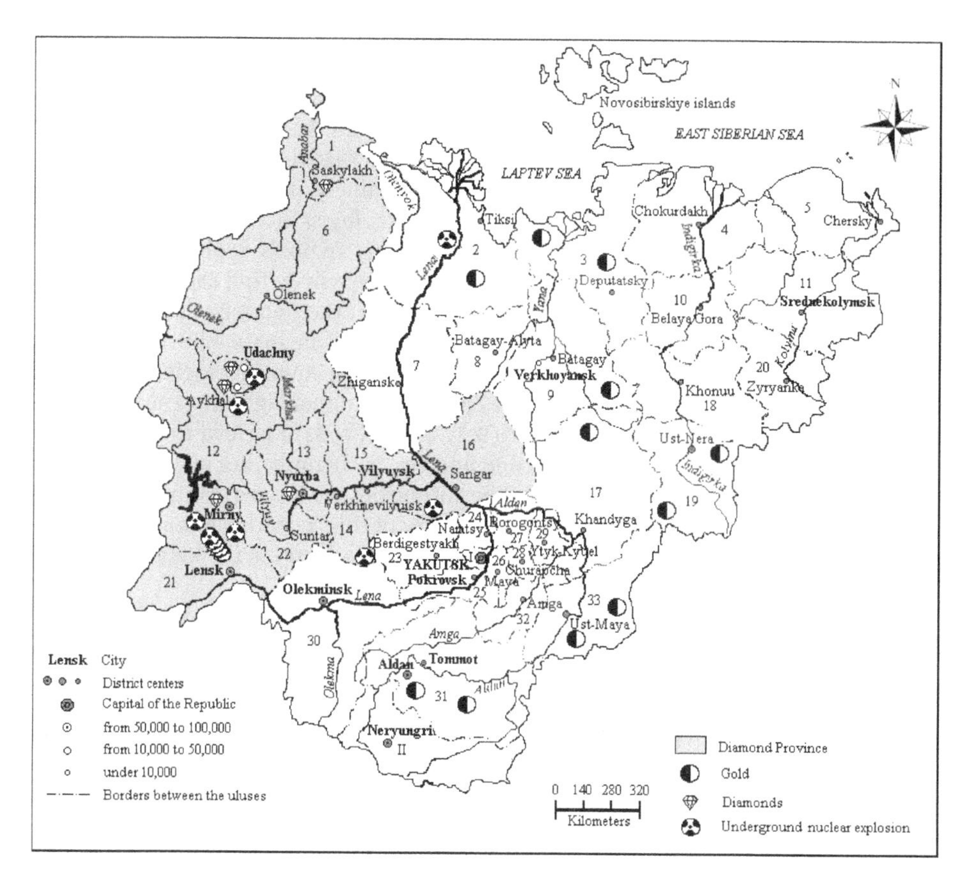

Districts of the Republic of Sakha: 1. Anabar; 2. Bulun; 3. Ust-Yana; 4. Allaikhovsky; 5. Nizhne-Kolymsky; 6. Olekminsk; 7. Zhigansk; 8. Eveno-Bytantaisky; 9. Verkhonyansky; 10. Abyi; 11. Srednekolymsky; 12. Mirny; 13. Nyurba; 14. Verkhnevilyuysk; 15. Vilyuysk; 16. Kobyai; 17. Tompon; 18. Momsky; 19. Oymyakon; 20. Verkhnekolymsky; 21. Lensk; 22. Suntar; 23. Gorny; 24. Namsky; 25. Khangalass; 26. Megino-Khangalass; 27. Ust-Aldan; 28. Churapcha; 29. Tatta; 30. Olenek; 31. Aldan; 32. Amga; 33. Ust-Maya; I. Territory of the city Yakutsk; II. Territory of the city Neryungri.

The socioeconomic development of the Vilyuy regions is directly linked to diamond mining in the Republic, which brings both economic prosperity and environmental and social problems (Crate 2006a; 2003a).

History of resource exploitation in the Republic of Sakha (Yakutia)

Historically, the area of the contemporary Republic of Sakha (Yakutia) was first exploited by Russian colonisers, beginning in the mid-1600s, for its 'soft gold'—sable pelts via *yasak*. The next exploitation was for minerals, starting with gold discoveries in

Aldan in 1851. However, the absence of roads and the remoteness from settlements made it economically unviable to develop these placer deposits. The establishment of Bolshevik rule in Soviet Russia prompted the 1920s gold rush, attracting large numbers of migrant labour from other parts of Russia (Sagers 1997). The central authorities quickly took control of gold mines and created Yakutzoloto (Yakut gold) (Aldanzoloto 1994), prompting the construction of energy stations, roads and the Aldan–Yakutskaya railroad, an extension of the Baikalo–Amurskaya railroad (Yakovleva 2005).

A second wave of mineral resource-related development in the Republic occurred shortly after World War II (Forsyth 1989). Following geologist Victor Sobolev's observation of similarities between South African and Vilyuy region's geological structure (Kharkiv *et al.* 1997), the Soviet government invested heavily to find the expected diamonds (Duval *et al.* 1996). Exploration for diamonds in the Republic near the Vilyuy River started in 1947 with the first diamonds discovered in 1949. Alluvial diamond deposits were discovered in 1950. On 21 August 21 1954, the young geologist Larisa Popugayeva discovered the first kimberlite pipe Zarnitza (Summer Lightning) near the Daldyn River, a tributary of the Vilyuy River (Belykh *et al.* 1984; Kiseleva 1999; Vecherin 1997). When geologists discovered a second and very rich kimberlite pipe in 1955 near the Irelyakh River, they sent a telegraph to the authorities in Moscow stating, 'We smoke the pipe of peace, excellent tobacco.' The discovery of this second diamond pipe, later called Mir (Peace) was the start of the diamond mining industry in the Republic (Tanin 2001).

In January 1957, the Soviet government established a state-owned enterprise Yakutalmaz (Yakut diamond) to conduct diamond-mining development in the Republic. Yakutalmaz extracted the first carats of gem-quality diamonds on 10 June 1957 and launched two enrichment plants working with ore from the Mir pipe in 1958 (Belykh 1993; Belykh *et al.* 1984). From 1957 to 1992, Yakutalmaz operated the Republic's alluvial and ore diamond mines. The industrial development in western Yakutia provided an impetus for new urban development for miners, including Mirny (1959), Chernyshevsky (1961), Aykhal (1962), Almazny (1963) and Udachny (1968) (Belykh *et al.* 1984). The majority of the labour force for diamond mining came from western parts of the former Soviet Union and resulted in a large migration to a sparsely populated area.

Following the fall of the Soviet Union, Yakutalmaz was liquidated in 1992 and a new joint-stock company ALROSA was formed. ALROSA currently operates diamond mines in Anabar, Mirny and Nyurba districts, all located within the 'diamond province' of Anabar, Kobyai, Lensk, Mirny, Nyurba, Olenek, Suntar, Verkhnevilyuysk and Vilyuysk (see Fig. 12.2).

Environmental impacts of diamond mining on the Vilyuy

Soviet diamond-mining activities disregarded environmental laws and regulations, resulting in (1) watershed-wide chemical contamination, (2) nuclear contamination, and (3) related human health and social effects (Crate 2003a, 2006a; Kopylov *et al.* 1993; Labutina and Smolenko 1990). There are three main sources of chemical contamination: phenols, thallium, used in the final stage of processing, and highly mineralised brine water. These sources, in turn, interact with the ecosystem and result in other contamination. For example, scientific examinations of the watershed in early 1990 found severe contamination of the environment with heavy metals immediately around the mines and at far distances. For instance, in Syuldukar (72 km [45 miles] from Mirny, the centre of diamond mining) the content of titanium in flood-plain sediments exceeded 5 times the normal levels; in Bordon (174 km [108 miles] from Mirny) 3.5 times; and in Djarkhan (265 km [165 miles] from Mirny) 1.5 times (Savvinov *et al.* 1992). Research teams have found high concentrations of aluminium, chromium, nickel, cobalt, copper, zinc, scandium, vanadium, titanium, manganese, barium and strontium in the bottom sediments of the watershed's rivers including the Markha, Malaya Botuobiya and Vilyuy (Savvinov *et al.* 1992).

The large water reservoir established in 1958 for the Vilyuy hydroelectric station (Vilyuy GES) to supply the nascent diamond industry was one source of surface water contamination (Belykh 1993). The reservoir, encompassing 148,000 ha (356,000 acres) of prime fields and woodlands containing haying, pasturing and hunting areas and economically valuable timberlands, disrupted the river's natural ebb and flow, 'softened' the local climate, inundated native settlements and valuable land resources, and contaminated the surface waters with phenols (Crate 2006a; 2003a). Two to three years after the flooding the water was significantly polluted, and more than ten years were needed for the stabilisation of the oxygen regime (Yakovleva *et al.* 2000). This artificial regulation of the Vilyuy River flow significantly changed the hydrological, hydro-chemical and hydro-biological characteristics of the flooded area, as well as the ice regime, and the reservoir itself became the source of hydrogen sulphide and phenols that contaminated the environment (Kirrilov 1990). Sakha elders describe the pre-hydrodam Vilyuy River as having crystal-clear waters abundantly teeming with sturgeon, freshwater salmon and other valuable fish species, which are now rarities if found at all (Crate 1995). The river otter (*Lutra lutra* L.) and black or hooded crane (*Grus monacha*), once common to the Vilyuy and its tributaries, are gone (Andreev 1987). A study carried out from 1975 to 1990 showed that in some sections of the river there had been a complete loss in natural productivity of fish (Savvinov *et al.* 1992; Savvinov 1993; Shipkov 1999), and both the volume and diversity of fish had been dramatically reduced (Egorov 1990).

As a part of a programme for seismic and geological prospecting, boosting oil and gas production, creating underground gas stores and excavating reservoirs and dams, the Soviet government performed 12 underground nuclear explosions in the Republic of Sakha (Yakutia) between 1974 and 1987. Two of the 12 explosions, Kristall and Kraton-3, had catastrophic above-ground fallout, releasing caesium-137, strontium-90, plutonium-239, plutonium-240, and americium-241 into the atmosphere (Burtsev 1993; Burtsev and Kolodeznikova 1997). Kristall was the first of eight explosions planned to free the subsoil of permafrost to build a dam for one of several waste filtration ponds of the Udachny diamond industry. In 1978, a charge of 1.7 kilotons was placed at a depth of 98 metres. The explosion resulted in the release of radionuclides into the atmosphere, killing 550,000 square metres of forest. After the catastrophic fallout of this first explosion, the remaining seven were cancelled. In the 1990s, levels of beta-radiation in the area of the nuclear explosion were five to six-and-a-half times above the natural background level, and only in 1992 was reclamation work undertaken around Kristall to cover the area with a protective sarcophagus (Burtsev and Kolodeznikova 1997; Yakovleva 2002).

At the time of the nuclear explosion the local Vilyuy populations were not given the information needed to protect themselves and their communities. To this day Vilyuy Sakha elders recall the dark cloud that passed over their community at the time of the explosion and left several children, who were outside playing at the time, fatally ill (Crate 2003a). The public only knew the facts after the release in the 1980s of information about the gamut of environmental impacts of diamond mining, including information about the nuclear explosions conducted in the Republic (Crate 2003a; Yakovleva and Alabaster 2001). Once the information did become public, the government paid no heed to citizens' appeal that the affected areas be properly analysed and rehabilitated and that their communities' health be appropriately protected. Despite the life-threatening nature of these accidents, there has yet to be a comprehensive study of the affected areas.

Human health effects

The deterioration of human health in the Vilyuy watershed is the most potent and well-studied environmental effect of diamond-mining activities. The Vilyuy watershed is inhabited by 192,000 people, 18% of the total population of the Republic of Sakha (Yakutia) and 35% of the total indigenous population. Morbidity rates in this area are higher than in the Republic of Sakha (Yakutia) overall (Krivoshapkin *et al.* 1990), with increases in viral hepatitis, gastrointestinal infections and dysentery linked to the poor quality of drinking water (Petrova and Volozhin 1996). Metal contamination is a pervasive problem. A high content of cadmium, copper and nickel is found in the blood of citizens of Syuldukar, a high content of chromium in blood tests of citizens from Djarkhan, and a high content of nickel in the hair of people from Egoldja settlement. The incidence of malignant tumours in the indigenous population in the Vilyuy region is considerably higher than the Republic average, as are congenital anomalies in the cardiovascular system, large vein and endocrine pathologies, and chronic inflammatory and allergic diseases of the respiratory organs and urogenital system (Petrova and Volozhin 1996).

The effects on livelihoods and social structures

Establishing the diamond-mining industry meant the relocation of local populations. Flooding for the hydro-electric reservoir displaced 600 people along the Chona River, a tributary of the Vilyuy (Crate 2006a, 2003b). These communities made up one of the last two 'outposts' of reindeer communities left in the Vilyuy regions at that time. Their relocation delivered the final blow to these two groups, the Shologinsky (meaning 'inhabitants of the upper river areas' (Tugolukov 1986: 188) and the Sadinksy, whose ancestors had settled in these more mountainous areas several centuries earlier (Crate 2006a, 2003b). Additionally, immigrants practising poaching and the intrusive effects and fragmentation resulting from uncontrolled transport links displaced local inhabitants from their hunting, foraging and fishing.

The discovery of diamonds on the Vilyuy reoriented many priorities of rural Sakha livelihoods. From the late 1950s' state farm consolidation until the late 1980s, diamonds for Vilyuy inhabitants were a symbol of regional pride and their local contribution to building the Soviet society, mainly through defence and armaments (Argunova-Low

2004: 259). Although most rural Vilyuy Sakha never held real diamonds, they knew that all work efforts on the state farm were in their name. The recently established state-owned farms were required to produce meat and milk for the diamond industry, essentially colonising Vilyuy Sakha as resource suppliers for the Soviet industrial complex (Crate 2006a, 2003c). In the change from collective to state farming, Vilyuy Sakha went from being part-owners of collective operations to being members of the working class in a vast state farm system. Vilyuy Sakha had a serf-like relationship with the diamond industry. Instead of earning for their cooperative's annual profit in meat and milk, workers punched the clock for bi-weekly wages.

The move to a working class, receiving pay cheques in place of locally produced food, changed the indigenous diet. The government rationed meat to one kilogram a month for state farm workers' household consumption in order to increase the supply of foodstuffs to the mining areas. Households could consume more meat if they kept their own cows. However, most did not have the time or energy for private cows since work on the state farm kept them busy for 10–12 hours a day. At this point most people had more money than time and so the tendency was for the natural Sakha diet to be replaced by imported store-bought foods. Dairy products, especially milk and cream, were also scarce consumer products. Inhabitants of Elgeeii village, the state farm centre, recall the daily helicopter that came to pick up a ton of crème fraiche for delivery to the diamond workers.

The state farm consolidation affected Vilyuy Sakha work regimes. In order to increase production, jobs became specialised with workers performing limited tasks with large numbers of animals. A milker, who formerly tended their cows' new calves, hauled water, hay and supplemental feed, cleaned stalls and hauled manure, now only milked. The other duties were assigned to new specialists: the calf watcher, the barn manure cleaner and hauler, and the feed and water carrier. In the course of a generation, the change to specialisation worked to erode much of Vilyuy Sakha's traditional knowledge of horse and cattle breeding.

Concomitantly, with the advent of the state farm, Vilyuy Sakha were forced to abandon much of their subsistence production, including their private herding, fishing, hunting and foraging. These activities, if rendered profitable by the government, were now assigned to work brigades. However, mining contamination of the Vilyuy River resulted in a loss of many prize fish species. The river water was also the main drinking source for Vilyuy Sakha villages located near the river. In the same way that the contamination of drinking water in the villages affected human health, it also affected the health of domestic horses and cattle, which in turn affected the human populations consuming those products (Crate 2006a; Petrova and Volozhin 1996).

Mining changed the demographics of the Vilyuy regions. The Soviet government 'imported' workers from Ukraine, Belorussia and European Russia, changing the proportion of indigenous (Sakha, Evenk, Even, Chukchi, Yukagir and Dolgan) to migrant (Crate 2006a, 2003a). In 1926, before the beginning of wide-scale industrial development in the Republic of Sakha (Yakutia), migrants comprised 12% of the total population as compared to 1989 when they made up 64% (Poiseev 1999).

The local indigenous inhabitants were not initially recruited to work in the diamond industry, mostly because their labour was considered more valuable producing the meat and milk necessary to fuel immigrant-worker populations (Crate 2006a). Diamond mining on the Vilyuy has resulted in increasing income disparity between agricultural (mostly indigenous) and mining (mostly immigrants) workers. Whereas in 1985 an agricultural worker received 82% of what an industrial worker received, in 1994 this was only 43% (Poiseev 1999; Yakovleva 2002). In conjunction with the stark contrast between living conditions in rural villages and mining centres, such income disparity created significant rural–urban tension (Crate 2006a).

Routes to addressing interests of indigenous people in the diamond province

Voice of the community in the diamond province

Since the late 1980s, on the heels of *glasnost* (openness) and *perestroika* (restructuring), when inhabitants across the Soviet Union first gained access to information about the environmental offences of the Soviet period, Vilyuy Sakha have actively voiced their concerns about the social and environmental issues of diamond mining. Concerned citizens in collaboration with representatives of the capital city Yakutsk's scientific intellectual community founded the Public Ecological Centre in early 1989 to research and disseminate information about the environmental legacies of their homelands. Later that autumn, the centre organised a Republic-wide conference to discuss ecological problems and to form regional watershed-based organisations to initiate local activism. One of these organisations, the Vilyuy Committee, formed by Pyoter Martinov and other concerned Nyurba residents, focused on the Vilyuy watershed.[4] Concomitantly the Vilyuy Committee and the Public Ecological Centre organised a conference in Mirny, the diamond-mining centre, to meet with representatives of ALROSA (Crate 2006a, 2003a). On 30 October 1990, the members of the conference signed an appeal to the Supreme Soviet of the USSR, the Supreme Soviet of the RSFSR (Russian Soviet Federative Socialist Republic, one of the 15 republics of the former Soviet Union and the territory now making up the Russian Federation) and the Supreme Soviet of the Yakutia-Soviet Socialist Republic (SSR) demanding immediate actions for environmental improvement in the Vilyuy regions (Yakovleva 2002).

Concerned citizens throughout the diamond province initiated local chapters of the Vilyuy Committee and began taking action by organising village meetings and discussing environmental concerns (Crate 2006a, 2003a). In 1992, heads of local authorities in Nyurba, Suntar, Verkhnevilyuysk and Vilyuysk districts and the Vilyuy Committee signed a similar appeal to the Supreme Soviet of the Russia Federation, the President of the Republic of Sakha (Yakutia) and the Supreme Soviet of the Republic of Sakha (Yakutia) (Pavlov and Afanasieva 1997). In addition, local authorities from Kobyai district, government agencies, associations and other organisations approached the Supreme Soviet of the Republic of Sakha (Yakutia) with propositions to introduce actions for environmental improvement and socioeconomic development in the Vilyuy River basin (Yakovleva 2002).

4 Personal communication with P.N. Martinov, Nyurba, 1997.

These citizen actions resulted in a number of unsuccessful environmental initiatives by the regional government. As early as 1990, the Supreme Soviet of the Yakutia-SSR commissioned the Council of Ministers (later the Government of the Republic of Sakha [Yakutia]) to develop a special programme for environmental improvements in the Vilyuy River basin. However, the programme failed due to lack of funding (Yakovleva 2002). During 1990–91, the Supreme Soviet of the Republic appealed to the Soviet and Russian authorities to assign the Vilyuy River basin with a status of 'environmentally unfavourable zone', which would have given a legal basis for developing programmes for rehabilitation of environmental damage caused by industrial developments in the diamond province. However, the Supreme Soviet and Council of Ministers of the USSR and Supreme Soviet and Council of Ministers of the RSFSR rejected these initiatives (Yakovleva 2002).

Following a lack of support from central authorities to recognise the significance of environmental problems in the diamond province, in 1992 the regional authorities decided to address the problems in the Vilyuy River basin using regional resources. On 20 March 1992, the Presidium of the Supreme Soviet of the Republic of Sakha (Yakutia) issued a Decree 'about a set of primary actions for improvement of the environmental situation in the Vilyuy river basin'.[5] This important document formulated the official position of the regional authorities towards the environmental problems in the Vilyuy regions (Pavlov and Afanasieva 1997). Following this decision, the government established a special fund (Fund of Vilyuy Ecology) for environmental rehabilitation of the Vilyuy River basin.[6] However, the plan for the implementation of the Fund of Vilyuy Ecology was soon overturned (Mironov 1999).

The citizen activism that brought such promise in the early 1990s similarly was severely hampered over the course of the years to follow. Initially, the Vilyuy Committee, having started its work within the membership of the Public Ecological Centre in Yakutsk, had substantial support from many government representatives and a lot of coverage in the press. 'Our press supported us and played a big role in getting the word out', commented Lyubov Yegorova, one of the original founders of the Vilyuy Committee:

5 The Presidium agreed in general with the proposed programme for environmental protection and socioeconomic development actions for 1992–95 in the Vilyuy River basin, brought by the local authorities and heads of administration in Kobyai, Vilyuysk, Verkhnevilyuysk, Nyurba and Suntar districts, ministries, associations and organisations of the Republic. This Decree stipulated that the Government of the Republic of Sakha: (1) define the status of the environmental situation in the Vilyuy River basin and define the volume of damage, caused by Yakutlamaz Corporation, Yakutenergo Corporation geological exploration, to the environment and health of the population in the Vilyuy River basin; (2) develop and adopt the programme for compensation for the population of the region affected by the damage using endowment from the Russian Federation, Republic of Sakha (Yakutia), Yakutalamaz Corporation, Yakutenergo Corporation, Lenaneftegasgeologia Corporation; (3) create a special fund for environmental rehabilitation of the Vilyuy River basin endowed by assignments from the Government of the Russian Federation, Government of the Republic of Sakha, Administration of districts, Yakutalmaz Corporation and Yakutskenergo Corporation.

6 The Government of the Republic of Sakha issued a Decree on 27 April 1992, No. 201 'About emergency socioeconomic and environmental protection actions for rehabilitation of environmental situation in the Vilyuy river basin'. The Decree was signed by the Vice President of the Republic of Sakha, Vyacheslav Shtyrov, who in 1993 became co-chairman of the Supervisory Council of ALROSA, in 1995 a president of ALROSA, and in 2002 the President of the Republic of Sakha.

> Soon the whole Republic knew about 'The Tragedy of the Vilyuy', and all the other things they called it. Everyone knew about what was going on. We kept talking about the problems, and then in 1991, with the help of the Committee to Save the World and the Sakha Minister of Ecology, and we shot a film about . . . the radiation problems and all the ecological problems. We were working, and we gathered a lot of money by showing the ecological problems throughout the Republic.[7]

Public and institutional support to ameliorate the ecological situation on the Vilyuy began to wane when, in 1994, the diamond company announced its discovery of 'the biggest diamond pipe in the Vilyuy regions', located near several indigenous communities in the Nyurba region. Local inhabitants openly voiced their opposition to these new mines, claiming that their settlements had already received their share of environmental havoc from previous mining activity. The Ministry for Nature Protection (the new post-Soviet name for the Ministry of Ecology of the Sakha Republic) adopted higher environmental standards from the mid-1990s, but what tangible changes in practices have resulted remains unclear. The Ministry promised extensive monitoring of the new diamond area. However, most inhabitants did not believe these promises,[8] and in fact the Ministry later curtailed monitoring owing to a lack of funds (Crate 2006a, 2003a).

Pyoter Martinov, a life-long resident of the Nyurba region experienced in the technology of diamond transport and passionate about citizen advocacy, travelled with most of the ecological expeditions, researching the nuclear accident sites in the Vilyuy and to Yakutsk to meet representatives in the state parliament. In 1996 Martinov spearheaded several referendums, demanding environmentally safe technology, to impede the diamond industry's plans to exploit new diamond pipes. The referendums only temporarily set back the company's plans. In 1997 Pyoter Martinov died an untimely death after a prolonged struggle with liver cancer. Martinov was the guiding vision and force behind the Vilyuy Committee and his loss was a major setback to its efforts. After Martinov's death, most of the original Vilyuy Committee members left the organisation. Concurrently, the diamond company started a propaganda campaign aimed at averting the remaining environmental concerns.[9] Between 1997 and 1999, Vilyuy citizens were told outright that, if they pushed their rights to a clean environment too far, they would risk losing their state salaries, subsidies and pensions. It was evident that these threats worked, as an active and concerned citizenry turned into a silent and apathetic one over the course of those two years (Crate 2006a, 2003a).

The regional representatives of the Vilyuy Committee also changed markedly. While the committee was preparing to celebrate its tenth anniversary in 2000, a complete turnover of membership produced new priorities that were anything but environmental. The new personnel, all key figures in regional economic development, transformed

7 Personal interview with L. Yegorova, Yakutsk, 2003. Former Sakha president Mikhail Nikolaev also played an advocacy role by writing an article in the Russian newspaper about the atomic explosions on the Vilyuy.

8 Personal interview with P.N. Martinov, Nyurba, 1996.

9 The company did conduct some citizens' forums before the opening of the new kimberlite pipes Botuobinskaya and Nyurbinskaya, but their significance and the scale of citizen inclusion remain unclear. ALROSA-Nyurba also made arrangements with the local administrations on training and recruitment of the local population to work in diamond-cutting factories, which are state-of-the-art and maintain standards of operation much higher than the old factories in Mirny and Aikhal.

the Vilyuy Committee from an environmental NGO focused on involving the citizenry in environmental activism, to a bureaucratic board of local officials who gather privately to discuss their plans. In short, the active environmental agenda of the original Vilyuy Committee had been successfully co-opted (Crate 2003a).

Industry and government policy towards communities in the diamond province

In tandem with the ebb and flow of activism among Vilyuy citizens, the industry has also made attempts, mostly failed, to ameliorate its social and environmental legacy on the Vilyuy. Table 12.1 charts industrial and community activity beginning in the 1980s.

During the Soviet time, the extractive industry in the Republic of Sakha (Yakutia) operated under a slogan 'Give more gold and diamonds' and issues of community and environmental protection were largely neglected. There was some improvement in environmental protection in the diamond-mining industry in the 1980s. On 15 January 1981, the Central Committee of the Communist Party of the Soviet Union and the Council of Ministers of the USSR issued a Decree on 'actions for enforcement of protection of seas, rivers and other watercourses of the Arctic basin from contamination'. It requested an improvement of effluent treatment facilities to eliminate the discharge of untreated waste-water into watercourses and to reduce the discharge of contaminants into the atmosphere. In 1982, Yakutalmaz developed the first plan in the Republic to eliminate the discharge of untreated waste-water into watercourses. In 1986, Yakutalmaz constructed closed-cycle tailings for diamond enrichment factories, aiming to solve the problem of direct discharge of underground water from quarries into the river system by storing them in specially designed reservoirs and constructed sewage treatment facilities in the cities of Mirny and Udachny and the settlement Chernyshevsky.[10] However investigations since that time have shown that the system is in many ways faulty, since the holding ponds overflow in the spring floods, and contaminants find their way into the watershed system (Crate 2003a).

The next major changes in environmental and community policy in the diamond-mining industry occurred during the 1990s, when the industry was privatised[11] (Yakovleva 2002). First, the regional and local authorities received greater economic power over diamond mining in the Republic of Sakha (Yakutia). When ALROSA was created in 1992, the company's shareholders were: the Government of the Russian Federation and the Government of the Republic of Sakha (Yakutia), both holding 32% of shares; employees of ALROSA holding 23%; the Administrations of eight districts of the diamond province, holding 8%; and a Fund for guarantees of servicemen, an organisation which

10 Interview with a representative of the Ministry for Nature Protection of the Republic of Sakha, 30 July 2001 (Yakovleva 2002).

11 The Yakutalmaz Corporation was a state-owned enterprise that was governed and funded by the Department in the Government of the Soviet Union. Yakutalmaz was liquidated as a legal entity, because there was no more Soviet Union. A new private enterprise was created, a joint-stock venture, with shares owned by various shareholder organisations, government departments, municipalities and individuals. However, the majority of shares were given to the Government of the Russian Federation and the Government of the Republic of Sakha (Yakutia). ALROSA is a private company working under the regulations of a private enterprise but controlled by ownership by the government. The formation of ALROSA happened during the privatisation reform of Russian industry.

TABLE 12.1 Industrial and community activity in the Vilyuy regions

Year	Industry actions	Result	Year	Community actions	Result
1981	Central Committee Decree on waste-water	Yakutalmaz developed plan			
1982	Yakutalmaz plan to defer waste-water				
1983–1988	Waste-water discharged into watercourses	River/drinking water contamination			
1988–1993	Waste-water discharged to reservoirs	Proved faulty; continued river/drinking water contamination	1989	Public ecological centre and Vilyuy Committee formed	
			1990	Vilyuy Committee formed	
1992	ALROSA created		1992	• Appeal to central government • Regional efforts to rehabilitate watershed resulting in establishment of Fund for Vilyuy Ecology	• Failed due to lack of funds • Failed due to change of political system
1993	SAPI formed		1994	Began speaking out on new diamond mines in Nyurba	
1993–1999	SAPI activity: economic development, construction, etc.	Did not address environmental issues adequately; money misused	1996	Vilyuy Committee submits referendums	Only temporarily stalled diamond company
			1997	Pyoter Martinev passes away	Vilyuy Committee lost activist focus
			1997–1999	Propagandising to inhabitants	Citizen activism squashed
2000	Target fund established	Addressed environmental issues as part of government programmes			

became a shareholder of ALROSA under the Government of Russian Federation, holding 5% (Yakovleva 2005). Since then some redistribution of shares has occurred. Currently the majority shareholder is the Government of the Russian Federation with 37% of shares; the Government of the Republic of Sakha (Yakutia) holds 32%, employees of ALROSA and other organisations hold 23%, and the Administrations of eight districts in the diamond province hold 8% (ALROSA 2006). Importantly, the inclusion of the Administrations of eight districts from the diamond province was a first step towards acknowledging that the local communities have a stake in natural resource exploitation in their territories.

Second, following political changes in the relationship between the Republic and the Russian Federation, the diamond-mining developments were considered to be the property of the Republic. After creating the private corporation ALROSA, the corporation and the Republic signed an agreement 'Lease of land sites, diamond deposits, other natural resources and fixed and current assets of Yakutalmaz Corporation'. This allowed ALROSA to lease diamond deposits for a 50-year period, stipulating several types of leasing payments, including payments for a specially created environmental fund for the rehabilitation of environmental damage in the diamond province, amounting to 2% of the value of ALROSA's diamond output.

Following the signing of the lease agreement, in 1993 the Sakha government established the Sakha Diamond Province Investment Company (SAPI) Financial Corporation to receive the 2% of ALROSA's revenue and fund remedial socioeconomic and environmental programmes in the diamond province (Crate 2006a, 2003a; Yakovleva and Alabaster 2001). The SAPI Foundation delivered a number of initiatives from 1993 to 1999.[12] However, over this period only 10% of the US$23 million yearly average income was allocated to environmental amelioration. More importantly, regional representatives revealed that SAPI was fraught with issues of state-level, regional and local theft and corruption, and that SAPI funds were often depleted at the regional level, before reaching village populations, those most affected by environmental contamination.[13] The Sakha Republic's president at that time, Mikhail Nikolaev, made direct references to suspicions of the company's mismanagement: 'We await good work from SAPI.'[14]

After disputes between the SAPI Foundation and the government of the Republic, in 2000 the Sakha government created a Target Fund which continued to deliver programmes in the diamond province[15] (Yakovleva and Alabaster 2003). These monies, intended for environmental rehabilitation, were used for various initiatives, most of which were not directly linked with the environment, but rather with economic development such as construction, support for diamond cutting, and oil and gas developments (Yakovleva and Alabaster 2003).

12 SAPI Foundation delivered various initiatives in the diamond province in relation to business development, community development, culture, arts and sports, economic development, education and training, environmental protection, health, job creation, research, agriculture development, social welfare and women.

13 Personal interview with A. Romanov, Nyurba, 1997.

14 Personal interview with V.G. Alekseev, Yakutsk, 1997.

15 The full name of the Target Fund is 'Target budget fund for implementation of socioeconomic and environmental programmes in *uluses* [districts] of the diamond province, established by the Decree of President of the Republic of Sakha (Yakutia) in 2000'.

The third change in environmental and community policy during the 1990s was that ALROSA itself undertook a range of initiatives targeting the population in the diamond province via a corporate social programme and corporate regional policies. To this day, ALROSA engages in community development initiatives in eight districts of the diamond province including training local community members for future employment in ALROSA, supporting social programmes targeting populations in the diamond province, provisioning corporate employees with products from the local agricultural industry, developing patronage relationships between ALROSA and local community organisations, and supporting sports and cultural organisations (Yakovleva 2005). However, ALROSA has to date not implemented the corporate social responsibility practices that are increasingly commonplace in corporations and that aim to maximise the corporation's productivity and efficiency, minimise impacts that will affect future generations, and address all environmental implications of operations, facilities and products, including the elimination of wastes. The company does have procedures in place for environmental protection, health and safety, personnel policy and community development, and it does allocate resources and staff, plans and implements programmes within each area and has a system of internal reporting and assessment. However, these measures are designed only to meet regulatory requirements, and so far ALROSA has not been willing to go beyond these requirements. The company has therefore adopted a reactive approach to environmental issues and limits its activities to conforming with, and at times failing to conform with, environmental legislation.

ALROSA also signed a series of agreements with the eight administrative districts in the diamond provinces supporting small business development, such as diamond cutting, food processing and agriculture and education and training. ALROSA provides support for school children, funding health expenses, sporting activities, recreation and social care and it also supports local voluntary organisations and contributes to rural communities in kind (transport and fuel) (Yakovleva 2002).

ALROSA also collaborates with governmental agencies of the Republic of Sakha (Yakutia) in two ways: via a specific agreement between ALROSA and governmental agencies and in response to specific events or issues. For example, according to the agreement between ALROSA and the Ministry for Nature Protection of the Republic of Sakha (Yakutia), the company undertook an initiative for nature conservation in the diamond province that included the creation of an educational zoological park featuring animals native to the Republic. ALROSA, as the leading enterprise of the region, is frequently approached by the government to respond to specific events in the Republic of Sakha (Yakutia). For example, when spring floods damaged many settlements in the Republic of Sakha (Yakutia) in 2001, the company spent about US$130 million for restoration of regional infrastructure (Yakovleva 2002).

The impetus for the enhancement of the diamond mining company's community relations came from the concerns of local communities, wider general public and local and regional authorities (Yakovleva 2005). It is impossible to deny that ALROSA has contributed towards social development in the region, and since the 1990s it has taken steps to rectify environmental damage. However, these efforts have not been proven to be effective. The changes in the operations of the diamond industry towards society and the environment were linked to social pressures that have affected the company, specifically: (a) increased public awareness about environmental and social impacts and risks of mining development, particularly since the late 1980s; (b) concerns about compensation for damage caused by the mining development to local communities; (c) employee concerns over their welfare; (d) expectations of local authorities for input into community development (Yakovleva 2005).

Bridging the gap: the human rights and environmental justice implications of Vilyuy diamonds

Despite ALROSA's impressive portfolio of activities and investments, in many ways the company, the SAPI Foundation, the Target Fund and the government are 'missing the target'. Real rehabilitation will proceed only when the environmental damages of the Soviet and post-Soviet periods are not only recognised but accounted for and ameliorated. One pervasive issue that illustrates how the many targets of environmental amelioration in the Vilyuy regions are missed is that of safe drinking water. In the context of a three-year community sustainability project, during focus group sessions discussing the hurdles to future village-level sustainability, local inhabitants were first and foremost concerned about the lack of safe drinking water (Crate 2006b). They offered countless testimonies about the serious health effects a lack of safe water has meant for them, their communities and their animals.

In the words of Vasili Alekseev, the now former Minister of Ecology of the Sakha Republic, the Target Fund was established 'to ensure that the money has a definite address'.[16] But, while Alekseev and his regional representatives all agree that the first and foremost need of Vilyuy inhabitants is safe drinking water, target fund monies are allocated to develop gas, oil and high-voltage electricity development on the Vilyuy (Crate 2006a, 2003a).

The issue of safe drinking water on the Vilyuy is complicated by the ecosystem's subarctic water regime. There is a lack of adequate water supply in the Vilyuy regions throughout the year. Due to low levels of precipitation, less than 200 mm per year and pervasive permafrost, the Republic of Sakha (Yakutia) is classified as an area with relatively low availability of drinking water. This is in part why settlements are located close to waterways (Petrova and Volozhin 1996). However, due to industrial and agricultural contamination of the Vilyuy River and other waters in the diamond province, there are practically no natural sources of water in the Vilyuy regions that would meet standards for drinking water. Therefore, there is a need to install water purification facilities in Vilyuy settlements. The urban areas of Lensk, Mirny, Aikhal, Udachny and Chernyshevsky have water-purification systems funded by the diamond-mining industry. However, significant financing is required to construct purification facilities in the rural areas of the diamond province. The municipalities themselves have no financial means to construct these facilities in every village. One solution to the problem could be a centralised water supply with transportation of water to smaller villages, though this would also require a significant investment, considering the huge distances between the settlements.[17]

16 Personal interview with V.G. Alekseev, Yakutsk, 2003.

17 Interview with a representative of Vilyuy Basin Committee for Nature Protection, a branch of the Ministry for Nature Protection of the Republic of Sakha (Yakutia), 20 August 2001 (Yakovleva 2002).

The SAPI Foundation had initially planned to construct water supply and sewage facilities in the rural areas. However, from 1993 to 1999 the Foundation did not allocate sufficient funds for the improvement of drinking water in the diamond province, apart from several operations for the construction of roads to water collection points in some districts and a sewage facility in Nyurba. Since the establishment of the Target Fund in 2000, a portion of its revenue goes towards a Republic-wide programme for the provision of drinking water, delivered by the Ministry for Nature Protection of the Republic of Sakha (Yakutia) and sourced from the Government of the Russian Federation, the Government of the Republic of Sakha (Yakutia) and the Target Fund. This means that the Target Fund, which was specifically set up to invest in the diamond province, could be dispersed in programmes for the entire Republic, with little benefit for the drinking-water situation in the diamond province.[18]

When asked why they were not more vocal about having safe drinking water and their other basic human rights addressed by their government, most citizens of the Vilyuy expressed concern that they would lose their salaries, pensions and subsidies.[19] Despite their extensive scientific and anecdotal knowledge that regional diamond-mining activities generate a complex of environmental issues that directly affect their health and the health of their local environment, residents continue to be reluctant to speak out for fear of the consequences.

For Vilyuy Sakha two key points should be emphasised. First, although they are able to participate in the government of their region, their lack of special status as an indigenous people within the Russian Federation limits their ability to realise greater concessions. According to legislation, the state provides support and protection to ethnic minorities who: consider themselves an independent ethnic entity; occupy the territories of traditional settlement of their ancestors; retain traditional lifestyles, economies and trades; and whose population is not greater than fifty thousand. Due to their large numbers, Sakha are not recognised within the Russian Federation as indigenous.

Second, the early 1990s initiatives for regional sovereignty in the Republic, and attempts to build regional power by capturing or partaking in the control over economic resources, were not fully developed. The battle for sovereignty of the Republic of Sakha (Yakutia) was dedicated to the acquisition of economic powers that would lift the Republic from the status of a resource colony. The Declaration of Sovereignty of the Republic of Sakha (Yakutia), adopted in 1990, established the Republic of Sakha (Yakutia) as a sovereign state within the RSFSR, and also attempted to establish the rights of the Republic's population to natural resources: 'Land, its minerals, water, forests, flora and fauna, other natural resources, air space and the continental shelf on the territory of the republic shall be its exclusive property.' The Republic of Sakha (Yakutia) has been one of the most active regions in Russia to establish special relationships with the federal centre, and it was able to secure important economic provisions such as the distribution of revenues from the regional production of diamonds and gold. Economic demands and opportunity to retain and manage the Republic's wealth to alleviate social and economic crisis in Sakha lie at the heart of political agenda of the Republic (Jackson and Lynn 2002).

18 Interview with a representative of the Ministry for Nature Protection of the Republic of Sakha (Yakutia), 22 August 2001 (Yakovleva 2002).

19 Interviews and field data, Suntar and Nyurba regions, Sakha, Russia, 1996–97.

However the development of centre–periphery relations is ongoing and the initial liberties given to the development of regions in Russia have been under scrutiny under President Putin and now President Medvedev. The Constitution of the Republic of Sakha (Yakutia) that was adopted on 27 April 1992 did not coincide fully with the Constitution of the Russian Federation that was adopted later on 12 December 1993. Article 72 of the Constitution of the Russian Federation stipulates that the following issues are under joint jurisdiction of the Russian Federation and of the subjects of the Russian Federation (i.e. republics, *krays, oblasts, okrugs*):

> c) issues of the possession, use and management of the land, mineral resources, water and other natural resources; d) delimitation of state property; e) management of natural resources, protection of the environment and ecological safety; specially protected natural reserves; protection of historical and cultural monuments.

President Putin later ordered the unification of principles of regional statutes within the Russian Constitution, which is the main law ruling the Russian Federation, and the Constitution of the Republic of Sakha (Yakutia) was accordingly altered.

Conclusion

Even in 2008, with the democratisation of Russia in process for over 15 years, the plight of Vilyuy Sakha, like that of other native peoples across Russia, has not improved, in part because of continued economic and political instability. For Vilyuy Sakha this is in part owing both to continuing centre–periphery tensions and to the fact that rural Sakha and urban/political Sakha appear to have different interests. There are many Sakha of the Vilyuy regions who have attained political power with at least an initial intention to better the life of the rural areas but, in the process of gaining a place in government, have ended up severely abusing their power. The voice of Sakha is not that united after all. Additionally, with President Putin's and now President Medvedev's emphasis on natural resource development as Russia's primary source of economic recovery and with much of that resource wealth found in indigenous areas, it appears that environmental issues will have a low priority (Peterson 2001, 2002).

This is clearly the case for Vilyuy Sakha. In recent years the Russian government has been assuming more control over ALROSA. This action has met with considerable citizen protests in the capital Yakutsk.[20] There are also rumours that the Sakha Republic may lose its status as a republic and become a district of lower status, so that the Federation can extract more of its resource wealth.[21] The significance of this move is the potential influence it would have on the ability of Sakha not only to reap benefits from their natural resources, but to actually control their development. An environmental justice

20 There are several websites that feature the citizen protests. See diaspora.sakhaopenworld.org/alrosa6.shtml, accessed 15 August 2008; www.regnum.ru/news/519845.html, accessed 15 August 2008.

21 Personal interview with I. Shamaev, Yakutsk, 2005. See also www.knia.ru/news/489.html, accessed 15 August 2008.

model fits Vilyuy Sakha's plight well if we understand its frame to include issues of racism, elitism and economic disparity as factors in the 'unequal siting of environmentally undesirable land uses, routine marginalization from environmental decision-making processes and denial of just compensation and informed consent in environmental matters' (Sandler and Pezullo 2007: 8). With the autonomy of the entire Sakha nation at stake, there are clear issues of justice and human rights that need to be brought to the fore (Crate forthcoming).

Vilyuy Sakha are not alone in their plight. Across Russia indigenous peoples are struggling to resolve the environmental issues of the vast territories they depend on. The contamination of indigenous lands by chronic oil spills, radioactive leaks and surface water pollution, in addition to infringements on land areas for resource extraction, continues to threaten the cultural survival of Russia's indigenous populations (Kohler and Wessendorf 2002; Forbes 1989, 1999; Wiget and Balalaeva 2000, 1997). If we look to other world areas where indigenous populations are confronted with similar issues but have been able to win some concessions to protect their environments, economies and health, we find they are in areas where indigenous peoples have land claims and a certain level of self-government and self-determination (Crate 2006a, 2006b). Additionally, there exist mechanisms for corporate social responsibility, which in Russia were absent until the 1990s.

Russia's indigenous peoples, like their counterparts worldwide, desire control over their lives, economies and local resources, and hope for their children's sustainable future and for that of the coming generations. Post-Soviet Russia's indigenous communities struggle daily with the reality of failing political, economic, ecological and social systems. This is further complicated both by the Soviet legacy that undermined local ecological knowledge, kinship settlement patterns, land and resource rights, and healthy natural ecosystems, and by the contemporary effects of globalisation and modernity, which further erode local efforts to build economies and maintain communities (Crate 2006b). In addition, northern Russia is known as a place where the government largely ignores the interests and needs of local indigenous populations when devising natural resource management strategies, the effects of which further exacerbate a loss of exclusive rights in historically based economic spheres (Langlais 1999: 65).

The Vilyuy Sakha study presented here and other case studies of indigenous peoples in Russia show that these inhabitants of northern Russia have adapted creatively to the conditions of the transition (Crate and Nuttall 2004; Crate 2006a, 2006b). Many researchers working with Russia's indigenous peoples are operating in part with the hope that Russia's northern inhabitants will eventually realise similar levels of property rights, material compensation, self-determination arrangements and environmental justice, as witnessed in other parts of the circumpolar north: for example, in Alaska, Canada and Greenland. Key to the success of such similar moves in Russia will be involvement with the international community, most notably with indigenous groups fighting similar issues, with research initiatives, and with governmental bodies elsewhere in the circumpolar north that can facilitate the flow of ideas, experiences and examples of human rights and environmental justice movements to cross international boundaries. However, considering the Vilyuy Sakha case represented here and the trends towards the future for Russia in place at this time, it seems highly unlikely that Russia's indigenous peoples will realise their circumpolar counterparts' rights anytime soon.

References

AHDR (*Arctic Human Development Report*) (2004) (Akureyri, Iceland: Stefansson Arctic Institute).

Aldanzoloto (1994) *70 Let Akzionernoi Kompanii Aldanzoloto 1924–1994* [*70 Years of Aldanzoloto Company 1924–1994*] (Moscow: Aldanzoloto).

ALROSA (2006) *Annual Report 2005* (Mirny, Russia: ALROSA Company Ltd; www.alrosa.ru, accessed 20 August 2008).

Andreev, B.N. (1987) *Ptitsy Viliuiskovo Basseina* [*Birds of the Vilyuy Basin*] (Yakutsk, Russia: Yakutskoe Knizhnoe Izdatel'stvo).

Argunova-Low, T. (2004) 'Diamonds: Contested Symbol in the Republic of Sakha (Yakutia)', in E. Kasten (ed.), *Properties of Culture—Culture as Property: Pathways to Reform in Post-Soviet Siberia* (Berlin: Dietrich Reimer Verlag): 257-65.

Belykh, Z.P. (1993) *Novye grani almaznogo kraya* [*New Facets of Diamond Province*] (Moscow: Soyuzreklamkultura).

——, M.I. Vybornov and M.I. Nepomnyashii (eds.) (1984) *Lyudi i almazy* [*People and Diamonds*] (Yakutsk, Russia: Yakutskoe Knizhnoe Izdatel'stvo).

Berkes, F. (1999) *Sacred Ecology: Traditional Ecological Knowledge and Resource Management* (Philadelphia, PA: Taylor & Francis).

Brown, M. (2003) *Who Owns Native Culture?* (Cambridge, MA: Harvard University Press).

Burtsev, I.S. (1993) *Yadernoye zagryazneniye Respublika Sakha: Problema yadernaya bezopasnosti* [*Nuclear Contamination of the Sakha Republic: The Problem of Nuclear Safety*] (Yakutsk, Russia: Polygraph).

—— and E.N. Kolodeznikova (1997) *Raioazionnaya obstanovka v almazonosnykh raionakh Yakutii* [*Radioactive Situation in the Diamond Regions of Yakutia*] (Yakutsk, Russia: YNZ SO RAN [Yakutsk Scientific Centre, Siberian Branch, Russian Academy of Sciences]).

Caulfield, R. (1997) *Greenlanders, Whales, and Whaling: Sustainability and Self-determination in the Arctic* (Hanover, NH: University of New England Press).

Chance, N.A., and E.N. Andreeva (1995) 'Sustainability, Equity, and Natural Resource Development in Northwest Siberia and Arctic Alaska', *Human Ecology* 23: 217-40.

Crate, S.A. (1995) *Kwek tiin* [*Green Spirit*] (Yakutsk, Russia: Bichik).

—— (2003a) 'Co-option in Siberia: The Case of Diamonds and the Vilyuy Sakha', *Polar Geography* 26.4: 289-307.

—— (2003b) 'The Legacy of the Viliui Reindeer Herding Complex', *Cultural Survival Quarterly* 27.1: 25-27.

—— (2003c) 'Viliui Sakha Adaptation: A Subarctic Test of Netting's Smallholder Theory', *Human Ecology* 31.4: 499-528.

—— (2006a) *Cows, Kin and Globalization: An Ethnography of Sustainability* (Walnut Creek, CA: Alta Mira Press).

—— (2006b) 'Investigating Local Definitions of Sustainability in the Arctic: Insights from Post-Soviet Sakha Villages', *Arctic* 59.3:115-31.

—— (forthcoming) 'Viliui Sakha of Sub-arctic Russia and their Struggle for Environmental Justice', in J. Agyeman and Y. Ogneva-Himmelberger (eds.), *Environmental Justice and Sustainability in the Former Soviet Union* (Cambridge, MA: MIT Press).

—— and M. Nuttall (2004) 'Russia in the Circumpolar North', *Polar Geography* 27.2: 85-96.

Dahl, J., J. Hicks and P. Jull (2000) *Nunuvat: Inuit Regain Control of their Lands and their Lives* (Copenhagen: International Work Group for Indigenous Affairs).

Duval, D., T. Green and R. Louthean (1996) *The Mining Revolution* (London: Rosendale Press).

Egorov, E.G. (1990) 'Ekonomicheskie aspekty ekologii basseina Reki Vilyuy' ['Economic Aspects of Ecology in the Basin of the River Vilyuy'] in I.Ya. Egorov and V.F. Chernyavsky (eds.), *Vorposy regional'noj gigieny, sanitarii i epidemiologii* [*Questions of Regional Hygiene, Sanitation and Epidemiology*] (Yakutsk, Russia: Minzdrav Yakutskoi ASSR [Ministry of Health of the Yakut Autonomous Soviet Socialist Republic]): 22-23.

Forbes, B. (1989) 'The Indigenous Peoples of Siberia in the 20th Century', in A. Wood and R.A. French (eds.), *The Development of Siberia* (London: Macmillan): 72-95.

—— (1999) 'The End of the Earth: Threats to the Yamal Region's Cultural and Biological Diversity', *Wild Earth* 9.3: 46–50; home.planet.nl/~innusupp/english/forbes2.html, accessed 20 August 2008.

Forsyth, J. (1989) 'The Indigenous Peoples of Siberia in the 20th Century', in A. Wood and R.A. French (eds.), *The Development of Siberia* (London: Macmillan): 72-95.

FSGS (Federalnaya Sluzhba Gosudarstvennoi Statistiki Territorialny Organ po Respublike Sakha [Yakutia]) [Federal Service of State Statistics Territorial Body in the Republic of Sakha (Yakutia)] (2005) *Statisticheski Ezhegodnik Respubliki Sakha (Yakutia): Statisticheskii sbornik* [*Statistical Year Book for the Republic of Sakha (Yakutia): Statistical Collection*] (Yakutsk, Russia: Sakhapoligraphizdat).

Goskomekologia (2000) *Gosudarstvenny Dolkad o Sostoyanii Prirodnoi Sredy v Rossiiskoi Federazii v 1999 Godu* [*State Report on the Environmental Situation in the Russian Federation in 1999*] (Moscow: Gosudarstvenny Komitet po Ekologii [State Committee for the Environment]).

Grant, B. (1995) *In the Soviet House of Culture: A Century of Perestroikas* (Princeton, NJ: Princeton University Press).

Habeck, J. (2003) *Sustainable Development of the Pechora Region in a Changing Environment and Society* (Rovaniemi, Lapland: University of Lapland; www.ulapland.fi/home/arktinen/spice/spice.htm, accessed 20 August 2008).

ILO (International Labour Organisation) (2003) *ILO Convention on Indigenous and Tribal Peoples 1989 (No. 169): A Manual* (Geneva: International Labour Office; www.ilo.org/global/What_we_do/Publications/ILOBookstore/Orderonline/Books/lang--en/docName--WCMS_PUBL_9221134679_EN/index.htm, accessed 20 August 2008).

Jackson, L., and N. Lynn (2002) 'Constructing Federal Democracy in Russia: Debates over Structures of Power in the Regions', *Regional and Federal Studies* 12: 91-125.

Jull, P. (2003) 'The Politics of Sustainable Development: Reconciliation in Indigenous Hinterlands', in S. Jentoft *et al.* (eds.), *Indigenous Peoples: Resource Management and Global Rights* (Delft, Netherlands: Eburon Academic Publishers): 21-44.

Kharkiv, A.D., N.N. Zinchuk and V.M. Zuyev (1997) *Istoriia almaza* [*The History of Diamonds*] (Moscow: Nedra).

Kirrilov, A.F. (1990) 'O vliyanii zaregulorovannogo rechnogo stoka Vilyua na ekosystemu pritoplyaemogo uchastka' ['About the Impact of Artificial Flow Regulation on the River Vilyuy to the Flooded Area'], in I.Ya. Egorov and V.F. Chernyavsky (eds.), *Vorposy regional'noi gigieny, sanitarii i epidemiologii* [*Questions of Regional Hygiene, Sanitation and Epidemiology*] (Yakutsk, Russia: Minzdrav Yakutskoi ASSR [Ministry of Health of the Yakut Autonomous Soviet Socialist Republic]): 223.

Kiseleva, G. (1999) 'Put' k almazam' ['Road to Diamonds'], *Yakutia* 29.29055 (17 February 1999); www.gazetayakutia.ru, accessed 20 August 2008.

Kohler, T., and K. Wessendorf (eds.) (2002) *Towards a New Millennium: Ten Years of the Indigenous Movement in Russia* (Copenhagen: International Workgroup on Indigenous Affairs).

Kopylov, R.N., V.K. Marshintsev and M.M. Tyaptirgyanov (1993) 'Obschaya ekologisheskaya situatsia Territorii Yakutii' ['General Ecological Situation on the Territory of Yakutia'], in I.S. Burtsev (ed.), *Radioazionnoe zagryaznenie Territorii Respubliki Sakha (Yakutia): Problemy radiazionnoi bezopasnosti. Sbornik Dokladov Respublikanskoi nauchno-prakticheskoi konferenzii (Yakutsk, 14–15 janvarya 1993)* [*Radioactive Contamination in the Territory of the Republic of Sakha (Yakutia): Problems of Radioactive Safety. Collection of Reports from the Republic Academic and Research Conference (Yakutsk, 14–15 January 1993)*] (Yakutsk, Russia: Poligrafist): 3-10.

Krivoshapkin, V.G., Neustroeva, G.A., Timofeev, G.A. (1990) 'Promyshlennoe Zagryaznenie Basseina Reki Vilyuy i Sostoyanie Zdorovya Naseleniya v Gruppe Vilyuyskikh Rayonov' ['Industrial Pollution of the Vilyuy River Watershed and the Health Status of the Viliyuy Region Populations'], in I.Ya. Egorov and V.F. Chernyavsky (eds.), *Vorposy regional'noi gigieny, sanitarii i epidemiologii* [*Questions of Regional Hygiene, Sanitation and Epidemiology*] (Yakutsk, Russia: Minzdrav Yakutskoi ASSR [Ministry of Health of the Yakut Autonomous Soviet Socialist Republic]): 91-92.

Ksenofontov, G.V. (1937) *Uraangkhai Sakhalaar* [*Points in Ancient History of the Yakut (Sakha)*] (Vol. 2; Yakutsk, Russia: Natsinal'noye Izdatel'stvo; 2nd edn [1992]).

Labutina, T.M., and L.I. Smolenko (1990) 'Vliyanie promyshlennogo kompleksa na khimicheskii sostav prirodnykh vod' ['The Impact of Industrial Complex on Chemical Composition of Natural Waters'], in I.Ya. Egorov and V.F. Chernyavsky (eds.), *Vorposy regional'noi gigieny, sanitarii i epidemiologii* [*Questions of Regional Hygiene, Sanitation and Epidemiology*] (Yakutsk, Russia: Minzdrav Yakutskoi ASSR [Ministry of Health of the Yakut Autonomous Soviet Socialist Republic]): 221-22.

Langlais, R. (1999) 'Envisioning a Sustainable Arctic: Nunavut in Contrast to the Russian North', in H. Petersen and B. Poppel (eds.), *Dependency, Autonomy, and Sustainability in the Arctic* (Aldershot, UK: Ashgate:): 65-78.

Matveev, A. (1998) 'Almazy Rossii-Sakha: Sostoyanie, Perspektiby, Problemy Kompanii' ['Almazy Rossii-Sakha: Situation, Perspectives and Problems of the Company'], *Problemy Teorii i Praktiki Upravleniya* [*Journal of Problems of Theories and Practice of Management*] 3; www.uptp.ru/articles-all/articles-all_3121.html, accessed 20 August 2008.

—— and E.D. Cherny (2000) 'AK ALROSA Strategiya Razvitiya' ['ALROSA Company: Strategy for Development'], *Mineral'no-syr'evye Resursy Rossii* [*Journal of Mineral Resources of Russia*] 5–6: 58-67.

Mironov, V. (1999) 'Pyat let suschestvovaniya SAPI: No nelzya vse spikhivat' na nego' ['Five Years of the SAPI Operations: Should not be Entirely Blamed'], *Yakutia* 29066 (4 March 1999); www.gazetayakutia.ru, accessed 20 August 2008.

Nuttall, M. (1998) *Protecting the Arctic* (Amsterdam: Harwood Academic Press).

Pavlov, N.P., and V.M. Afanasieva (1997) *Tragediya i bol' sedogo Vilyuya* [*Tragedy and Pain of Grey Vilyuy*] (Yakutsk, Russia: SAPI-Torg-Kniga).

Peterson, D.J. (2001) 'The Reorganization of Russia's Environmental Bureaucracy: Implications and Prospects', *Post-Soviet Geography and Economics* 42.1: 65-76.

—— (2002) 'Russia's Industrial Infrastructure: A Risk Assessment', *Post-Soviet Geography and Economics* 43.1: 13-25.

Petrova, P.G., and L.I. Volozhin (1996) *Ekologiya cheloveka v usloviyakh severa: Yakutia, Respublika Sakha* [*Human Ecology in Conditions of the North: Yakutia, the Republic of Sakha*] (Moscow: Tipografiya Voenno-inzhinernoi Akademii).

Poiseev, I.I. (1999) *Ustoichivoe razvitie severa* [*Sustainable Development of the North*] (Novosibirsk, Russia: Nauka).

Sagers, M. (1997) 'Regional Trends in Russian Gold Production', *Post-Soviet Geography and Economics* 38.6: 315-56.

Sandler, R., and P. Pezzullo (2007) *Environmental Justice and Environmentalism: The Social Justice Challenge* (Cambridge, MA: MIT Press).

Savvinov, D. (ed.) (1993) *Ekologiya r. Vilyuy* [*Ecology of the River Vilyuy*] (Yakutsk, Russia: RAN Institut Prikladnoi Ekologii Severa [Russian Academy of Sciences, Institute of Applied Ecology of the North]).

——, M.M. Teptirgyanov and V.K. Marshintsev (1992) *Ekologiya basseina r. Vilyuy: promyshlennoe zagryaznenie* [*Ecology of the Basin of the River Vilyuy: Industrial Contamination*] (Yakutsk, Russia: YNZ RAN SO [Yakutsk Scientific Centre, Russian Academy of Sciences, Siberian Branch]).

Sejersen, F. (2002) *Local Knowledge, Sustainability, and Visionscapes in Greenland* (Eskimologis Skrifter No. 17; Copenhagen: University of Copenhagen).

Shipkov, R.Ju. (1999) 'Ekologicheskaya obstanovka almaznoi provinzii: problemy I puti esheniya: doklad zamestitelya predsedatelya Praviletstva RS (Y) na parlamentskikh slushaniyakh 26 marta 1999', ['Environmental Situation in the Diamond Province: Problems and Solutions: Report of the Vice Prime Minister of the Government of RS (Y) on Parliament Hearings on 26 March 1999'] (unpublished report; Yakutsk, Russia).

Shnirelman, V. (1999) 'Introduction: North Eurasia', in R. Blee and R. Dalys (eds.), *The Cambridge Encyclopedia of Hunters and Gatherers* (Cambridge, UK: Cambridge University Press): 119-73.

Sirina, A.A. (2005) 'Clan Communities among the Northern Indigenous Peoples of the Sakha (Yakutia) Republic: A Step to Self-determination?', in E. Kasten (ed.), *Rebuilding Identities: Pathways to Reform in Post-Soviet Siberia* (Berlin: Dietrich Reimer Verlag): 197-216.

Slezkine, Y. (1994) *Arctic Mirrors* (Ithaca, NY: Cornell University Press).

Tanin, S. (2001) 'Trubka mira: legendarny gorod stal garantom politicheskogo spokoistviya Respubliki Sakha' ['Mir Pipe: The Legendary City has Become a Guarantee of Political Stability in the Republic of Sakha'], *Nezavisimaya Gazeta—Regiony* 11.75 (3 July 2001); regions.ng.ru/far/1999-11-16/5_peace_pipe.html accessed 20 August 2008.

Tugolukov, V.A. (1985) *Tungusy Srednei I Zapadnoi Sibiri* [*Tungus of Middle and Western Siberia*] (Moscow: Nauka).

Vecherin, P.P. (1997) *Ot tresta do kompanii, ot palatok do gorodov: Khronologiya 'Yakutalmaza' 1957–1992 gg* [*From Trust to Company, from Tents to Cities: Chronology of 'Yakutalmaz' 1957–1992*] (Mirny, Russia: Mirninsky Rabochy).

Wiget, A., and O. Balalaeva. (1997) 'Black Snow: Oil and the Eastern Khanty'; www.nmsu.edu/~english/hc/IMPACTOIxL.html, accessed 13 February 2008.

—— and O. Balalaeva (2000) 'Alternative to Genocide: The Yuganskii Khanty Biosphere Preserve'; www.nmsu.edu/~english/hc/hcbiosphere.html, accessed 20 August 2008.

Wilson, R. (1999) 'Corporate Citizenship in a Multinational Business', paper presented to *Global Values, Global Difference Conference*, London, 1 July 1999 (London: Centre for Tomorrow's Company).

Yakovleva, N. (2002) 'Environmental and Social Responsibility in the Extractive Industry: A Case Study of Precious Metals and Minerals in the Republic of Sakha' (PhD thesis, University of Sunderland, UK).

—— (2005) *Corporate Social Responsibility in the Mining Industries* (Aldershot, UK: Ashgate).

—— and T. Alabaster (2001) 'Ecological Modernisation: Critical Analysis of the SAPI Foundation in the Republic of Sakha (Yakutia)', in I. Massa and V. Tynkkynen (eds.), *The Struggle for Russian Environmental Policy* (Helsinki: Kikimora Publications): 107-22.

—— and T. Alabaster (2003) 'Tri-sector Partnership for Community Development in Mining: A Case Study of the SAPI Foundation and Target Fund in the Republic of Sakha (Yakutia)', *Resources Policy* 29.3–4: 83-98.

——, T. Alabaster and P.G. Petrova (2000) 'Natural Resource Use in the Russian North: A Case Study of Diamond Mining in the Republic of Sakha', *Environmental Management and Health* 11.4: 318-36.

13
Conclusion

Saleem H. Ali
University of Vermont, USA

This volume has attempted to consolidate the diffuse literature on indigenous people and corporate social responsibility (CSR) in the extractive industries that has emerged over the past decade from various social science and humanities disciplines. A critical part of the context for the study is the demand for raw materials to propel the development booms in the BRIC (Brazil, Russia, India and China) economies, which is causing tremendous growth in the extractive sector. Exhaustion of deposits in more accessible regions focuses attention on remoter areas which often intersect with indigenous lands. Companies may also be more attracted to deposits that can most easily be granted approval for development, and, historically, remote indigenous lands had relatively lax regulatory enforcement. Given the constraints of the market and capricious pricing, companies are often in a rush to develop these deposits while the price of a commodity is high. The result of this 'rush', coupled with incipient post-colonial conflicts, has been a growing rise in tensions between indigenous communities, development ventures, governments and certain environmental organisations.

These are times of transition for the extractive industries as well as for indigenous populations worldwide. On the corporate side, there is increasing scrutiny of company activities by governments, civil society networks and financial institutions leading to a push towards CSR. While the manifestation of CSR practices has multiple interpretations (as discussed in the Introduction), a shift in corporate culture is clearly occurring. Whether or not this shift has yet brought palpable benefits to communities is widely contested. At the same time, indigenous movements are also gaining strength in terms of asserting their sovereignty at multiple levels by challenging established nation-states. The assertion of indigenous sovereignty is a particular conundrum for multinational corporations who see themselves as agents of globalisation. While cross-national barriers to commerce are eroded by global agreements and institutions such as the World Trade

Organisation, internal assertions of indigenous sovereignty are arising within nation-states.

We are thus witnessing a simultaneous push towards harmonised governance regimes at the international level, while also encountering a fracturing of state sovereignty through indigenous movements. Extractive corporations find themselves in the midst of these fault lines and are trying to establish their own norms of coping with these simultaneous shifts. Among the mechanisms employed for this purpose have been transnational global policy networks to voluntarily assess corporate performance and set benchmarks for improved conduct (Khagram and Ali 2008). This book has attempted to understand how CSR is configured within these transnational global policy networks, through relevant legislative requirements, negotiated agreements and other mechanisms that affect relationships between companies, communities and governments.

Polarised perspectives?

In 2002, the mining industry concluded an effort to consider its environmental and social responsibility through a transnational global policy network called the Mining, Minerals and Sustainable Development initiative, and launched its report at the United Nations World Summit on Sustainable Development. Issues pertaining to indigenous people were also to be discussed as part of this effort, and alliances were declared between industry and conservation organisations including the International Union for the Conservation of Nature. However, numerous indigenous groups rejected this effort and issued a joint statement which they referred to as 'United Outcry Against Mining Greenwash'. One of the more strident activists in this movement, Joji Carino from the Tebtebba Foundation and the Indigenous Peoples International Centre for Policy-Research and Education in the Philippines, declared that: 'Entering a partnership on Mining and Biodiversity with the World Conservation Union, while marginalizing indigenous peoples and local communities, who are most severely impacted, is a gross cynicism and non-accountability on the part of these global organizations' (quoted in Larsen 2003). These critics were not just 'outlier' activists, but included groups that gained international prominence through the United Nations Permanent Forum on Indigenous Peoples, which was also very active at the time and whose work resulted, for instance, in the adoption of the UN Declaration on the Rights of Indigenous Peoples in 2007, with the only dissenting votes coming from Australia, Canada, New Zealand and the USA.

In 2003, the World Bank published the results of an introspective review of its financing policies for extractive industries projects which had been led by an Indonesian academic and former environment minister Emil Salim. In his attempt to perhaps placate many of the concerns emanating from the Mining, Minerals and Sustainable Development initiative, Dr Salim openly engaged with numerous indigenous organisations and gave them a platform to voice their concerns in an almost cathartic way. One result of this effort was a 2003 workshop held by the Forest Peoples Programme and Tebtebba Foundation in April 2003 in which an 'Indigenous Peoples' Declaration on Extractive Industries' was issued, which declared that:

> We, indigenous peoples, reject the myth of 'sustainable mining': we have not experienced mining as a contribution to 'sustainable development' by any reasonable definition. Our experience shows that exploration and exploitation of minerals, coal, oil, and gas bring us serious social and environmental problems, so widespread and injurious that we cannot describe such development as 'sustainable'. Indeed, rather than contributing to poverty alleviation, we find that the extractive industries are creating poverty and social divisions in our communities, and showing disrespect for our culture and customary laws (quoted in Colchester *et al.* 2003).

While many indigenous groups have expressed these feelings of antipathy towards any extractive venture, there are many others who are inclined to develop their natural resources and often require technical help in order to do so. Organisations such as the Canadian Aboriginal Mining Association or the Council of Energy Resource Tribes, which consider resource development to be an assertion of their sovereignty, have also existed for several decades and are gaining strength. Differentiation through this perspective is viewed as a mark of independence, similar to casino development compacts among US tribes. Many indigenous groups resent external involvement by environmental groups or civil society as interference in their prerogatives. Fortunately, some environmental groups are beginning to realise the limits of their persuasion power and respecting alternative views towards development (Coumans, Chapter 3 this volume).

The discourse on corporate social responsibility among indigenous people has emerged out of this highly polarised environment, and the chapters in this book have tried to grapple with the diverse attitudes evident within indigenous communities and in the relationship between them and governments as well as corporations. Underlying these chapters has been a clear recognition for the role of regulatory mechanisms and other means of accountability. As Weitzner (2002) reminds us: 'Corporate social responsibility should not be confused with or substituted for government social responsibility. Governments need to uphold and implement their national and international legal obligations to indigenous peoples, and strengthen legal, regulatory and judicial frameworks where these are weak.'

Recognising this distinction, chapter authors have nevertheless considered the impact of corporate regimes on governments and the potential for mutual co-optation. Our findings reveal the range of indigenous experiences around the world, but also show some coalescence of indigenous communities around key principles of action such as free, prior and informed consent as well as a desire for sustainable livelihoods. To synthesise the findings and goals of this book, let us consider some of the key initial questions that we started with and the lessons that can be gleaned from the research presented in these pages.

What is the nature and extent of CSR initiatives in the extractive industries and how should they be understood in the context of indigenous people?

CSR activities in the indigenous domain are often characterised by an acknowledgement of cultural deference and respect as well as some recognition of past injustices by settler societies. Rarely does this take the form of an apology, unless the same company has been involved in past actions. The extent to which such respect and recognition is manifest in actual agreements is highly variable. Some companies, especially in Latin America, are continuing with consultation mechanisms rather than direct negotiations. Dialogue tables and consultation forums, while useful in fostering some degree of cultural connectivity, start with a premise of asymmetric negotiating power and are often perceived to be unjust by communities. Increasingly, impact–benefit agreements are becoming a way to contractually mark this respect in Australia and Canada. However, their implementation may also affect political relations between indigenous peoples, the state and civil society, and thus deserve more integrated analysis.

What motivates companies to pursue CSR policies that focus on indigenous people?

CSR in the extractive industries has largely been spurred as a result of direct action campaigns by indigenous people and their allies or through regulatory enforcement. In some cases the emergence of CSR norms has involved litigation such as with BHP Billiton's mining in Papua New Guinea and Chevron's oil development in Ecuador. Even if the outcome of the litigation is in favour of the company, the level of negative publicity generated by such litigation is enough to cause voluntary change in behaviour by many prominent companies. The impact of this effect is likely to be greater with publicly traded companies based in developed countries, and particularly those that might not have the resources to have protracted litigation appeals. More broadly, the rise of transnational policy networks and institutions such as the International Council on Metals and Mining suggests that an irrevocable cultural shift in corporate behaviour may be under way.

What is the relation between indigenous political action and CSR?

CSR activities have accentuated some of the political aspirations of indigenous communities in constructive ways by allowing for greater scrutiny of corporate–indigenous interaction. Indigenous people can use CSR forums as an opportunity to interact with the state but often governments can also use such forums to shield inaction on their own part. Interactions between non-indigenous company employees and indigenous populations has produced some constructive confrontation as well as some healing. However, much of this process, as analysed in Scandinavian and Australian cases, is ephemeral in part because of high turnover rates among company employees of the extractive sector. Thus the political capital that may be developed with particular relationships though CSR forums can be very easily lost unless there is greater consistency and more systemised ways of delineating obligations. Networks of devolved governance also need to be clearly articulated between the state, the indigenous community and the corporation to avoid conflict as a result of misperceived expectations.

Under what conditions, if any, can CSR help bring about a fundamental change in the distribution of benefits and costs from large-scale resource exploitation?

Under current market mechanisms, the efficacy of CSR largely depends on either creating incentive mechanisms or instituting punitive action for non-compliance of certain standards. If we are to move beyond the cynicism concerning CSR, particularly among indigenous populations, we must accept the reality of political opposition and come to grips with the nuances involved in dealing with indigenous populations, rather than trying to co-opt resistance. In other words, we need to move from conflict *management* to conflict *resolution* or from *consultation* to *negotiation*. This process may initially seem less efficient to companies since it will probably take longer to implement and may also require specific attention to power relations, such as the role of gender or tribal hierarchies. In addition, CSR activities from the private sector must be complemented with public sector involvement from various tiers of government. Civil society groups should play an epistemic role where possible by providing relevant technical information and acting as a means of redressing power imbalances between settler societies and indigenous communities. However, they must also be sensitive to the limitations of their role in this context and allow space for constructive engagement rather than positional entrenchment.

Further study

Any narrative concerning indigenous communities may be highly evanescent, given the rapidity of changes in the legal standing, economic trajectories and cultural exclusivity of these populations. At the same time, there are also key features of indigenous communities that are timeless and firmly engrained in the fabric of culture, such as a profound association with the land. Over time, indigenous people are likely to exert further influence on development ventures but will also be held to a higher standard in terms of their own self-reliance.

Several of the authors have made specific suggestions for further research to augment the work of this volume. Trebeck suggests that further research should consider how to best 'harness the business case' for indigenous interests. Haley and Magdanz highlight the need for 'new primary data on social networks and well-being for indigenous people of common cultural heritage representing the full spectrum of degrees of market integration . . . to tease out these hypothesised relationships'. O'Faircheallaigh concludes that contractual agreements will need to be evaluated over the long term to consider if they are indeed the most effective way for meeting Aboriginal aspirations for development. Gibson and Kemp exhort us to more fully explore the power relations between women, indigenous and non-indigenous, and the large-scale mining economy by monitoring these interactions over time. Echoing the need for such long-term studies, Barker also concludes that Aboriginal employment levels within the industry need to be carefully monitored and studied to consider nuances of attrition and attraction to such vocations. In addition to these needs for further enquiry emanating from specific chapters, there are also some overarching themes that we have not been able to adequately address in this volume but could be addressed in future work.

This book has dealt with industrial-scale mining, given the primary topic of *corporate* social responsibility. However, there are also many indigenous small-scale miners all over the world whose lives and norms of conduct regarding extraction need to be better understood in the context of individual entrepreneurship on the one hand and corporate employment on the other. For example, the Kankanaey and Ibaloi of Benguet province in the Philippines have engaged in both small-scale pocket mining and panning for gold, and wage labour in commercial mines for several decades (UN 2002: 24). Such cross-cutting cases deserve further study to appreciate the evolution of indigenous world-views about extraction of non-renewable materials.

Some scholars have suggested that accountability must be considered as a reciprocal phenomenon and that notions of 'culturally appropriate accountability' detract from functional political development and improvement of relations between indigenous and settler societies (Rowse 2000). This post-colonial approach suggests that the playing field has now been 'levelled' to the extent that objective factors of impact such as environmental pollution and occupational health should be considered without cultural allowances. Such assertions deserve further study in terms of the performance of extractive ventures that may be managed by indigenous communities in the near future. There is, however, a danger that such an approach may dilute the authenticity of indigenous claims. Any research in this domain must be considered with some trepidation and within international norms as it may reify the elusive definition of an 'indigenous' person and affect the actual power that they may exercise in particular development efforts (Ooft 2006). Definitions in the international community can be empowering insofar that laws can subsequently be formulated with clarity and persuasion. However, they are also confining in their coverage of ethnicity and identity which are in constant flux among modern societies. The discourse on definitions thus tends to be inherently political in scope, and researchers must be mindful of how such approaches can be co-opted by power brokers and the resultant impact on indigenous aspirations.

Finally, this book aims to transcend an academic audience and reach policy-makers among governments, corporations, civil society and indigenous communities themselves. Evaluating the long-term impact of CSR policies by all these stakeholders must continue. Academics must endeavour to verify claims and refine metrics of performance in order to provide more clear prescriptive guidance. In a world of increasing scarcity of resources, there is likely to be more tension around any extractive industry projects as stakes for rewards are raised by the markets. It is essential that the cost of such a rush to riches does not neglect those who have already been most marginalised.

References

Colchester, M., A.L. Tomayo, R. Ravillos and E. Caruso (eds.) (2003) *Extracting Promises: Indigenous Peoples, Extractive Industries and the World Bank* (Moreton-in-Marsh, UK: Forest Peoples Programme; Baguio Cuty, Philippines: Tebtebba Foundation).

Khagram, S., and S.H. Ali (2008) 'Transnational Transformations: From Government-Centric Inter-State Regimes To Multi-Actor, Multi-Level Global Governance?', in J. Park, K. Conca and M. Finger (eds.), *The Crisis of Global Environmental Governance* (London/New York: Routledge).

Larsen, P. (2003) *Mining and Indigenous People: An Analysis from IUCN's Social Policy Perspective* (Gland, Switzerland: International Union for Conservation of Nature).

Ooft, M. (2006) *UNDP and Indigenous Peoples towards Effective Partnerships For Human Rights and Development* (Oslo: UNDP Governance Centre).

UN (United Nations) (2002) *Report of the Workshop on Indigenous Peoples, Private Sector Natural Resource, Energy and Mining Companies and Human Rights* (E/CN.4/Sub.2/AC.4/2002/3; New York: United Nations Publications).

Weitzner, V. (2002) *Cutting-edge Policies on Indigenous Peoples and Mining: Key Lessons for the World Summit and Beyond* (Ottawa: North-South Institute).

Acronyms and abbreviations

AHA	Aboriginal Heritage Act
AHDR	*Arctic Human Development Report*
AI	Amnesty International
ALROSA	Almazy Rossii-Sakha Company Ltd
ARCO	Atlantic Richfield Company
BBT	Brethren Benefit Trust
BC	British Columbia
BCL	Bougainville Copper Ltd
BRIC	Brazil, Russia, India and China
CAO	Compliance Advisory Ombudsman
CEPE	Ecuadorian State Petroleum Corporation
CERD	Committee on the Elimination of Racial Discrimination
CGC	Compañia General de Combustibles
CIDA	Canadian International Development Agency
CLC	Central Land Council
CMT	culturally modified tree
CNCA	Canadian Network on Corporate Accountability
CONACAMI	National Coordinating Body of Communities Affected by Mining in Peru
CONAIE	Confederation of Indigenous Nationalities of Ecuador
CORECAMI	Regional Coordinator of Communities Affected by Mining
CRA	Conzinc Riotinto Australia
CSR	corporate social responsibility
CSRM	Centre for Social Responsibility in Mining
DCMI	DIOPIM Committee on Mining Issues
EIA	environmental impact assessment
EIR	Extractive Industries Review
ERA	Energy Resources of Australia
FEROCAFENOP	Federacion de Rondas Campesinas Femeninas del Norte Del Peru
FOE	Friends of the Earth
FOEI	Friends of the Earth International

FONCODES	Fondo Común de Desarrollo/Compensation Fund for Social Development
FPIC	free, prior and informed consent
GMI	Global Mining Initiative
GRUFIDES	Grupo de Formación e Intervención para el Desarrollo Sostenible
HCA	Heritage Conservation Act (Canada)
IBA	Impact and Benefit Agreement
ICCR	Interfaith Center on Corporate Responsibility
ICMM	International Council on Mining and Metals
IFC	International Finance Corporation
IFI	international financial institution
ILO	International Labour Organisation
ILUA	Indigenous Land Use Agreement
IPPP	Indigenous Peoples Partnership Program
IRMA	Initiative for Responsible Mining Assurance
LRC-KsK	Legal Rights and Natural Resources Center-Kasama sa Kalikasan
MCA	Minerals Council of Australia
MCEP	Mining Certification Evaluation Project
MIGA	Multilateral Investment Guarantee Agency
MMSD	Mining Minerals and Sustainable Development
NAFA	National Aboriginal Forestry Association
NGO	non-governmental organisation
NNTT	National Native Title Tribunal
NSI	North-South Institute
NWA	Northwest Arctic
OAS	Organization of American States
OHCHR	Office of the High Commissioner for Human Rights
OMAN	Ontario Mining Action Network
OPIC	Overseas Private Investment Corporation
PDAC	Prospectors and Developers Association of Canada
PDAP	Placer Dome Asia Pacific
PIPLinks	Philippine Indigenous Peoples Links
PNG	Papua New Guinea
PNGDOM	Papua New Guinea Department of Mining
PNGSDP	Papua New Guinea Sustainable Development Program Ltd
PWA	Porgera Women's Association
RAN	Rainforest Action Network
RCMP	Royal Canadian Mounted Police
RNC	Rhéébù Nùù Committee
RSFSR	Russian Soviet Federative Socialist Republic
RTZ	Rio Tinto Zinc
SAPI	Sakhaalmazproinvest Financial Corporation
SEC	Securities and Exchange Commission
SIA	social impact assessment
SLiCA	Survey of Living Conditions in the Arctic
SNIP	National System of Public Investment Peru
STAN	shareholder transnational advocacy network
SWB	subjective well-being

TAN	transnational advocacy network
UNEP	United Nations Environment Programme
WCCCA	Western Cape Communities Co-existence Agreement
WCED	World Commission on Environment and Development
WMAN	Western Mining Action Network
WSDP	Western Shoshone Defense Project

About the contributors

Saleem H. Ali is associate professor of environmental planning at the University of Vermont and on the adjunct faculty of Brown University's Watson Institute for International Studies. He is the author of *Mining the Environment and Indigenous Development Conflicts* (University of Arizona Press, 2004) and the editor of *Peace Parks: Conservation and Conflict Resolution* (MIT Press, 2007). Dr Ali has a bachelor's degree in Chemistry from Tufts University, a master's in Environmental Studies from Yale University and a doctorate in environmental planning from MIT. Further details about his research can be found on www.uvm.edu/~shali.

Bill Angelbeck is a doctoral candidate in archaeology at the University of British Columbia in Vancouver, Canada. He has conducted research in the Coast Salish region and the broader Northwest Coast and Interior Plateau of North America. Focal interests include the development of social inequality, archaeological heritage, ideation and ideology, and the role of warfare in the past. He is the editor of *The Midden*, a quarterly about the archaeology and heritage of the Northwest Coast published by the Archaeological Society of British Columbia.

Isabelle Anguelovski is a doctoral student in Environmental Policy and Planning at the Department of Urban Studies and Planning at MIT. Her research focuses on innovative forms of planning and participation processes for the sustainable development of poor communities impacted by extractive industries and resource management concerns. She is particularly interested in the interaction between environmental justice and community engagement initiatives of corporations conducting oil and mining extraction and delivering water services in Latin America.

Glenn Banks is a human geographer specialising in studies of the social and economic impact of major mining projects in Papua New Guinea and Indonesia. He has also published on the use of GIS technologies in monitoring the impact of large-scale mines and the practice of corporate social responsibility in the mining industry. He has worked as a consultant on socioeconomic issues to a number of mining corporations, as well as contributing to the Mining, Minerals and Sustainable Development project and the production of a sustainable development policy for the PNG Department of Mining.

Tanuja Barker is a social researcher and has undertaken several indigenous employment projects at the University of Queensland's Centre for Social Responsibility in Mining. She holds a Master of Resource Management from the University of Queensland, Australia, and a Bachelor of Science from the University of Auckland, New Zealand.

John Burton specialises in social mapping, landowner identification and land ownership issues in Melanesia, the social impacts of mining on traditional owners, and Native Title research in Australia. He was previously lecturer in Anthropology and Sociology at the University of PNG, Senior Anthropologist at the Torres Strait Regional Authority, and has worked on social mapping and social impact projects at the Lihir, Porgera, Freeport and Ok Tedi mines among others.

Jamie Cerretti graduated from Boston University with a degree in biology and went on to earn her master's in Natural Resources Planning at the University of Vermont. Her thesis examined the organising strategies of indigenous groups in the Ecuadorian Amazon. Currently, she is engaged in environmental advocacy work in the extractive industries around Denver, Colorado.

Catherine Coumans holds an MSc (London School of Economics) and a PhD (McMaster University) in Cultural Anthropology. She carried out postdoctoral research at Cornell University and taught at Cornell and McMaster. She is Research Coordinator and responsible for the Asia-Pacific Program at MiningWatch Canada. Catherine has worked with NGOs and mining-affected communities in India, Burma, Thailand, Indonesia, the Philippines, Papua New Guinea and New Caledonia. Her work has focused on indigenous peoples affected by Canadian mining companies. Catherine has published numerous peer-reviewed articles and reports on mining and co-authored *Framework for Responsible Mining: A Guide to Evolving Standards* (2005).

Susan Crate is an applied anthropologist who has conducted research with Vilyuy Sakha of north-eastern Siberia, Russia, since 1991. She specialises in human–environment interactions, cultural/political ecology, sustainability studies, circumpolar peoples and climate change. Crate's most recent research focuses on understanding the perceptual and cultural implications of global climate change which is causing unprecedented change in the Arctic/sub-Arctic and other climate-sensitive world regions. Her 2007–2010 NSF project is aimed to locate research approaches to both tap into anthropology's particular ways of knowing and move anthropologists, conducting research with indigenous communities confronting global climate change, from impartial observers into the realm of action-oriented researchers, at home and abroad.

Colin Filer is an anthropologist by training and an expert on the social context and impact of resource development and conservation projects in Melanesia. He has engaged with the mining sector as a teacher, researcher, consultant and policy-maker since taking up an appointment at the University of PNG in 1983. He was also a member of the Assurance Group for the Mining, Minerals and Sustainable Development project in 2001–2002 and drafted a number of policy papers for the PNG government under the terms of the Mining Sector Institutional Strengthening Project funded by the World Bank in 2002–2003.

Ginger Gibson works as an anthropologist with communities affected by extractive industries projects. As a Trudeau Scholar, she completed her PhD research in Mining Engineering studying the negotiation of change by Dene miners, families and communities in the diamond-mining economy of northern Canada. Her research for communities affected by mining in Latin America and Canada has focused on negotiation, capacity building, engagement and risk assessment and communication. She is currently working for indigenous governments on social and mining and mineral policy issues, and, together with Ciaran O'Faircheallaigh, writing a manual on negotiation of Impact and Benefit Agreements for communities.

Sharman Haley is a professor of economics and public policy at the Institute of Social and Economic Research, University of Alaska Anchorage. She has been studying rural development and socioeconomic impacts of extractive industries in remote regions of Alaska and the circumpolar Arctic for more than a decade. She is a contributing author to the *Arctic Human Development Report*, the *Assessment of Poten-*

tial Effects of Oil and Gas Activities in the Arctic, and *Arctic Oil and Gas: The Challenges of Sustainable Development*, and is active in the International Arctic Social Science Association (IASSA) working group on extractive industries.

Richie Howitt's research focuses on indigenous rights and the interface between indigenous communities, natural resource development, governments and corporations at the scales of the project, the community, the landscape and the nation.

Deanna Kemp is a Senior Research Fellow at the Centre for Social Responsibility in Mining, which is part of the Sustainable Minerals Institute at The University of Queensland (UQ), Australia. For the past 12 years her work and research has concentrated on the social and community impacts of large-scale industry (primarily mining). Much of this work has been international at large-scale projects in India, Pakistan, Bangladesh, Indonesia, Laos, Canada, the United States, Australia and New Zealand. At CSRM she focuses on relationships between mining companies and local communities, including issues such as company–community disputes, community participation in resource development and industry responses to community relations and development challenges. Deanna holds a PhD from UQ and a Master of Social Science from RMIT University.

Rebecca Lawrence is an Adviser to the Saami Council and is completing her PhD in the Department of Sociology, University of Stockholm, and the Department of Human Geography, Macquarie University, New South Wales. Her research focuses on the interplay between indigenous rights and corporate social responsibility in Australia and Scandinavia.

James Magdanz is a subsistence resource specialist for the Alaska Department of Fish and Game. He is responsible for designing, conducting and analysing subsistence harvest surveys in north-west Alaska.

Emily McAteer is a research analyst with RiskMetrics Group's Climate Risk Management Team, where she conducts research and analysis on corporate strategies to address climate change for institutional investors and other clients. Emily is a co-author of the January 2008 report *Corporate Governance and Climate Change: The Banking Sector.* Prior to joining RiskMetrics, Emily completed a bachelor's degree in Environmental Studies at Brown University, where she graduated Phi Beta Kappa.

Ciaran O'Faircheallaigh is Professor of Politics and Public Policy at the Griffith Business School, Griffith University, Brisbane. He is the author of *A New Approach to Policy Evaluation: Mining and Indigenous People* (Ashgate, 2002) and numerous monographs on indigenous people and resource development, negotiation, social impact assessment and public management. He has worked as a negotiator and advisor for many of Australia's leading Aboriginal organisations.

Katherine Trebeck is a Research and Policy Executive at the Wise Group, a Glasgow-based social enterprise. Before commencing at the Wise Group, Katherine was a Research Fellow at the University of Glasgow from 2005 to 2008. The empirical research that informs Chapter 1 was carried out as part of a PhD thesis in Political Science at the Centre for Aboriginal Economic Policy Research at the Australian National University, 2002–2005.

Natalia Yakovleva is a Research Associate in the ESRC-funded Research Centre for Business Relationships, Accountability, Sustainability and Society (BRASS) at Cardiff University, UK. She has a PhD in Environmental Studies and her main research interests relate to CSR, community development and reporting in the extractive industry. Her current research projects include investigation of governance and sustainability challenges in the small-scale mining sector in Ghana; and corporate–community interactions in the gold-mining industry in Argentina. She is specifically interested in local community impacts of extractive projects and methods of community investment and participation. She continues to conduct research in Russia; her current project examines responses of indigenous and regional communities in southern Yakutia to construction of an oil pipeline in Eastern Siberia. In addition, Natalia studies the sustainability of food supply chains with the development of indicators.

Index

Note: page numbers in *italic figures* refer to figures and tables.

For Product Safety Concerns and Information please contact our EU representative GPSR@taylorandfrancis.com Taylor & Francis Verlag GmbH, Kaufingerstraße 24, 80331 München, Germany

Printed and bound by CPI Group (UK) Ltd, Croydon, CR0 4YY

08/05/2025

01864348-0001